D0635780

quanta

Photograph by Lloyd E. Saunders courtesy of The University of Chicago

Gregor Wentzel

Essays in Theoretical Physics Dedicated to Gregor Wentzel

Edited by P. G. O. Freund, C. J. Goebel, and Y. Nambu

The University of Chicago Press, Chicago and London

Standard Book Number: 226-26280-4
Library of Congress Catalog Card Number: 70-108268

The University of Chicago Press, Chicago 60637
The University of Chicago Press, Ltd., London

Printed in the United States of America

CONTENTS

PREFACE

Professor Gregor Wentzel has decided to retire from the University of Chicago at the end of this academic year. His contributions to the development of Quantum Theory are, by now, classical. His gift as a teacher is widely known. Some of Gregor Wentzel's colleagues and former students would like to offer him these Quanta as an expression of their admiration.

The generous assistance of Mrs. Judith Reiffel in the realization of this project is gratefully acknowledged.

--The Editors
Chicago, May, 1969

INTRODUCTION

This collection of papers is dedicated to Gregor Wentzel on the occasion of his retirement from the University of Chicago --to honor the scientist, teacher, and friend.

Gregor Wentzel was born in Düsseldorf, Germany, on February 17, 1898. He attended the Universities of Freiburg, Greifswald, and finally Munich, where Sommerfeld was his teacher. There he received his doctorate in 1921. He stayed in Munich until 1926 (becoming "Privatdozent" in 1922), and moved to the University of Leipzig as Professor of Mathematical Physics.

In 1928 he became Schrödinger's successor at the University of Zurich, at about the same time when his close friend Pauli, another Sommerfeld student, was called to the Swiss Federal Institute of Technology across the street. For many years the team Pauli-Wentzel represented theoretical physics in Zurich, creating an atmosphere of intellectual stimulus and warm "camaraderie" which will remain unforgettable to anyone who was fortunate enough to witness it.

During the war years, which Pauli spent at the Institute for Advanced Study in Princeton, Wentzel had to carry the burden of the theoretical physicist at both Zurich institutions. More than that, as a senior Swiss physicist he had a major share in leading Swiss physics--which was then largely isolated--through an extremely difficult time. Nevertheless, this was one of Wentzel's most productive scientific periods, as a glance at the bibliography will show.

In 1948 he left Zurich for the University of Chicago.

Wentzel's rich scientific work spans now almost fifty years, and the bibliography contains more than 100 items. It starts in 1921 with the Bohr-Sommerfeld theory of atomic electrons, the "elementary particles" of those days, and leads to the analysis of mesons and hyperons, the elementary particles of a different epoch. Wentzel's work is too well known to need a detailed commentary. But it might be useful--if only for the benefit of a younger generation of physicists--to point out that a good number of the ideas and methods which we use today as a matter of course originated in his lucidly and beautifully written papers. One item is of course the W-K-B method.[1]

Another point in case is the study of what Wentzel himself called "aperiodic phenomena". Scattering processes in the widest sense of the word form a recurring theme in the first decade of his scientific activity. A paper of 1922[2] contains probably the most careful and comprehensive treatment of scattering processes in pre-quantum mechanical times. The criteria for the occurrence of multiple scattering processes derived there have been frequently used ever since to analyze experimental data. The first wave mechanical treatment of the photo electric effect in atoms was given by Wentzel.[3] This paper contains already the separation of outgoing waves b complex integration along suitable paths and other good friends of the present day physicist. A number of other papers followed, until the appearance of the classical Handbuch article on collision and radiation phenomena[4] for a long time unmatched in depth, scope, and mathematical elegance.

Lastly we wish to emphasize that Wentzel is the author of the first serious and comprehensive papers on the "strong coupling" theory of nuclear forces[5], a subject to which he returned time and again.

Wentzel came to Chicago at a time when the first post-war crop of physicists was about to emerge. This whole generation had learned their quantum field theory from his remarkable book.[6] The first true textbook on the subject, it is still a model of clarity and pedagogical skill. Wentzel brought to Chicago in his first set of lectures a deep interest in and critical knowledge of Heisenberg and Møller's early work on S-matrix theory. It was a rare privilege for those of us who learned from him these developments and had his guidance in the progress being made at that time in quantum electrodynamics.

Another aspect of the Chicago era over and above teaching and research was the role Wentzel played in the weekly informal seminars held at the Enrico Fermi Institute. His wisdom in guiding the discussion over an extremely broad spectrum of all of science made these occasions most valuable for his colleagues. In many cases he had either worked on specific problems that arose or had through his research profoundly influenced them.

Gregor Wentzel has made deep fundamental contributions to physics. He has influenced several generations of physicists through his book and Handbuch articles. For over twenty years at the University of Chicago he has taught superbly, inspired students and contributed immeasurably to the quality and style of physics at that University.

Both of us were students and later colleagues of Gregor Wentzel. It has been a pleasure and an honor to have been associated with a man of such scientific and moral integrity. We look forward to reading many more of his papers and to learning from him at future conferences and celebrations.

--V. Bargmann and M. L. Goldberger

References

1. Eine Verallgemeinerung der Quantenbedingungen fur die Zwecke der Wellenmechanik, Zeitz. f. Physik 38, 518 (1926).

2. Zur Theorie der Streuung von Korpuskularstrahlen, Physik. Zeits. 23, 435 (1922).

3. Zur Theorie des photoelektrischen Effekts, Zeits. f. Physik 40, 574 (1926).

4. Wellenmechanik der Stoss- und Strahlungsvorgaenge, Handbuck der Physik, Geiger-Scheel, 24, 695 (1933).

5. Zum Problem des statischen Mesonfeldes, Helv. Phys. Acta 13, 269 (1940).

Zur Hypothese der hoeheren Proton-Isobaren, Helv. Phys. Acta 14, 3 (1941).

Zum Problem des statischen Mesonfeldes. Nachtrag, Helv. Phys. Acta 14, 633 (1941).

6. Einfuhrung in die Quantentheorie der Wellenfelder, Vienna, Franz Deuticke 1943; reprinted by Edwards Brothers, Inc., Ann Arbor, Michigan, 1946; English translation: Quantum Theory of Fields, translated by Charlotte Houtermans and J. M. Jauch, with an appendix by J. M. Jauch, New York, Interscience Publishers, Inc., 1949.

quanta

Elementary Derivation of General Multipole Moments for a Classical Charge and Current Distribution

by

N. Kemmer

University of Edinburgh

The occasion for the issue of this volume is one which at last gives me the opportunity to acknowledge the great debt of gratitude that I owe to Gregor Wentzel. My first encounter with the man later to become my teacher and friend will always remain a vivid memory. It occurred in Zurich in the autumn of 1931 at the first of his course of lectures on "Optics." (My lecture notes on the course are still to hand!) I remember how a slim and boyish figure appeared before us and how I turned to my neighbour with the words "I wonder why the Professor has sent his assistant to lecture to us?".

However, the lucid, superbly organized exposition, of which we had our first taste on that day, soon convinced us that this young man rightly deserved the august position to which he had risen.

I have no doubt that the future direction of my interests was decisively determined by this experience of really inspired teaching. Gregor Wentzel's influence on me certainly did not end there; he gave me my first research problem and guided me to a doctorate, and in the subsequent critical phase of my career, I owed more to his advice and friendship than I can express. But on this occasion I feel that the best way I can contribute in honoring him is to remember particularly his great gifts as a teacher. Other colleagues will have important and topical contributions to today's problems in physics to expound. I feel I cannot do so. Nevertheless perhaps it is not entirely unseemly for me to submit a piece of work which I believe to have some interest as an elementary presentation of a well explored classical subject. The pages that follow certainly could have been written over thirty years ago and perhaps they might have appealed to Gregor Wentzel then--even enough to have received mention in those great lectures.

1. Introduction

There exist in the literature[1] several derivations of the complete expansion of the radiation field of an arbitrary distribution of electric charge and current into a series of multipole fields. In most of this work the emphasis is on the multipole fields rather than on the multipole moments that give rise to the fields and where the moments are explicitly discussed it is always only for charges and currents varying harmonically with time. Although this is not a serious restriction in practice, it will be shown in the following that the problem of expressing the electric and magnetic 2^{ℓ}-moments can be solved by quite elementary methods without any such restriction. The procedure here developed follows very closely the lines familiar from the theory of static fields and shows directly how the "effective" electric and magnetic 2^{ℓ}-moments, i.e. the exact quantities that determine, by their retarded values, the 2^{ℓ}-pole field, are related to the corresponding moments of static theory. The full expressions obtained are equivalent and closely related in form to those of P. R. Wallace, whose work is quite similar to the present in the form of its results, but the method of derivation owes rather more to the work of Bouwkamp and Casimir.

An additional feature stressed in the present work is that the expressions obtained are equally valid for entirely continuous charge-current distributions and for those involving surface discontinuities. When, for instance, the currents are taken to be confined within the sharp boundaries of a conductor, the equations derived, unlike those of some of the earlier work, require no modification or special interpretation.

2. Potentials and Hertz vectors

As a preliminary to the main problems, let us consider the familiar expressions

$$\phi(\underset{\sim}{r}) = \int \rho(\underset{\sim}{r}',t-|\underset{\sim}{r}-\underset{\sim}{r}'|/c)\ \frac{dV'}{|\underset{\sim}{r}-\underset{\sim}{r}'|} \tag{1a}$$

$$c\underset{\sim}{A}(\underset{\sim}{r}) = \int \underset{\sim}{j}(\underset{\sim}{r}',t-|\underset{\sim}{r}-\underset{\sim}{r}'|/c)\ \frac{dV'}{|\underset{\sim}{r}-\underset{\sim}{r}'|} \tag{1b}$$

for the retarded potentials of the localised distribution characterised by charge density ρ and current density $\underset{\sim}{j}$, with these quantities satisfying

$$\nabla\cdot\underset{\sim}{j} + \dot{\rho} = 0\,. \tag{2}$$

The notation should be sufficiently familiar as not to need further explanation. Choosing an origin of coordinates in the vicinity of the distribution, we consider the field only at points distant from the origin, i.e. for which $r >> r'$. We can then expand under the integrals in powers of r'/r and, using the circumstance that $\underset{\sim}{r}$ enters (1) only in the combination $\underset{\sim}{r}-\underset{\sim}{r}'$ this expansion can be most briefly written as

$$\phi(\underset{\sim}{r}) = \frac{1}{r}\int\rho dV' - \frac{1}{1!}\partial_i\,\frac{1}{r}\int x_i'\rho dV' + \frac{1}{2!}\partial_i\partial_k\,\frac{1}{r}\int x_i'x_k'\rho dV' -+\ldots \tag{3a}$$

$$cA_i(\underset{\sim}{r}) = \frac{1}{r}\int j_i dV' - \frac{1}{1!}\,\partial_k\,\frac{1}{r}\int x_k'j_i dV' + \frac{1}{2!}\,\partial_k\partial_\ell\,\frac{1}{r}\int x_k'x_\ell'j_i dV'-+\ldots \tag{3b}$$

In (3) and subsequently the following conventions of notation have been used. Derivatives $\frac{\partial}{\partial x_i}$ are abbreviated to ∂_i and summation over repeated suffixes is implied. More important, the variables $\underset{\sim}{r}$, t on the left hand side and the variables $\underset{\sim}{r}'$, $t-r/c$ under the integral signs on the right are omitted. Throughout this work no integrals of quantities taken at non-retarded times will appear so that no ambiguity will arise from the short notation, but it must be borne in mind that differentiations such as ∂_i, while, of course, not acting on $\underset{\sim}{r}'$, do act on densities such as ρ and j_i via the retardation in their time dependence.

A useful transformation of (3b) can be performed with the use of (2): We have

$$\partial_n(x_i j_n) = j_i - x_i\dot{\rho}$$

$$\partial_n(x_i x_k j_n) = j_i x_k + j_k x_i - x_i x_k \dot{\rho} \tag{4}$$

and generally

$$\partial_n(x_i x_k \cdot\cdot x_m j_n) = j_{(i} x_k \cdot\cdot x_{m)} - x_i x_k \cdot\cdot x_m \dot{\rho} \; .$$

Here and in the following the notation of bracketed suffixes is used to denote symmetrization. The symmetrised quantity denoted by the bracketed term is understood to mean the sum of unequivalent terms only, i.e.

$$x_{(i}x_{k)} = x_i x_k$$

but $\qquad\qquad$ (5)

$$x_{(i}y_{k)} = x_i y_k + x_k y_i \; .$$

On taking volume integrals on both sides of (4) the left hand side reduces to a surface integral over any surface discontinuities that there may be and/or a surface integral over a large surface enclosing the distribution. Since $\underset{\sim}{j}$ is necessarily tangential at any discontinuity and the distribution is localised the left hand side clearly vanishes. It follows that (3) can be rewritten as:

$$\phi = \frac{e}{r} - \nabla\cdot\underset{\sim}{\Pi}^{(e)} \tag{6a}$$

$$\underset{\sim}{A} = \frac{1}{c}\dot{\underset{\sim}{\Pi}}^{(e)} + \nabla\times\underset{\sim}{\Pi}^{(m)} \tag{6b}$$

$$\left[e = \int \rho dV'\right]$$

with the electric and magnetic Hertz vectors defined as

$$\Pi_i^{(e)} = \frac{1}{1!}\frac{1}{r}\int x_i' \rho dV' - \frac{1}{2!}\partial_k\frac{1}{r}\int x_i' x_k' \rho dV' + \frac{1}{3!}\partial_k\partial_\ell\frac{1}{r}\int x_i' x_k' x_\ell' \rho dV' - + \ldots \tag{7a}$$

$$\Pi_i^{(m)} = \frac{2\cdot 1}{2!}\frac{1}{r}\int \xi_i dV' - \frac{2.2}{3!}\partial_k \frac{1}{r}\int \xi_i x_k' dV' + \frac{2.3}{4!}\partial_k \partial_\ell \frac{1}{r}\int \xi_i x_k' x_\ell' dV' - + \ldots \tag{7b}$$

with

$$\underset{\sim}{\xi}(\underset{\sim}{r}') = \frac{1}{2}\underset{\sim}{r}' \times \underset{\sim}{j}(\underset{\sim}{r}') \tag{8}$$

The expansions (7) have some of the features of true multipole expansions, but not all, because the tensors appearing in the integrals are not traceless nor, in the case of (7b), symmetrical so that they are not irreducible and hence the radiation field from the ℓ-th term is not a pure 2^ℓ-pole field.

However, it is known that the Hertz potentials $\underset{\sim}{\Pi}^{(e)}$ and $\underset{\sim}{\Pi}^{(m)}$ can be subjected to a wide range of gauge transformations[2] that do not alter the field intensities and by means of such transformations it is possible to go from (7) to a true multipole expansion. For example, if in the integrand of the second term of (7a) the quantity $\frac{1}{3} r'^2 \delta_{ik}$ is subtracted, this alters $\Pi_i^{(e)}$ only by the gradient $\frac{1}{6}\partial_i \frac{1}{r}\int r'^2 \rho dV'$ which satisfies the homogeneous wave equation and the addition of such a vector to $\Pi_i^{(e)}$ does not change the fields.

In symmetrising and removing traces from other terms one is led also to utilise the fact that the simultaneous replacements

$$\underset{\sim}{\Pi}^{(e)} \rightarrow \underset{\sim}{\Pi}^{(e)} + \frac{1}{c}\dot{\underset{\sim}{F}}$$

$$\underset{\sim}{\Pi}^{(m)} \rightarrow \underset{\sim}{\Pi}^{(m)} + \nabla \times \underset{\sim}{F}$$

where $\underset{\sim}{F}$ is any vector field satisfying the homogeneous wave equation, leaves the fields unaltered. Using such transformations a true multipole expansion can be developed step by step starting from (7), but the process is tedious and only interesting as being probably the most direct method of procedure to the desired end. However we shall show in the following that much of the complication can be avoided by a modified approach, suggested by the work of Bouwkamp and Casimir.

3. Debye potentials and radial fields

Let us begin by assuming that the required multipole expansion has been found in the form

$$\Pi_i^{(e)} = \frac{1}{r} p_i - \frac{1}{2!} \partial_k \frac{1}{r} q_{ik} + \frac{1}{3!} \partial_k \partial_\ell \frac{1}{r} r_{ik\ell} - + \ldots \tag{9a}$$

$$\Pi_i^{(m)} = \frac{1}{r} m_i - \frac{1}{2!} \partial_k \frac{1}{r} n_{ik} + \frac{1}{3!} \partial_k \partial_\ell \frac{1}{r} o_{ik\ell} - + \ldots \tag{9b}$$

The form of (9) is the same as of (7) with the important difference that the integrals of (7) have been replaced by irreducible tensors (taken at the time $t - r/c$), of which only the first approximation can be directly read off from (7). As already remarked, the complete transition from (7) to (9) is rather an involved process; instead we shall pass from (9) to directly related expansions of other quantities for which the comparison corresponding to that of (7) with (9) is a much simpler matter. The first step in this process depends on a transformation of every term of (9) separately which we illustrate on a typical term. Utilising the symmetry and tracelessness of the tensor involved, we derive the identities:

$$\begin{aligned} x_i \partial_k \partial_\ell \partial_m (\tfrac{1}{r} o_{klm}) &= \partial_k \{x_i \partial_\ell \partial_m (\tfrac{1}{r} o_{k\ell m})\} - \partial_\ell \partial_m (\tfrac{1}{r} o_{i\ell m}) \\ &= \partial_k \partial_\ell \{x_i \partial_m (\tfrac{1}{r} o_{k\ell m})\} - \partial_k \partial_m (\tfrac{1}{r} o_{kim}) - \partial_\ell \partial_m (\tfrac{1}{r} o_{i\ell m}) \\ &= \partial_k \partial_\ell (\frac{x_i x_m}{r} \frac{\partial}{\partial r} (\tfrac{1}{r} o_{k\ell m})) - 2\partial_k \partial_\ell (\tfrac{1}{r} o_{ik\ell}) \\ &= \partial_k \partial_\ell \{x_m \partial_i (\tfrac{1}{r} o_{k\ell m})\} - 2\partial_k \partial_\ell (\tfrac{1}{r} o_{ik\ell}) \\ &= \partial_i [\partial_k \partial_\ell (\frac{x_m}{r} o_{k\ell m})] - 3\partial_k \partial_\ell (\tfrac{1}{r} o_{ik\ell}). \end{aligned} \tag{10}$$

Performing analogous calculations on all other terms we see that we can put

$$\Pi_i^{(e)} = x_i U + \partial_i F \tag{11a}$$

$$\Pi_i^{(m)} = x_i V + \partial_i G \tag{11b}$$

where

$$U = -\partial_i\left(\frac{p_i}{r}\right) + \frac{1}{2}\frac{1}{2!}\partial_i\partial_k\left(\frac{q_{ik}}{r}\right) - \frac{1}{3}\frac{1}{3!}\partial_i\partial_k\partial_\ell\left(\frac{r_{ik\ell}}{r}\right) + - \ldots \tag{12a}$$

and

$$V = -\partial_i\left(\frac{m_i}{r}\right) + \frac{1}{2}\frac{1}{2!}\partial_i\partial_k\left(\frac{n_{ik}}{r}\right) - \frac{1}{3}\frac{1}{3!}\partial_i\partial_k\partial_\ell\left(\frac{o_{ik\ell}}{r}\right) + - \ldots \tag{12b}$$

Now, the relations between the Hertz vectors and the field intensities are, as is well known:

$$\underset{\sim}{E} = e\,\frac{\underset{\sim}{r}}{r^3} + \left(\nabla^2 - \frac{1}{c^2}\frac{\partial^2}{\partial t^2}\right)\underset{\sim}{\Pi}^{(e)} + \nabla \times (\nabla\times\underset{\sim}{\Pi}^{(e)}) - \nabla\times\dot{\underset{\sim}{\Pi}}^{(m)} \tag{13a}$$

$$\underset{\sim}{B} = \nabla\times(\nabla\times\underset{\sim}{\Pi}^{(m)}) + \nabla\times\dot{\underset{\sim}{\Pi}}^{(e)} \tag{13b}$$

and it is evident that the term $\partial_i G$ can be dropped from the expression for $\underset{\sim}{\Pi}^{(m)}$ without altering the fields at all. The same is not directly true for the corresponding term $\partial_i F$ in $\underset{\sim}{\Pi}^{(e)}$, because F does not necessarily satisfy the wave equation, even outside the localised distribution, but since $\underset{\sim}{\Pi}^{(e)}$ itself does, it is legitimate in this region to ignore $\partial_i F$ also, provided the term $(\nabla^2 - \frac{1}{c^2}\frac{\partial^2}{\partial t^2})\,\underset{\sim}{\Pi}^{(e)}$ is first dropped from (13a). We can therefore write

$$\underset{\sim}{E} = e\,\frac{\underset{\sim}{r}}{r^3} + \nabla \times \left\{\nabla \times (\underset{\sim}{r}U)\right\} - \nabla \times (\underset{\sim}{r}\dot{V}) \tag{14a}$$

$$\underset{\sim}{B} = \nabla \times \left\{\nabla \times (\underset{\sim}{r}V)\right\} + \nabla \times (\underset{\sim}{r}\dot{U}) \,. \tag{14b}$$

Equations (12) and (14) give the explicit representation of the field intensities in terms of the multipole moments via the <u>Debye potentials</u> U and V. The decisive step taken by Bouwkamp and Casimir consists in going over to the <u>radial components</u> of the field intensities. In terms of these the equations involving U and V separate, giving

$$\underset{\sim}{r}\cdot\underset{\sim}{E} = \frac{e}{r} - (\underset{\sim}{r}\times\nabla)^2 U \tag{15a}$$

$$\underset{\sim}{r}\cdot\underset{\sim}{B} = -(\underset{\sim}{r}\times\nabla)^2 V \,. \tag{15b}$$

It is now an easy step to write down the explicit multipole expansions for $\underset{\sim}{r}\cdot\underset{\sim}{E}$ and $\underset{\sim}{r}\cdot\underset{\sim}{B}$.

Returning to equations (12) we note that owing to the traceless nature of the moment tensors we can also write*

$$U = -x_i \left(\frac{1}{r}\frac{\partial}{\partial r}\right)\left(\frac{p_i}{r}\right) + \frac{1}{2}\frac{1}{2!} x_{(i}x_{k)} \left(\frac{1}{r}\frac{\partial}{\partial r}\right)^2 \left(\frac{q_{ik}}{r}\right) - + \ldots \quad \text{(1}$$

$$V = -x_i \left(\frac{1}{r}\frac{\partial}{\partial r}\right)\left(\frac{m_i}{r}\right) + \frac{1}{2}\frac{1}{2!} x_{(i}x_{k)} \left(\frac{1}{r}\frac{\partial}{\partial r}\right)^2 \left(\frac{n_{ik}}{r}\right) - + \ldots \quad \text{(1}$$

and moreover the tensors $x_{(i \ldots\ldots} x_{n)}$ occurring in (16) can be made traceless without altering U and V. In this way it is directly exhibited that the dependence on angles of the ℓ-th term in (16a,b) is that of an ℓ-th order spherical harmonic. Hence the action of the operator $(r \times \nabla)^2$ on it is just to multiply it by $-(\ell + 1)\ell$. Therefore we have

$$\mathbf{r} \cdot \mathbf{E} = \frac{e}{r} - 2\partial_i \left(\frac{p_i}{r}\right) + \frac{3}{2!}\partial_i\partial_k \left(\frac{q_{ik}}{r}\right) - + \ldots \quad \text{(1}$$

$$\mathbf{r} \cdot \mathbf{B} = \quad - 2\partial_i \left(\frac{m_i}{r}\right) + \frac{3}{2!}\partial_i\partial_k \left(\frac{n_{ik}}{r}\right) - + \ldots \quad \text{(1}$$

and this is the final form which we shall use to determine the explicit effective moments.

4. Direct calculation of the expansion for the radial fields.

The following equations are simple to verify. They are substantially due to Bouwkamp and Casimir.

$$\left(\nabla^2 - \frac{1}{c^2}\frac{\partial^2}{\partial t^2}\right)(\mathbf{r}\cdot\mathbf{E}) = -4\pi\left[\rho - \nabla \cdot (\mathbf{r}\rho) - \frac{1}{c^2}(\mathbf{r} \cdot \dot{\mathbf{j}})\right] \quad \text{(1}$$

$$\left(\nabla^2 - \frac{1}{c^2}\frac{\partial^2}{\partial t^2}\right)(\mathbf{r}\cdot\mathbf{B}) = -4\pi[-2\nabla \cdot \boldsymbol{\xi}] . \quad \text{(1}$$

By the usual method of integration we obtain

$$\mathbf{r}\cdot\mathbf{E} = \int [\rho - \nabla'\cdot(\mathbf{r}'\rho) - \eta] \frac{dV'}{|\mathbf{r}-\mathbf{r}'|} \qquad \left(\eta(\mathbf{r}) = \frac{1}{c^2}(\mathbf{r}\cdot\dot{\mathbf{j}}(\mathbf{r}))\right) \quad \text{(1}$$

$$\mathbf{r} \cdot \mathbf{B} = -2 \int (\nabla' \cdot \boldsymbol{\xi}) \frac{dV'}{|\mathbf{r}-\mathbf{r}'|} \quad \text{(1}$$

---

*The transformation embodied in (16) was suggested to me by Mr. A. Nisb where again the functions under the integral are taken at $\underset{\sim}{r}'$ and $t - \frac{1}{c}|\underset{\sim}{r}-\underset{\sim}{r}'|$, but the ∇' differentiations apply only to the direct dependence on $\underset{\sim}{r}'$, not the dependence via the retardation. Because the source functions entering (19) involve derivatives of ρ and $\underset{\sim}{j}$ the expressions are valid as they stand only for everywhere differentiable functions ρ and $\underset{\sim}{j}$. When there are sharp boundaries to the distribution the right hand sides must be augmented by surface terms. However, these surface terms can be avoided by the use of identities

$$\frac{\nabla'\cdot(\underset{\sim}{r}'\rho)}{|\underset{\sim}{r}-\underset{\sim}{r}'|} = \nabla'\cdot\left(\frac{\underset{\sim}{r}'\rho}{|\underset{\sim}{r}-\underset{\sim}{r}'|}\right) + \nabla\cdot\left(\frac{\underset{\sim}{r}'\rho}{|\underset{\sim}{r}-\underset{\sim}{r}'|}\right) \qquad [\rho = \rho(\underset{\sim}{r}')] \tag{20a}$$

$$\frac{\nabla'\cdot\underset{\sim}{\xi}}{|\underset{\sim}{r}-\underset{\sim}{r}'|} = \nabla'\cdot\left(\frac{\underset{\sim}{\xi}}{|\underset{\sim}{r}-\underset{\sim}{r}'|}\right) + \nabla\cdot\left(\frac{\underset{\sim}{\xi}}{|\underset{\sim}{r}-\underset{\sim}{r}'|}\right) \qquad [\underset{\sim}{\xi} = \underset{\sim}{\xi}(\underset{\sim}{r}')] \tag{20b}$$

On integration the surface terms arising from the first terms on the right hand side of (20) just compensate the ones referred to before, so that for any localised distribution the following alternative to (19) is correct.

$$\underset{\sim}{r}\cdot\underset{\sim}{E} = \int[\rho-\eta]\frac{dV'}{|\underset{\sim}{r}-\underset{\sim}{r}'|} - \nabla\cdot\int\underset{\sim}{r}'\rho\frac{dV'}{|\underset{\sim}{r}-\underset{\sim}{r}'|} \tag{21a}$$

$$\underset{\sim}{r}\cdot\underset{\sim}{B} = -2\nabla\cdot\int\underset{\sim}{\xi}\frac{dV'}{|\underset{\sim}{r}-\underset{\sim}{r}'|}\,. \tag{21b}$$

Now the expansion analogous to (3) is readily written down:

$$\begin{aligned}\underset{\sim}{r}\cdot\underset{\sim}{E} = {}& \frac{e}{r} - \frac{2}{1!}\partial_i\frac{1}{r}\int\rho x_i dV' + \frac{3}{2!}\partial_i\partial_k\frac{1}{r}\int\rho x_i' x_k' dV' - + \ldots \\ & - \frac{1}{r}\int\eta dV' + \frac{1}{1!}\partial_i\frac{1}{r}\int\eta x_i' dV' - \frac{1}{2!}\partial_i\partial_i\frac{1}{r}\int\eta x_i' x_k' dV' + - \ldots\end{aligned} \tag{22a}$$

$$\begin{aligned}\underset{\sim}{r}\cdot\underset{\sim}{B} = {}& -2\partial_i\frac{1}{r}\int\xi_i dV' + \frac{2}{2!}\partial_i\partial_k\frac{1}{r}\int\xi_{(i}x'_{k)}dV' \\ & - \frac{2}{3!}\partial_i\partial_k\partial_\ell\frac{1}{r}\int\xi_{(i}x'_k x'_{\ell)}dV' + - \ldots\end{aligned} \tag{22b}$$

In (22b) an obvious symmetrisation has been performed.

A similar process to the one described on p. has now to be performed, but here it is very considerably simpler both because now electric and magnetic multipoles are completely separate and because there are no further symmetrisations required:

We define

$$x_{[i_1} x_{i_2} \cdots x_{i_\ell]} = x_{i_1} x_{i_2} \cdots x_{i_\ell} - \frac{r^2}{(2\ell-1)} \delta_{(i_1 i_2} x_{i_3} \cdots x_{i_\ell)}$$

$$+ \frac{r^4}{(2\ell-1)(2\ell-3)} \delta_{(i_1 i_2} \delta_{i_3 i_4} x_{i_5} \cdots x_{i_\ell)} - + \cdots$$

The series terminates with the term

$$\frac{r^\ell}{(2\ell-1)\ldots(\ell+1)} \delta_{(i_1 i_2} \cdots \delta_{i_{\ell-1} i_\ell)}$$

for even ℓ and with

$$\frac{r^{\ell-1}}{(2\ell-1)\ldots(\ell+2)} \delta_{(i_1 i_2} \cdots \delta_{i_{\ell-2} i_{\ell-1}} x_{i_\ell)}$$

for odd ℓ and the tensors so defined are traceless and are of course just solid harmonics of order ℓ. If in (22a) we insert these polynomials everywhere in place of the products $x_i' \ldots\ldots x_\ell'$ the additional terms introduced will involve expressions of the form $\partial_i \partial_k \cdot \frac{1}{r} \delta_{ik} f(t-\frac{r}{c})$ and each such term can be replaced by $\frac{1}{c^2} \frac{1}{r} \ddot{f}(t-r/c)$. This leads to integrals involving second time derivatives of the original sources and polynomials $x_{i_1}' \ldots\ldots x_{i_{\ell-2}}' r'^2$.

These can be treated in turn just as the original integrals and so forth.

It is fairly easy to verify that the final result of the procedure is the following

$$\begin{aligned} \underset{\sim}{r} \cdot \underset{\sim}{E} &= \frac{1}{r} \int [X_o(\tfrac{r'}{c} \tfrac{\partial}{\partial t})\rho - Y_o(\tfrac{r'}{c} \tfrac{\partial}{\partial t})\eta] dV' \\ &- \partial_i \frac{1}{r} \int x_i' [X_1(\tfrac{r'}{c} \tfrac{\partial}{\partial t})\rho - Y_1(\tfrac{r'}{c} \tfrac{\partial}{\partial t})\eta] dV' \\ &+ \frac{1}{2!} \partial_i \partial_k \frac{1}{r} \int x_{[i}' x_{j]}' [X_2(\tfrac{r'}{c} \tfrac{\partial}{\partial t})\rho - Y_2(\tfrac{r'}{c} \tfrac{\partial}{\partial t})\eta] dV' - + \cdots \end{aligned} \tag{24}$$

where $Y_n(u) = 3\cdot5\ldots(2n+1)\ (-\frac{1}{u}\frac{d}{du})^n\ \frac{rmhu}{u}$ (25)

and $X_n(u) = (1 + n + u\frac{d}{du})\ Y_n(u)$. (26)

These functions can also be simply given in terms of Bessel functions of imaginary argument and half odd integral order. (Not surprisingly; in standard presentation with harmonic time dependence the corresponding Bessel functions of $\frac{r\omega}{c}$ are a well known feature). The first integral in (24) is capable of being greatly simplified with the aid of the identity

$$r^{2n}\ddot{\rho} - 2nc^2r^{2n-2}\eta = -\nabla\cdot(r^{2n}\dot{j}). \tag{27}$$

Using it, the n-th term in the power series for $X_o\rho$ is readily seen to cancel the (n-1)st term in the power series for $-Y_o\eta$ (apart from vanishing surface integrals), thus leaving only the constant term of the first series, i.e. just ρ. In this way the expected spherically symmetric term $\frac{e}{r}$ is reproduced. The 2^ℓ pole terms, on the other hand, cannot be substantially simplified without destroying the desirable feature that the leading term (i.e. the term involving no time derivatives) in the 2^ℓ moment is just $\int x'_{[i_1}\ldots x'_{i_\ell]}\,\rho dV'$.

We define further the irreducible tensors

$$\xi_{[i_1}x_{i_2}\cdots x_{i_\ell]} = \xi_{(i_1}x_{i_2}\cdots x_{i_\ell)} \frac{-r^2}{(2\ell-1)}\delta_{(i_1i_2}\xi_{i_3}x_{i_4}\cdots x_{i_\ell)}$$

$$+ \frac{r^4}{(2\ell-1)(2\ell-3)}\delta_{(i_1i_2}\delta_{i_3i_4}\xi_{i_5}x_{i_6}\cdots x_{i_\ell)} - + \cdots$$

with the final terms

$$\frac{r^{\ell-1}}{(2\ell-1)\ldots(\ell+2)}\delta_{(i_1i_2}\delta_{i_3i_4}\cdots\delta_{i_{\ell-2}i_{\ell-1}}\xi_{i_\ell)}$$

for odd ℓ, and

$$\frac{r^{\ell-2}}{(2\ell-1)\ldots(\ell+3)}\delta_{(i_1i_2}\delta_{i_3i_4}\cdots\delta_{i_{\ell-3}i_{\ell-2}}x_{i_{\ell-1}}\xi_{i_\ell)}$$

for even ℓ. The same procedure as before thus leads to the result

$$(\underset{\sim}{r} \cdot \underset{\sim}{B}) = 2\Big(-\partial_i \frac{1}{r} \int Y_1\left(\frac{r'}{c}\frac{\partial}{\partial t}\right) \xi_i dV'$$
$$+ \frac{1}{2!} \partial_i \partial_k \frac{1}{r} \int Y_2\left(\frac{r'}{c}\frac{\partial}{\partial t}\right) \xi_{[i} x'_{k]} dV' \tag{29}$$
$$- \frac{1}{3!} \partial_i \partial_k \partial_\ell \frac{1}{r} \int Y_3\left(\frac{r'}{c}\frac{\partial}{\partial t}\right) \xi_{[i} x'_k x'_{\ell]} dV' .$$

The comparison of (24) and (29) with (17a,b) now determines all effective 2ℓ moments explicitly. As an illustration we give a table of the first few terms in the expansion of the lowest moments.

Electric 2^ℓ pole moments

ℓ =

1 $$\int x'_i \left[\rho - \frac{1}{2}\eta + \frac{r'^2}{5c^2}\ddot{\rho} - \frac{r'^2}{20c^2}\ddot{\eta} + \frac{3}{280}\frac{r'^4}{c^4}\rho^{(iv)}\right]dV'$$

2 $$\int x'_{[i} x'_{k]} \left[\rho - \frac{1}{3}\eta + \frac{5}{42}\frac{r'^2}{c^2}\ddot{\rho} - \frac{r'^2}{42c^2}\ddot{\eta} + \frac{r'^4}{216c^4}\rho^{(iv)}\right]dV'$$

3 $$\int x'_{[i} x'_k x'_{\ell]} \left[\rho - \frac{1}{4}\eta + \frac{r'^2}{12c^2}\ddot{\rho} - \frac{r'^2}{72c^2}\ddot{\eta} + \frac{r'^4}{352c^4}\rho^{(iv)}\right]dV'$$

Magnetic 2^ℓ pole moments

1 $$\int \left[\xi_i + \frac{r'^2}{10c^2}\ddot{\xi}_i + \frac{r'^4}{280c^4}\xi_i^{(iv)} + \ldots\right]dV'$$

2 $$\int x'_{(i}\left[\frac{2}{3}\xi_{k)} + \frac{r'^2}{21c^2}\ddot{\xi}_{k)} + \frac{r'^4}{756c^4}\xi_{k)}^{(iv)} + \ldots\right]dV'$$

3 $$\int x'_{[i} x'_k \left[\frac{1}{2}\xi_{\ell]} + \frac{r'^2}{36c^2}\ddot{\xi}_{\ell]} + \frac{r'^4}{1584c^4}\xi_{\ell]}^{(iv)} + \ldots\right]dV'$$

References

1) P. R. Wallace, Canad. J. Phys. 29, 393-402 (1951).
C. J. Bouwkamp and H. B. G. Casimir, Physica 20, 539-554 (1951).

2) A. Nisbet, Physica 21, 799-802 (1955); Proc. Roy.Soc. (A) 231, 250-263 (1955).

Received 3/10/69

Ueber eine Ungleichung von E.P. Wigner und M.M. Yanase

by

Res Jost

Eidg. Technische Hochschule, Zürich

1. Einleitung

Wir behandeln im folgenden immer Operatoren über einem endlichdimensionalen (komplexen) Hilbertraum H. Die Verallgemeinerung auf unendlich-dimensionale Hilberträume interessiert uns hier nicht. Fur $P \geq 0$ und $K^* = K$ definieren Wigner und Yanase[1] als schiefe Information von P relativ zu K den Ausdruck

$$I(K,P) = 1/2 \, Sp[K,\sqrt{P}][\sqrt{P},K]. \tag{1}$$

Dabei bedeutet $\sqrt{P}$ immer die positive Quadratwurzel aus P. Offenbar gilt für $\lambda \geq 0$:

$$I(K,\lambda P) = \lambda I(K,P). \tag{2}$$

Viel tiefer liegt der Beweis der folgenden Aussage (A)[1]:

<u>Aussage (A)</u>: $\forall\alpha,\ 0 \leq \alpha \leq 1,\ \forall P_1 \geq 0,\ \forall P_2 \geq 0$

gilt

$$I(K,\alpha P_1 + (1-\alpha)P_2) \leq \alpha I(K,P_1) + (1-\alpha)I(K,P_2).$$

Ich diskutiere in dieser Note den höchst originellen Wigner-Yanase'schen Beweis von (A) und versuche dabei, das mir wesentlich Erscheinende etwas klarer herauszuarbeiten als dies in der ursprünglichen Arbeit vielleicht geschehen ist. Veranlasst sehe ich mich zu dieser Untersuchung durch den Umstand, dass ich mich mit Fritz Baumann vor kurzer Zeit mit Ungleichungen über positive Operatoren befasst habe, die in ihrer Struktur sehr ähnlich zur Aussage (A) sind, die wir aber nur in 2 Dimensionen beweisen konnten[2]. Ausserdem lassen sich nach F. J. Dyson (siehe [1a)]) unendlich viele weitere Aussagen (A^θ) aufstellen, indem man statt (1)

$$I^\theta(K,P) = 1/2\ Sp[K,P^\theta][P^{1-\theta},K], \tag{1'}$$

$0 < \theta \leq 1/2$, setzt. Die Aussagen (A^θ) sind in 2 Dimensionen wieder richtig (ebenso wie der Grenzfall $\theta \to 0$), doch ist mir der allgemeine Beweis nicht gelungen.

Die hier wiedergegebenen Ueberlegungen habe ich im Herbst 1968 am Institute for Advanced Study in Princeton ausgeführt. Ich danke dem Direktor Carl Kaysen für die mir gewährte Gastfreundschaft, meinem Freund Freeman Dyson dafür, dass er mich auf die Ungleichung von Wigner und Yanase aufmerksam gemacht hat und der N.S.F. für ihre finanzielle Unterstützung.

Eine besondere Freude ist es mir, diese Note meinem verehrten Lehrer Gregor Wentzel widmen zu dürfen. Gerne hätte ich ihm eine gehaltvollere Arbeit dediziert aber verschiedene, auch äussere, Umstände haben das verunmöglicht. Mein Dank an Gregor Wentzel soll dadurch nicht vermindert sein.

2. Umformung der Aussage (A):

Setzt man in der Aussage (A) $\alpha = 1/2$ dann erhält man die Aussage ($A_{1/2}$). Es ist wohlbekannt, dass ($A_{1/2}$) und (A) äquivalent sind. ($A_{1/2}$) lässt sich mit (2) zur Aussage (B) umformen.

Aussage (B): $\forall P_1 \geq 0,\ \forall P_2 \geq 0$ gilt

$$I(K,P_1+P_2) \leq I(K,P_1) + I(K,P_2).$$

Lemma 1: (A) ist äquivalent zu (B).

Unter $P > 0$ verstehen wir im folgenden einen Operator, für den $(u,Pu) > 0$ oder $u = 0$ gilt. Die Menge $T = \{P;\ P > 0\}$ ist ein offener, konvexer Kegel. $\overline{T}$ sei seine Abschliessung. Da $I(K,\cdot)$ in $\overline{T}$ stetig ist, genügt es, (B) für $P_1 > 0$, $P_2 > 0$ zu beweisen.

Aussage (B_o): $\forall P_1 > 0, \forall P_2 > 0$ gilt

$$I(K,P_1+P_2) \leq I(K,P_1) + I(K,P_2).$$

Lemma 2: (B) ist äquivalent zu (B_o).

Nun folgt aus (B_o) für $\lambda \geq 0$

$$\int_0^\lambda d\lambda' \frac{d}{d\lambda'} I(K,P_1+\lambda' P_2) \leq \int_0^\lambda d\lambda' \; I(K,P_2)$$

und durch Ableitung nach λ an der Stelle $\lambda = 0$ die

Aussage (C): $\forall P_1 > 0, \forall P_2 > 0$ gilt

$$\frac{d}{d\lambda} I(K,P_1+\lambda P_2)\Big|_{\lambda=0} \leq I(K,P_2).$$

Durch Integration folgt daraus

Lemma 3: (C) ist äquivalent zu (B_o).

Definieren wir für beliebiges A und $P_1 = Q^2$, $Q > 0$, S durch

$$\frac{d}{d\lambda} \sqrt{P_1+\lambda A}\,\Big|_{\lambda=0} = S \tag{3}$$

dann erhalten wir für S die Gleichung

$$QS + SQ = A. \tag{4}$$

da S durch (3) eindeutig bestimmt ist, ergibt sich die erste Aussage von

Satz 1: Falls $Q > 0$ und A beliebig ist, hat die Gleichung

$$QS + SQ = A \tag{4}$$

genau eine Lösung. Aus $A > 0$ folgt $S > 0$.

Bemerkung: Dieser Satz ist im wesentlichen das Lemma 2 in [1b].

Der Beweis für die zweite Aussage folgt etwa aus der Monotonie von $\sqrt{\cdot}$, welche man wie folgt einsehen kann. Auf Grund der Resolventendarstellung von $\sqrt{\cdot}$ ergibt sich leicht als Integraldarstellung für S

$$S = \frac{1}{\pi} \int_0^\infty d\lambda \sqrt{\lambda} (\lambda + P)^{-1} A (\lambda + P)^{-1}, \tag{5}$$

woraus $S > 0$ für $A > 0$.
Substituiert man (1) in die Aussage (C) und verwendet (3) so erhält man die

Aussage (D): $\forall Q > 0,\ \forall R > 0,\ \forall K = K^*$ gilt

$$\mathrm{Sp}\ 2KQKS - \mathrm{Sp}\ KRKR \geq 0, \text{ wobei}$$

$$QS + SQ = R^2 = P_2 \quad \text{ist.}$$

Lemma 4: (D) ist äquivalent zu (C).

Zur Vereinfachung von (D) führen wir schliesslich ein

$$\Psi = Q^{1/2} K Q^{1/2}, \quad s = Q^{-1/2} S Q^{-1/2}, \quad r = Q^{-1/2} R Q^{-1/2} \tag{6}$$

und erhalten die

Aussage (D'): $\forall \Psi = \Psi^*,\ \forall Q > 0,\ \forall r > 0$ gilt

$$\mathrm{Sp}\ 2\Psi s \Psi - \mathrm{Sp}\ \Psi r \Psi r \geq 0, \text{ wobei}$$

$$Qs + sQ = rQr \quad \text{ist.} \tag{7}$$

Lemma 5: (D') ist äquivalent zu (D).
Damit sind wir mit der Umformung von (A) am Ende.

3. Beweis der Aussage (D')

Die selbstadjugierten Operatoren $\{\Psi,\chi,\ldots\}$ über H bilden einen reellen Vektorraum, den wir mit dem (reellen) Skalarprodukt

$$\langle\Psi,\chi\rangle = \mathrm{Sp}\Psi\chi \tag{8}$$

metrisieren können. In dieser Metrik ist T selbstdual.
Aus Satz 1 und aus (7) folgt, dass $S > 0$ ist. Also ist

$$A = \{\Psi;\ \Psi=\Psi^*,\ 2\mathrm{Sp}\Psi s\Psi = 1\} \tag{9}$$

kompakt. Daher nimmt $\mathrm{Sp}\Psi r\Psi r$ auf A sein Maximum an und zwar für ein Ψ, welches der Eigenwertgleichung

$$\lambda(s\Psi + \Psi s) = r\Psi r \tag{10}$$

genügt und zu maximalem Eigenwert λ gehört.
Ausserdem ist

$$\lambda = \sup_{\Psi\varepsilon A} \Psi r\Psi r. \tag{11}$$

Wir werden zeigen, dass $\lambda = 1$ ist. Daraus folgt dann die Richtigkeit von (D').

Wieder nach Satz 1 wird durch die Gleichung

$$s\chi + \chi s = r\Psi r \tag{12}$$

eindeutig eine lineare Transformation

$$\chi = L(\Psi) \tag{13}$$

definiert und (10) schreibt sich jetzt

$$\lambda\Psi = L(\Psi). \tag{14}$$

Die zu L bezüglich (8) adjungierte lineare Transformation L^* erfüllt

$$(L^*)^{-1}(\Psi) = sr^{-1}\Psi r^{-1} + r^{-1}\Psi r^{-1}s. \tag{15}$$

Substituieren wir in (15) für Ψ den Ausdruck rQr, so erhalten wir mit (7)

$$(L^*)^{-1}(rQr) = sQ + Qs = rQr, \tag{16}$$

also ist rQr Eigenfunktion von L^* zum Eigenwert 1. Nun definieren wir die Ebenen

$$E_\alpha = \{\Psi; \langle rQr,\Psi\rangle = \alpha\}. \tag{17}$$

Lemma 6: L lässt jede Ebene E_α invariant. Ausserdem bildet L den Durchschnitt $E_\alpha \cap T = D_\alpha$ in sich ab.*

Beweis:

1. $\langle rQr, L(\Psi)\rangle = \langle L^*(rQr),\Psi\rangle$

   $= \langle rQr,\Psi\rangle = \alpha$

   falls $\Psi \varepsilon E_\alpha$.

2. Falls $\Psi\varepsilon T$ dann $r\Psi r \ \varepsilon \ T$ und nach Satz 1 und (12) auch $\chi = L(\Psi)\varepsilon T$.

Jetzt fassen wir die Geraden des Raumes der selbstadjungierten Operatoren als projektiven Raum auf und führen E_o als unendlich ferne Ebene ein. Da L die Ebene E_o invariant lässt, induziert L eine affine Abbildung in E_1, die ausserdem die Kompakte, konvexe Menge $\overline{D}_1$ in sich transformiert. Daraus folgt nach elementaren Sätzen, dass L in $\overline{D}_1$ einen Fixpunkt hat. Tatsächlich erfüllt Q die Eigenwertgleichung (10) mit $\lambda = 1$, wie aus (7) unmittelbar folgt, und daher ist $Q/\mathrm{Spr}QrQ\varepsilon D_1$ Fixpunkt der affinen Abbildung $E_1 \to E_1$. Nach Lemma 6 ist dieser

---

*Vergleiche für das folgende den Abschnitt 5.6.3 in [3])

Fixpunkt stabil, also ist kein Eigenwert von L dem Betrag nach grösser als 1. Damit haben wir den Beweis von

Satz 2: Die Aussage (D') und damit auch die Aussage (A) sind richtig.

Bemerkung: Weniger anschaulich, aber kürzer, lässt sich Satz 2 nach Wigner und Yanase wie folgt einsehen:
Wir suchen das Minimum von

$$2 Sp\, s\Psi^2 - Sp\Psi r\Psi r \tag{18}$$

auf $\langle\Psi,\Psi\rangle = 1$. Das führt auf ein selbstadjugiertes Eigenwertproblem

$$s\Psi + \Psi s - \Psi s\Psi = \lambda\Psi, \tag{19}$$

wovon $\Psi = Q$ nach (7) eine Lösung zum Eigenwert 0 ist. Könnte (18) nun strikte negativ werden, dann besässe (19) einen kleinsten strikte negativen Eigenwert, und für den zugehörigen Eigenvektor χ wäre $\langle\chi,Q\rangle = 0$. Jetzt suchen wir bei gegebenem Ψ^2 das Maximum von $Sp\Psi r\Psi r$. Wählt man Ψ diagonal mit Diagonalelementen Ψ_i, dann ist

$$Sp\Psi r\Psi r = \sum_{i,k} |r_{ik}|^2 \Psi_i \Psi_k \leq \sum_{i,k} |r_{ik}|^2 |\Psi_i|\, |\Psi_k|. \tag{20}$$

Das Maximum von $Sp\Psi r\Psi r$ und damit das Minimum von (18) wird also für $\Psi_o = \sqrt{\Psi^2} \geq 0$ erreicht. Es gibt also zum kleinsten Eigenwert von (19) einen Eigenvektor $\chi_o \geq 0$ im Widerspruch zu $\langle\chi_o,Q\rangle = 0$.

Literatur:

1a) Eugene P. Wigner und Mutsuo M. Yanase, Proc. Nat. Acad. of Sciences 49, 910 (1963).

1b) Canadian Journal of Mathematics 16, 397 (1964).

2) Res Jost und Fritz Baumann, "Remarks on a Conjecture of Robinson and Ruelle Concerning the Quantum Mechanical Entropy."

3) D. Ruelle, Statistical Mechanics, New York Benjamin, 1969.

Ich danke David Ruelle nicht nur dafür, dass er mir sein ausgezeichnetes Buch vor dem Druck zugänglich gemacht hat, sondern auch für die Liebenswürdigkeit, mit der er immer bereit ist, mit mir zu diskutieren.

Received 2/17/69

Strong Coupling in Static Models

by

Charles J. Goebel
University of Wisconsin

I. Introduction

1. Historical introduction

Yukawa proposed his meson theory of nucleon forces in 1935; within a year the meson was found. Unfortunately, it was the wrong meson (the muon, now no longer called a meson). The right meson (the pion) was not found until 1947, and so there was a long uncomfortable period for field theorists in which they were expected to find a meson theory which yielded strong nucleon-nucleon forces but small meson-nucleon cross sections. The situation was summarzed by Wentzel (1947) in a review article written just before the discovery of the pion.

There were two important early theoretical observations. The one was by Heitler (1940) and Bhabha (1941), who noted, independently, that the cross section for the scattering of a neutral scalar meson on a nucleon would be quite small, because the direct and crossed Born terms (direct channel and crossed channel one nucleon exchange, in modern terminology) would very nearly cancel. For a charged meson the cross section is large, because only one of the Born terms exists. But if, in addition to the nucleons there existed nucleon isobars [i.e. low lying excited states of the nucleon] with charges of +2 and -1 units, then there would be two Born terms which could nearly cancel, as in the neutral meson case.

The other observation was by Heisenberg (1939) who calculated the scattering of a neutral vector meson coupled to the spin of a nucleon (a la magnetic dipole coupling of a photon; this is equivalent to the p-wave coupling of a neutral pseudoscalar meson). He used a classical approximation, and found that the large Born approximation result was greatly suppressed by reaction effects. As he put it, the self field of the nucleon (i.e. the part of the meson field which it carried around) leads to an increased inertia of the spin motion. The scattering cross section was found to be of the order of the size of the nucleon, a,

$$\sigma \sim a^2 k^4/\omega^4 < a^2 \quad . \tag{1.1}$$

Wentzel (1940) initiated the strong coupling theory by showing that the simplest static model, namely a charged scalar meson field coupled to a fixed scatterer, could be solved quantum mechanically in the limit of large coupling constant, i.e., the strong coupling limit. [He remarked that the alternative approach to a solution, the weak coupling expansion, was hopelessly complicated from the start; this was before Feynman diagrams, as well as renormalization.] An important discovery was the existence of isobars in the solution. But they did not result in a particularly small cross section because their excitation energy Δ was not small enough; Δ was of the order of μ/g^2, reducing the Born amplitude from the order of g^2/ω to

$$g^2/\omega + \frac{g^2}{\Delta-\omega} \approx \frac{g^2\Delta}{\omega^2} \sim \frac{\mu}{\omega^2} \tag{1.2}$$

so that the low energy cross section was of the order of μ^{-2}, i.e. 20 mb.

Oppenheimer and Schwinger (1941) treated the strong coupling limit of a p-wave model, namely neutral pseudoscalar, and found that Heisenberg's classical result for the scattering was reproduced. They pointed out that there were isobars in the solution [strangely, these had been overlooked by Heisenberg, who had failed to observe that a large moment of inertia implies low lying rotational states] and that the cross section could be said to have been reduced by the Heitler-Bhabha mechanism: the isobar excitation Δ is of the order of $a\mu^2/f^2$ where a is the "radius of the nucleon" [so that the meson-nucleon coupling is cut off at a momentum $k \simeq a^{-1}$] and f is a dimensionless p-wave coupling constant, and so the Born amplitude is reduced from the order of $f^2/\omega \ k^2/\mu^2$ to

$$\frac{f^2k^2}{\omega\mu^2} \cdot \frac{\Delta}{\omega} = a\, k^2/\omega^2 \ , \qquad (1.3)$$

which is Heisenberg's result. [There is virtually no unitarity correction, at least for $k \ll a^{-1}$, because ak^2/ω^2 is much smaller than the unitary limit $1/k$].

So, a p-wave model was found to give small "meson-nucleon cross sections, $\sigma \lesssim a^2$, independently of the nuclear force strength ($V \approx f^2 e^{-\mu r}/r$). But since the observed "mesons" were charged, it was necessary to treat the p-wave charged, or p-wave symmetric (i-spin 1), models. This was accomplished by Pauli and Dancoff (1942) and by Wentzel (1943), independently. The isobars in the p-wave symmetric model were found to have I=J, and formed the sequence (I,J)=(1/2,1/2) (the nucleon), (3/2,3/2), (5/2,5/2).... The meson-nucleon cross section was found to be of same order as in the neutral model, $\sigma \lesssim a^2$.

As summarized by Wentzel (1947) the only thing lacking was the observation of the low lying nucleon excitations, the isobars. What was found, instead, at that point was the right Yukawa meson, the pion, which was found to have a reasonably large nuclear scattering cross section, just as one would suppose in the first place. But the strong coupling theory did not become irrelevant, because the meson turned out to be pseudoscalar, and hence coupled to the nucleon in p-wave in the static limit; in p-wave models the condition for the validity of the strong coupling theory is that the isobar excitation energy Δ be small compared to the cutoff energy a^{-1}, which could be guessed to be $\approx m_N$; so the isobar need not be very low-lying. In terms of the coupling, the condition $\Delta \ll a^{-1}$ means $f^2 \gg a^2\mu^2 \approx \mu^2/m_N^2 \approx .02$, so that f^2 need not be very large for strong coupling to be valid. In fact, $f^2 \approx .08$ and the isobar turned up as the (3/2,3/2) resonance (now called the Δ) at an excitation of $\Delta \approx 300$ MeV. [The scattering data were first so interpreted by Brueckner (1952); see also Wentzel (1952).]

Since the isobar mass depends on the cutoff a, which is an ad-hoc parameter, the only quantitative thing predicted about the (3/2, 3/2) resonance is its reduced width, which is proportional to f^2. The strong coupling theory has had no further quantitative successes in describing the pion nucleon system: It has no applicability to nuclear forces, because of the largeness of the isobar excitation energy compared to nuclear binding energies. The next nucleon isobar, (I,J) = (5/2,5/2), is predicted at the mass $m_N + 8/3\ \Delta \tilde{\sim} 1700$ MeV; it has not been found. Further, a difficulty which was noticed early [Pauli-Dancoff (1942); see also Houriet (1945) and Miyazawa (1956)] was that the strong coupling theory predicted the neutron and proton magnetic moments to be equal in magnitude, i.e. a pure isovector nucleon moment. But it now appears that this difficulty is removed if one uses SU_3 symmetry [Goebel (1966)], extending the pion to the octet of 0^- mesons, and the nucleon to the octet of 1/2+ baryons. The strong coupling theory continues to describe moderately well the p-wave meson-baryon interaction, the lowest isobar being the decuplet 3/2+ baryon. In fact, compared to the pion-nucleon system, there are now many more quantities which are predicted, for example the f/d ratios of the meson-nucleon coupling, the baryon magnet moment, and the (octet) baryon mass splitting; they are given correctly to within $\tilde{\sim}$ 20%.

It was found very difficult to make a perturbation expansion in g^{-2} around the strong coupling limit. [Kaufman (1954); Nickle and Serber (1960)]. This stimulated the development of other non-perturbative methods of calculation in quantum field theory. Tomonaga (1947) in his "Intermediate Coupling Theory" made a variational determination of an approximate wave function. But in its region of validity, the simplest form of wave function is equivalent to strong coupling theory, and the use of a more accurate wave function leads to a rapid growth of complexity. However, in somewhat the same spirit, a direct numerical approach [Schwartz (1965)] seems possible.

Many non-perturbative approaches have been through integral equation formulations of renormalized perturbation theory [Tamm (1945) - Dancoff (1950), Dyson et al. (1954), Chew (1954)] The simplest approximation [Edwards-Matthews (1957) amounted to calculating exactly the g^2 (Born) and g^4 terms of the meson-nucleon scattering amplitudes, and then assuming the entire series to be a geometric one [this is the primitive version of the Padé method]; this yielded the (3/2,3/2) resonance correctly.

More recently, methods based on S-matrix properties have been developed; in particular the elegant Chew-Low formalism [Chew and Low (1956); Wick (1955)] is the S-matrix formulation of the static model. Here emphasis is laid on the correct form of the S-matrix at low energies, in particular the correct poles, cuts and crossing symmetry; as a result it provides an excellent description of low energy amplitudes.

2. Static models. S-matrix vs Hamiltonian methods

The p-wave meson-baryon interaction appears to be the only part of hadronic physics where the strong coupling theory is applicable; all other interactions appear to violate at least one of the conditions for its validity, which are that the coupling must be strong enough [it must also be attractive in at least one meson-baryon state; examples not satisfying this are the p-wave interaction of a neutral scalar meson, and the El scattering of a photon on a magnetic monopole] and the static model must be applicable. This last requires that a) the baryon (or other scatterer) must be sufficiently massive [this requirement is apparently not very important, at least qualitatively], and b) meson exchange between the meson and the baryon must be negligible, or at least of sufficiently short range.

But aside from any direct applications, static models have the interest of being non-trivial field theories, without pathologies [unless truncated, as in the Lee model, Lee (1954)], and yet much simpler than fully relativistic models. In static models one can almost prove conjectures concerning renormalizable field theory such as the vanishing of renormalization constants, the zero range of convergence of perturbation expansions in renormalized coupling constants, and the absence of non-field theory solutions of the Chew-Low equations [for the last two in the one-meson approximation, see Wilson (1964) and Huang and Mueller (1965), respectively]. One good "handle" on static models lies in their solvability in the two limits weak coupling and strong coupling ["interpolation is better than extrapolation"], by both field theoretical (Hamiltonian) and S-matrix (Chew-Low) methods. In both limits, the solvability is due to the same thing, the absence of meson production. It might also be remarked that both limits are non-uniform with respect to energy: For coupling arbitrarily small (resp. large) but fixed, the scattering amplitudes $f(\omega)$ in high energy limit $\omega \to \infty$ are not correctly given by the weak (resp. strong) coupling approximation. [As a consequence, strong coupling solutions do not satisfy Levinson's Theorem or the static model sum rules: Chew-Low (1956), Miyazawa (1956), and see Eq. A.14 below.]

Presumably the correspondence between solutions of the Chew-Low equations and amplitudes of a Hamiltonian lies in their freely variable parameters. For instance the Hamiltonian, and hence the amplitudes, of the π-N static model has one free parameter, namely the π-N-N coupling constant [we choose not to vary $a\mu$]. Conversely then, we expect that solutions of the Chew-Low equations of the ($SU_2 \times SU_2$, $3 \cdot 3$) p-wave model which have one free parameter are just the amplitudes of the π-N Hamiltonian; solutions with more free parameters must be the amplitudes of Hamiltonians with more elementary particles. Such solutions can always be obtained by "turning on" new poles [say by the mechanism of CDD poles] in some sheet or other of ω-space. [The example of K. Wilson's solutions of the charge-symmetric scalar model, (SU_2,3) in the one-meson approximation shows that a free parameter can easily be overlooked, even in an explicitly given solution.] On the other hand, a solution of a model with <u>no</u> free parameters would be a non-Hamiltonian "true bootstrap" solution. No such solution is known; as remarked above it has been proven impossible in the one-meson approximation [Huang-Mueller (1965)], and also in the strong coupling approximation, if one excepts the trivial solution in which all isobars are degenerate and all scattering amplitudes vanish.

If it is true that the two classes of amplitudes, Hamiltonian solutions and Chew-Low solutions, are identical, it then becomes a matter of taste or convenience whether to investigate one or the other. Two points might be noted: If one wishes to vary the free parameters of a solution over a wide range, the Hamiltonian will relate the weak and strong coupling limits of the solution. We cannot continue Chew-Low solutions between these limits because we cannot solve the Chew-Low equations for intermediate couplings; but it is trivial to continue the Hamiltonian in the coupling, and we can solve it in both limits. On the other hand, if one is using a static model in a phenomenological manner, to approximate experiment, there is no reason to start with a Hamiltonian.

By avoiding a Hamiltonian, one avoids introducing bare coupling constants and masses (or even a cutoff, in the renormalizable s-wave models). These bare quantities are presumably unobservable and irrelevant in the limit of no cutoff, although there is no known systematic demonstration of this, even in the strong coupling limit, unlike renormalized perturbation theory in weak coupling. However, the following is known about renormalization in the strong coupling limit.

As long as there is a cutoff, $a > 0$, the ratio of bare to physical couplings, $Z^{-1} = g_o/g$, is finite in the strong coupling limit $g \to \infty$. This is because transitions to isobars dominate over transitions to scattering states in the sum rule Eq. (4.13) which describes the distribution of the coupling strength over physical states; the result [Eq. (A.27)] is of the form

$$g_o^2 = c_1 g^2 + c_2 \ln(c_3/a\mu), \quad c_i = O(1),$$

so

$$g_o/g \to \sqrt{c_1} \quad \text{as} \quad g \to \infty \; .$$

Hence isobar excitation energies Δ, for example, which are $\sim g_o^{-2}$ in the strong coupling limit, are finitely expressed in terms of g as well, i.e. $\sim g^{-2}$, in leading order.

But in the limit of no cutoff, $a \to 0$, $Z \to 0$. So if g is held fixed, $g_o \to \infty$; but g_o is irrelevant to physical quantities and in particular Δ remains $\sim g^{-2}$, as made clear by the Chew-Low solution. On the other hand, if g_o is held fixed, g becomes small as $a \to 0$, and so strong coupling doesn't hold. A simple exhibition of this is the following: If Δ has an expansion in terms of the physical coupling, $\Delta = \frac{A}{g^2} + \frac{B}{g^4} + \ldots$, this expansion in terms of g_o

$$\frac{c_1 A}{g_o^2} + \frac{B + A\, c_2 \ln(c_3/a\mu)}{g_o^4} + \ldots$$

blows up as $a \to 0$. This is just what was found by Pais and Serber (1957) in the charged scalar model; Nickle and Serber (1960) carried out the calculation of the corrections to both Δ and g of relative order g_o^{-2}, and found that the coefficient of g^{-4} in the expansion of Δ was indeed finite and so independent of cutoff in the limit of no cutoff. As remarked above, no systematic expansion scheme in powers of g_o^{-2} is known; a Chew-Low expansion scheme, directly in powers of g^{-2}, might be possible. An interesting question is whether the expansion in g^{-2} is only semi-convergent, as seems to be the case for the weak coupling expansion in g^2 [Wilson (1964)].

3. The strong coupling limit of static models and its algebra

In the strong coupling limit, the scatterer has an infinite number of excited states, isobars; this is easily demonstrated by S-matrix methods (see Section IV). Their excitation energies Δ are of order g^{-2}. They form either a rotational or a vibrational band, or in complex cases, both. In the rotational case, the strong coupling limit is $g \to \infty$, and hence $\Delta \to 0$; in the vibrational case the strong coupling limit can be reached for g finite and Δ non-vanishing. The scattering amplitudes are directly determined by the isobars, because the scattering is purely "orthogonality scattering", i.e. due to the orthogonality of the propagating part of the meson field (frequencies $> \mu$) to the part which is bound to the scatterer [see Appendix E]. There is no meson production.

The bound part of the meson field clothes the bare scatterer to make the physical scatterer, which is a system with a small number of degrees of freedom ($\leq$ the number of components of the meson wave which is coupled to the scatterer); it either rotates or vibrates, or in complex cases both, and the isobars are the energy levels of these motions.

From the S-matrix (phenomenological) point of view, the isobars are characterized by the observable quantities, masses M_s and couplings, g_{s,A_s} which can be considered as matrix elements between isobar states of operators, M and $\underset{\sim}{g}_A$:

$$< s'|M|s> = M_s \delta_{s's} \quad , < s'|\underset{\sim}{g}_A|s> = g_{s',A_s} \tag{1.4}$$

As can be shown either from field theory or Chew-Low theory, these operators satisfy the fundamental equations [Eqs. (II.51), (II.52), (III.46), (III.53), (IV.24), (IV.26) or (IV.34).]

$$[\underset{\sim}{g}_A, \underset{\sim}{g}_B] = 0 \tag{1.5}$$

$$\Lambda_{CB}\, \Lambda_{BA} = \Lambda_{CA} \; , \tag{1.6}$$

where $\Lambda_{BA}(g) = [\underset{\sim}{g}_B, [M/m, \underset{\sim}{g}_A]]$.

m is a constant giving the mass scale. The reduced scattering amplitude is given by

$$f_{BA}(\omega) = \Lambda_{BA}\, f(\omega) \tag{1.7}$$

where $f(\omega)$ is a universal function; see Eqs. (II.38), (IV.27), (IV.35). From the S-matrix point of view these are "bootstrap" equations involving the observables M_s and $g_{s',As}$; the first was remarked by Chew (1962). From the operator point of view, the first means that $\underset{\sim}{g}_A$ are coordinates (i.e.,of the physical scatterer); the second means that M is a "non-relativistic particle Hamiltonian" quadratic in momentum:

$$M = m[\, \tfrac{1}{2}\, p_B\, \Lambda_{BA}(\underset{\sim}{g}) p_A + W_A(\underset{\sim}{g})\, p_A + V(\underset{\sim}{g}) \,] \tag{1.8}$$

where p_A is the canonical conjugate to g_A,

$$[g_A, p_B] = i\ \delta_{AB}$$

Further conditions (see Section IV) require that to a good approximation Λ_{BA} be constant, W_A at most linear, and V at most quadratic in g. The finding of the masses M_s and couplings $g_{s',As}$ is thus the fairly trivial problem of solving the single particle Hamiltonian (8).

Internal symmetry is a very important property of physically interesting models. For our purposes, the internal symmetry of a model is specified by the group C of transformations which leave the Hamiltonian invariant, and by the way the meson field transforms under C. For instance, the charge symmetric scalar model (i.e. the meson has $I = 1$ and $J^P = 0^+$, coupled in s-wave) is described as $(SU_2,3)$, the i-spin group being SU_2, and the meson transforming as a triplet under it. The phenomenological operators g_A and M transform like Φ_A (the coupled part of the field) and H respectively. Under the continuous (Lie) part of the symmetry group C, the static current g_A transforms as

$$[J_i, g_A] = g_B\ B_{BiA} \tag{1.9}$$

where the B_{BiA} are constants; the generators J_i satisfy

$$[J_i, J_j] = C_{ij}^{\ \ k}\ J_k \tag{1.10}$$

[The B_{BiA} are a representation of the J_i, of course, and the J_i can be written in terms of g_A and p_B, $J_i = i g_B B_{BiA} p_A$.] The mass operator M, like the Hamiltonian, is a scalar under C,

$$[J_i, M] = 0 \tag{1.11}$$

which means that the isobars form degenerate multiplets, transforming as irreducible representations of the symmetry group C.

In the simplest models the scatterer is a rigid rotator, which means that $M(P, \underset{\sim}{g})$ depends on the momenta P_A only through the generators J_i, $M = M(J, \underset{\sim}{g})$; the scalars of $\underset{\sim}{g}_A$ commute with M and are thus constants [this is the rigidity of the rotator]. Eqs. (5), (9) and (10) define a Lie algebra; the corresponding group G is a "rotation-translation" group, the J_i and $\underset{\sim}{g}_A$ generating rotations and translations respectively. Thus the problem of finding solutions of these equations, i.e. explicit matrices, is that of finding (matrix) representations of a Lie algebra, or loosely speaking of the group G. The rigidity means that these solutions are <u>irreducible</u> representations of G. The finding of irreducible representations of G is a standard group representation theory problem; the matrices can be given explicitly in terms of Clebsch-Gordan coefficients of C [Goebel (1966)]. The remaining step in finding a solution is to find mass operators $M(J, \underset{\sim}{g})$ which satisfy (6) and (11). These imply that M must be scalar and quadratic in J_i, and so is of the form

$$M = \frac{1}{2} A_{ji}(\underset{\sim}{g})\, J_j J_i + B_i(\underset{\sim}{g})\, J_i + C(\underset{\sim}{g}) \,. \qquad (1.12)$$

where A_{ji}, B_i and C transform appropriately under rotations to make M scalar; there are only a limited number of such forms. The constraint (6) is then to be applied.

Whether one has started with a Hamiltonian or not, these purely algebraic methods involve considerably less effort than writing down and solving rotator wave equations, especially in complex models.

The free parameters of a solution are the labels of the irreducible representations, i.e. a complete independent set of invariants formed from $\underset{\sim}{g}_A$ and J_i, plus the additional free parameters (if any) which occur in M. If desired, these free parameters can be related to the free parameters of Hamiltonians of the model. For instance in the model $(SU_2, 3)$, J_i and $\underset{\sim}{g}_A$ are both 3-vectors; there are two independent invariants, say $\vec{\underset{\sim}{g}}^2 \equiv g^2$ and $\vec{\underset{\sim}{g}} \cdot \vec{J} \equiv gJ_0$. J_0 plays the role of the minimum spin occurring in the representation, which has the rotational band of states $J = J_0, J_0 + 1, \ldots$. These parameters g and J_0 correspond to the coupling constant g_0 and bare isobar spin J_0, respectively, of a Hamiltonian in which the meson field $\vec{\Phi}$ has the coupling $-g_0 \vec{S}\cdot\int d^3r\, u(r)\vec{\Phi}(r)$ where $\vec{S}$ are $(2J_0+1)\times(2J_0+1)$ spin matrices. M has the form

$$M = \frac{a}{2}\vec{J}^2 + b \tag{1.13}$$

(any term $\vec{g}\cdot\vec{J}$ is lumped into b, because $\vec{g}\cdot\vec{J}$ is an invariant); a is fixed by the condition (6), giving

$$M = \frac{\vec{J}^2}{2g^2}\, m + b, \quad \overleftrightarrow{\Lambda} = 1 - \overleftrightarrow{gg}\ . \tag{1.14}$$

(Note that the form of $\vec{\Lambda}$ did not need to be known a priori). So M contains no additional free parameters, aside from the trivial additive constant b, which is always irrelevant in static models.

It is possible for the strong coupling solution to have more symmetry than the Hamiltonian. For instance, the isobars of the ($SU_2 \times SU_2$, 3.3) Hamiltonian with $\pi\Lambda\Sigma$ coupling, but no $\pi\Sigma\Sigma$ coupling, have the symmetry $SU_2 \times O_9$ [Wentzel (1962, 1963)]. Again, the isobars of the (SU_3, 8) Hamiltonian in which the meson is coupled to an octet scatterer with pure d coupling have the symmetry O_8 [Wentzel (1965)]. Similarly, it is possible to find an (SU_2, 3) Hamiltonian whose isobars have the symmetry SU_3 (3-dimensional harmonic oscillator). These are cases in which the form of the Hamiltonian does not constrain all the scalars of g_A under C to be fixed, in the low-lying states. Only certain functions of the (original) scalars of g_A are fixed; these are the scalars under the higher symmetry. The representations are generally reducible under the original symmetry. The phenomenon of higher symmetry can generally occur whenever g_A transforms as a representation not only of the original symmetry groups but also of a larger group.

The invariants, whose values label the irreducible representations of (5), (9) and (10), fall into two classes: Those formed from g_A alone, i.e. the scalars of g_A, and those formed from both g_A and

J_i. These last invariants are functions of the generators of the "little group" [L.G.] (Wigner's terminology), i.e. the subgroup of the symmetry group C which leaves g_A invariant, and they label irreducible representations of the L.G. [This statement, that the irreducible representations of the strong coupling groups are completely labeled by the scalars of g_A and the irreducible representations of the L.G., is the content of a theorem of Mackey; see Cook-Sakita (1967).] For instance, in (SU_2, 3) the L.G. is the $U_1[=0_2]$ group of rotations with $\vec{g}$ as axis, generated by $\vec{g}\cdot\vec{J}$, the body axis spin projection, which thus is a representation label (= is constant within each representation = is a constant of the motion).

In physical terms, g_A can be taken as the coordinates of a body; the L.G. is then the intrinsic symmetry group of the body, the "body symmetry group". It is important to realize that the L.G. is not completely fixed by the model, i.e. by the given coefficients $C_{ij}^{\ k}$ and B_{BiA} of Eqs. (9) and (10); not all representations of a model have the same L.G., in general. The L.G. will be larger when the scalars of g_A are particularly related; in physical terms, the body symmetry will be larger for particular shapes of the body. [Consider (SU_2,5) where g_A can be written as a symmetric traceless 3x3 tensor $\overleftrightarrow{g}$. In general, the L.G. is here the Viergruppe of 180° rotations around the principal axes of $\overleftrightarrow{g}$; but if two of its eigenvalues are equal, $\overleftrightarrow{g}$ has axial symmetry and so its L.G. contains all rotations around the axis of symmetry.] The consequence of a higher body symmetry is that the representation has fewer states (isobars), because more operators are invariants.

We elaborate on this. The two classes of invariants (constants of the motion) discussed above, have the interpretation of the intrinsic shape of the body and the intrinsic spins of the body, respectively. The rotational states of the body can be labeled by total angular momentum j, magnetic quantum number m in a laboratory (space fixed) frame, and magnetic quantum number m_o in an intrinsic (body fixed) frame. [Speaking generally, j,m, and m_o are each a set of numbers. In group theory terms, j labels irreducible representations of the symmetry group, and m (likewise m_o) labels states of a representation.] Generally, by choosing the definition of m_o and/or the orientation of the intrinsic frame relative to the body, the set of numbers m_o will contain the invariants constructed from the L.G.; <u>these</u> components of m_o are constants of a rotational band. This puts a restriction on the $|j,m,m_o>$ states which occur in the band (=irreducible representation.) The higher the body symmetry, the greater will be the number of invariants contained in the m_o, and hence the fewer the number of $|j,m>$ states in the band.

For example, in 3-dimensional space, where j,m, and m_o are each single numbers: a) An arbitrary body (e.g. polyatomic molecule) has no L.G. [aside from T^2, the square of time reversal] so it has rotational states with all values of j [more precisely, all integral (half-integral) values if $T^2 = +1(-1)$], each occurring 2j+1 times, being labeled by m_o, the projection of $\vec{J}$ on an arbitrary body axis. b) A rod-like body (e.g. diatomic molecule) has an $O_2[\sim U_1]$ L.G.; choosing m_o to be the projection of J on the rod, m_o is constant, hence each j state, $j = |m_o|, |m_o| + 1, \ldots$occurs once. c) A point-like body (e.g. monatomic molecule) has an $O_3[\sim SU_2]$ L.G., so j is fixed and m_o is irrelevant.

In the model (SU_3, 8), m, and likewise m_o, can be taken to be the set $(Y; I_3, I)$; $\underset{\sim}{g}_A$ is an "octet vector" $\underset{\sim}{g}_a$, a=1,2,...,8. We may choose the intrinsic frame so that in it only the components g_3 and g_8 are nonvanishing; their values g_3 and g_8 are scalars (hence invariants of the strong coupling group). The L.G. is $U_1 \times U_1$ with generators I_3 and Y in the intrinsic frame, and so the components I_{30} and Y_o of m_o are constant. The remaining component I_o is not constant, and so is a label of SU_3 multiplets in a band, labeled by g_3, g_8, Y_o and I_{30}. Of course the values of I_o are restricted by the requirement that (Y_o, I_{30}, I_o) must be a possible magnetic state of the multiplet [just like "$|m_o| < j$" in the O_3 example above]. For instance, a band with Y_o, $I_{30} = 0, 0$ has multiplets $1, 8_0, 8_1, 10, \overline{10}, 27_0, 27_1, 27_2, \ldots$where the subscript is I_o; a band with $Y_o, I_{30} = +1, +1/2$ has states $8, 10, \overline{10}, 27_1, 27_2, \ldots$ But if there is an intrinsic frame in which only g_8 is nonvanishing, the L.G. generators are $\vec{I}$ and

Y in this frame, so that all the components of m_o are constant (and I_{30} is irrelevant). Thus, for instance the band labeled by Y_o, I_o = +1, 1/2 has the states 8,10,27,..., with no duplication of multiplets.

The "physical" parameters of a strong coupling solution of a Hamiltonian [the invariants of the strong-coupling group representation to which the solution belongs plus any additional parameters which occur in the mass operator] are functions of the bare parameters of the Hamiltonian; they may be discontinuous functions. This was discovered by Wentzel (1962) in the (SU_2,3) model in which the scatterer has integral i-spin (iso-vector scalar meson "δ" coupled to "Λ" and "Σ"). Taking the Hamiltonian to have two coupling terms representing δΛΣ and δΣΣ vertices, with coefficients (bare coupling constants) g_o^Λ, g_o^Σ, respectively, he found that in the strong coupling solution one of the corresponding physical coupling constants g^Λ,g^Σ vanishes [stated more fully, g^Σ/g^Λ is a monotonic function of g_o^Σ/g_o^Λ, jumping from 0 to ∞ at a critical value of the latter]. This result follows simply from consideration of the representations of (SU_2,3), which as mentioned above [paragraph of Eq. (13)] are characterized by one coupling constant and by I_o, the minimum i-spin. Given that the Hamiltonian describes elementary states of the scatterer with I=0 ("Λ") and I=1 ("Σ"), there are two possibilities for the solution, I_o=0 or I_o= 1. In the former case the δΣΣ coupling vanishes [a consequence of a higher body symmetry, the rotation of g to -g in the body frame; all diagonal couplings vanish, and the selection rule can be expressed as the conservation of a parity, $(-)^I$ for the isobars and -1 for the mesons] and in the latter case the πΛΣ coupling vanishes, since the Λ does not belong to the representation. It should be added that when g_o^Σ/g_o^Λ has precisely the critical value, the Hamiltonian has doublet symmetry [see Wentzel (1962), *loc. cit.*] and this symmetry is obeyed by the solution, which then belongs to a reducible representation.

Another example, investigated by Wentzel's student Ramachandran (1965) is given by the (SU_2 x SU_2, 9+3) model in which p-wave π and η mesons are coupled to a "nucleon"; there are two bare couplings g_o^π and g_o^η. It was found that if g_o^η/g_o^π is less than a critical value, g^η vanishes; if g_o^η/g_o^π is greater, the ratio g^η/g^π is directly proportional to it. In terms of representations, in the first case the body symmetry is the large O_3 one (see below, the paragraph of Eq. (15), also in Section II, Eqs. (II.67)-(II.74)) in which $\langle\sigma_i\rangle=0$, so that g^η vanishes. In the second case the body symmetry is $U_1 \times U_1$, and the forms of the π and η currents are $g_{\alpha i}^{(\pi)} = g^\pi m_\alpha n_i$, $g_i^{(\eta)} = g^\eta n_i$, where m_α and n_i are unit vectors. [The mass formula M is derived in Appendix C, Eqs. (C.10) et seq. for this case.] The reason for discontinuity of the solution is that in a pure π case, the static energy for a given g_o is much lower for the first band $\{O_3\}$ (negative self-energy greater by a factor of three), so that it is stable with respect to the coupling of the η, up to the critical value of g_o^η/g_o^π. Beyond this value the static energy of the second band $\{U_1 \times U_1\}$ is lower, the η contributing more to the (negative) self energy than was lost by the change of the manner of coupling the π.

The physically interesting representations of the models with known physical relevance, $(SU_2 \times SU_2, 3\cdot 3)$ and $(SU_2 \times SU_3, 3\cdot 8)$, have high body symmetry, namely O_3 and $U_1 \times O_3$ respectively. In the former model, this occurs when the eigenvalues g_r of $\mathfrak{g}_A = \mathfrak{g}_{\alpha i}$ (a 3x3 matrix: α = i-spin index = 1,2,3; i=spin index= 1,2,3) are equal, so that $\mathfrak{g}_{\alpha i}$ is of the form

$$\mathfrak{g}_{\alpha i} = \sum_r A_{\alpha r} B_{ir} g_r = g \sum_r A_{\alpha r} B_{ir} = g\, O_{\alpha i} \qquad \text{(I.15)}$$

where $A_{\alpha r}$, B_{ir}, and therefore $O_{\alpha i}$ are orthogonal matrices. Consequently, $\mathfrak{g}_{\alpha i}$ commuted with $Y_r = I_\alpha A_{\alpha r} + J_i B_{ir}$, which are the generators of an O_3 L.G. An intrinsic frame can be defined by transformation with $A_{\alpha r}$ and B_{ir} on i-spin and spin indices respectively; in this frame I_α, J_i and $g_{\alpha i}$ become respectively $A_{\alpha r} I_\alpha \equiv I_r$, $B_{ir} J_i \equiv J_r$ and $g_r \delta_{rr'}$, $= g\delta_{rr'}$ in the case L.G.$=O_3$. Since Y is an invariant, its length y labels bands (and y_3 is irrelevant); a representation labeled by g,y contains one each of states i,j which satisfy

$|i-j| < y < i+j$.

Since the lowest π-N p-wave resonance has $(i,j)=(3/2,3/2)$, it and the nucleon should belong to a y=0 representation. A consequence is that the matrix elements between isobars of this band of any current operator are unique up to an overall factor [Goebel (1966 a,b)] because the spin and i-spin of the operator must add up to 0 in the body frame in order to couple to the y=0 of both the initial and final isobars. In particular, all matrix elements vanish if the spin and i-spin of the operator are unequal, as for example the η current discussed above (Ramachandran's model) and the isoscalar magnetic moment, both with J=1, I=0.

Similarly, in $(SU_2 \times SU_3, 3\cdot 8)$, when the body symmetry is $U_1 \times O_3$ there is a frame in which $\underset{\sim}{g}_{ai}$ (a=1,2,...,8; i=1,2,3) has the form $g\,\delta_{ai}$, and $y = |Y| = |I + J|$ and Y (hypercharge) in this frame are invariants. Since the lowest $(8,0^-)$-$(8,1/2^+)$ p-wave resonance is the $\Delta(10,3/2^+)$, it and the $(8, 1/2^+)$"ground state" baryon should belong to a $Y_o=1$, y=0 representation. Again, current matrix elements in this band are unique [Goebel (1966a,b)] and in particular, the f/d ratios of octet operators are fixed.

The solutions of the simplest field theory Hamiltonians of these last three models are as follows. The simplest $(SU_3,8)$ Hamiltonian has two bare coupling constants f_o and d_o which are the coefficients of f and d type couplings, respectively, of the meson to an octet bare scatterer. The scalars of $\underset{\sim}{g}_a$, e.g. g_3 and g_8 [or equivalently, the physical couplings f and d] are smooth functions of f_o and d_o, and the L.G. of the solution generally is $U_1 \times U_1$. The special kind of solution with the larger L.G. $SU_2 \times U_1$ occurs for a special value of f_o/d_o. When $f_o = 0$, the solution has the higher symmetry $(O_8,8)$ with an O_7 L.G.

The simplest $(SU_2 \times SU_2, 3\cdot 3)$ "π-N" Hamiltonian has just one bare coupling constant; the L.G. of the solution is O_3, and y=0 so the isobars have i=j [One cannot argue that if $\underset{\sim}{g}_{r'r} = g_r\,\delta_{r'r}$ depended on only one continuously variable parameter it had to be of the form $g\,\delta_{r'r}$; it could have been of the form $g\,\delta_{r'r}\,\delta_{r3}$, for instance.]

The simplest ($SU_2 \times SU_3$, 3·8) Hamiltonian has two bare coupling constants, f_o and d_o, the coefficients of the f and d type couplings of the octet 0^- meson to an octet $1/2^+$ scatterer. The solution has the large $U_1 \times O_3$ L.G. over the entire range of the ratio f_o/d_o [Dullemond (1967)]. The representation has $(Y_o, y) = (+1,0)$, $(0, 1/2)$ or $(-1,0)$ depending on the value of f_o/d_o; the representation is a discontinuous function of f_o/d_o.

It is perhaps significant that in both models ($SU_2 \times SU_2$, 3·3) and ($SU_2 \times SU_3$, 3·8) the simplest Hamiltonian suffices to fit experiment; this is especially striking in the latter case where the simplest Hamiltonian allows the physical f-d ratio of couplings to have only three discrete values, one of which agrees well with experiment [Goebel (1966a)]. From the S-matrix point of view, the same result follows from a minimality principle, namely to choose the representation with the least number of isobars; this singles out the $(Y_o, y) = (\pm 1, 0)$ representations.

II. The Classical Method in Strong Coupling

We present here in general terms the classical calculation of meson scattering in the strong coupling limit, done originally by Heisenberg (1939) [See also Pauli (1946)]. Because of the identity of the equations of motion of classical quantities and of the corresponding quantum mechanical operators, this can be described as the equations of motion method. Its usefulness is based on the linearity of the equation of motion of the scattered wave, in the strong coupling limit. Its solution is thus trivial, although left in terms of static ("bound") operators whose form and matrix elements can be found only by other methods.

The Hamiltonian of a meson field, Yukawa coupled to a fixed scatterer is of the form

$$H = H_\phi - g_{a\alpha} \bar{\Phi}_{a\alpha} \qquad (II.1)$$

where H_ϕ is the free meson Hamiltonian

$$H_\phi = \int d^3r\left(\frac{1}{2}\pi_\alpha^2 + \frac{1}{2}\Phi_\alpha\,\Omega^2\Phi_\alpha\right), \quad \Omega^2 = \mu^2 - \nabla^2 . \tag{11.2}$$

For simplicity we have assumed the multiplet of mesons to all have the same mass, μ. $\mathcal{J}_{a\alpha}$, the current, is a spinor matrix in the space spanned by the "elementary" states of the scatterer; for simplicity we take them to be degenerate in energy. Φ_α is likewise a spinor matrix in the Heisenberg picture, which we shall use. $\overline{\Phi}_a$ is an "average" of the meson field over the scatterer,

$$\overline{\Phi}_{a\alpha} = \int d^3r\ u_a\,\Phi_\alpha \tag{11.3}$$

where $u_a(r)$ is a multiple gradient of a fixed "cutoff function", $u(r)$; this couples particular orbital angular momentum parts of the meson field.

The equation of motion of the field is readily solved,

$$\Phi_\alpha = \Phi_\alpha^{in} + \frac{1}{\Omega^2 + (\partial_t - \varepsilon)^2}\,u_a\,\mathcal{J}_{a\alpha} \qquad \varepsilon = 0+ , \tag{11.4}$$

where Φ_α^{in} is a solution of the free wave equation,

$$(\Omega^2 + \partial_t^2)\,\Phi_\alpha^{in} = 0 ; \tag{11.5}$$

Φ_α^{in} is a scalar in the spinor space. For later use, we note that multiplying by u_a and integrating over space gives

$$\Phi_{a\alpha} = \Phi_a^{in} + I_{ab}(i\partial_t)\,\mathcal{J}_{b\alpha} \tag{11.6}$$

where

$$I_{ab}(\omega) = \int d^3r\ u_a\,\frac{1}{\Omega^2 - (\omega - i\varepsilon)^2}\,u_b . \tag{11.7}$$

Generally, this has the form $I_{ab}(\omega) = I(\omega)\,\delta_{ab}$, so Eq. (11.6) simplifies to

$$\Phi_A = \Phi_A^{in} + I(i\partial_t)\,\mathcal{J}_A \tag{11.8}$$

where we have replaced the two indices $a\alpha$ by one index A.

The non-trivial problem is to find the current; its equation of motion is

$$[H, \mathcal{J}_A] = [\mathcal{J}_A, \mathcal{J}_B]\, \Phi_B \,. \tag{11.9}$$

We now separate the different frequency parts of the operators (the frequency ω part has matrix elements only between states differing in energy by ω) and start drawing consequences from the assumed strongness of the coupling. The crucial point is that the low frequency part of $\mathcal{J}_A$ is large, because matrix elements of $\mathcal{J}_A$ between states with the same meson content are coupling constants; but the higher frequency parts of $\mathcal{J}_A$, whose matrix elements are scattering amplitudes, are of limited size because of unitarity. [The distinction between low and high frequency parts, in the case of a s-wave coupled meson, is simply whether $\omega < \mu$ or $\omega > \mu$; but in the case of higher waves the distinction is blurred. The centrifugal barrier makes metastable states (resonances), so that there are parts of $\mathcal{J}_A$ with frequencies $> \mu$ which are coupling constants. But at the same time, $I(\omega)$ is only weakly dependent on ω, precisely because of the centrifugal barrier, so that it is proper to include these in the "low frequency" part of $\mathcal{J}_A$.] If we assume that the low "bound" frequencies are small enough to be neglected against μ [this is valid in the strong coupling limit], the large part of Φ_α is the low frequency, "static" part $\Phi_\alpha^{\approx o}$,

$$\Phi_\alpha^{\approx o}(r) \approx \frac{1}{\Omega^2}\, u_a(r)\, \mathcal{J}_{a\alpha}^{\approx o} \,. \tag{11.10}$$

$\Phi_\alpha^{\approx o}$ can be taken to be a scalar constant (c-number), ϕ_α^o, since it commutes with H and its components commute with one another (this last, because $\Phi_\alpha^{\approx o}$ is the large part of Φ_α). From Eq. (10), $\mathcal{J}_A^{\approx o}$ is also a c-number, call it g_A; so Eq. (10), and correspondingly the static part of Eq. (8), read

$$\phi_\alpha^o = \frac{1}{\Omega^2}\, u_a\, g_{a\alpha} \,, \quad \bar{\phi}_A^o = I(0)\, g_A \,. \tag{11.11}$$

The static part of Eq. (II.9) implies

$$0 = [\mathcal{g}_A, \mathcal{g}_B]^o g_B \ . \tag{II.12}$$

We now consider the nonstatic parts. Eq. (II.8) gives

$$\bar{\Phi}^{\omega}_A = \bar{\Phi}_A^{in,\omega} + I(\omega)\ \mathcal{g}^{\omega}_A \ . \tag{II.13}$$

Eq. (II.9) linearizes in the frequency ω parts when we use the principle that static parts are the largest, giving

$$\omega\ \mathcal{g}^{\omega}_A = [\mathcal{g}_A, \mathcal{g}_B]^o\ \bar{\Phi}^{\omega}_B + [\mathcal{g}_A, \mathcal{g}_B]^{\omega}\ \phi^o_B \tag{II.14}$$

$$= [\mathcal{g}_A, \mathcal{g}_B]^o\ \bar{\Phi}_B^{in,\omega} + I(\omega)\ [\mathcal{g}_A, \mathcal{g}_B]^o\ \mathcal{g}^{\omega}_B + I(0)[\mathcal{g}_A, \mathcal{g}_B]^{\omega}\ g_B \ . \tag{II.15}$$

This linearization is the reason for the solubility of the strong coupling limit, and also implies the vanishing of meson production [Φ^{in} has matrix elements only between states differing by one meson].

Equation (II.15) for $\mathcal{g}^{\omega}_A$ has the complication that it also involves $[\mathcal{g}_A, \mathcal{g}_B]^{\omega}$. In general, $[\mathcal{g}_A, \mathcal{g}_B]$ can be expanded in a complete set of Hermitian nxn "spinor" matrices, some of which are the currents $\mathcal{g}_A$ themselves. Let the currents $\mathcal{g}_A$ be expressed in terms of a complete set of spinor matrices σ_A,

$$\mathcal{g}_A = g_A\ \sigma_A \quad \text{(no sum)} \tag{II.16}$$

where the σ_A have the commutation relation

$$[\sigma_A, \sigma_B] = C_{ABC}\ \sigma_C \ , \quad C_{ABC} \text{ totally antisymmetric} \tag{II.17}$$

[The σ_A are generators of U_N.]

The equation of motion for σ_A

$$[H, \sigma_A] = [\sigma_A, \mathcal{g}_B]\ \bar{\Phi}_B \tag{II.18}$$

has the static and nonstatic parts

$$0 = \sum_{B,C} C_{ABC}\, \sigma_C^{o}\, g_B^{2}\, \sigma_B^{o} \tag{II.19}$$

$$\omega\sigma_A^{\omega} = \sum_{B,C} C_{ABC}\, \sigma_C^{o} [g_B \bar{\Phi}_B^{in,\omega} + \{g_B^2\, I(\omega) - g_C^2\, I(0)\}\, \sigma_B^{\omega}\,] \tag{II.20}$$

respectively [when multiplied by g_A, these equations become (12) and (15) respectively, but they hold even when $g_A = 0$.] Defining

$$\sum_C C_{ABC}\, \sigma_C^{o} \equiv P_{AB} \tag{II.21}$$

$$\sum_C C_{ABC}\, \sigma_C^{o}\, (g_B^2\, I(\omega) - g_C^2\, I(0)) \equiv Q_{AB}, \tag{II.22}$$

the equations of motion (19) and (20) are

$$0 = P_{AB}\, g_B^2\, \sigma_B^{o} \tag{II.23}$$

$$\omega\sigma_A^{\omega} = P_{AB}\, g_B \bar{\Phi}_B^{in,\omega} + Q_{AB}\, \sigma_B^{\omega}\, . \tag{II.24}$$

The last has the formal solution

$$\sigma_A^{\omega} = (\omega - Q)^{-1}_{AB}\, P_{BC}\, g_C\, \bar{\Phi}_C^{in,\omega} \tag{II.25}$$

with the result

$$4\pi f_{AC}(\omega) = g_A (\omega - Q)^{-1}_{AB}\, P_{BC}\, g_C \tag{II.26}$$

for the reduced scattering amplitude ("reduced" means the omission of the centrifugal and cutoff factors belonging to each external meson, contributed by u_a). This amplitude, Eq. (26), will not obey crossing unless

$$Q_{AB} = P_{AB'}\, R_{B'B}\, , \quad \text{where } R_{B'B} \text{ is symmetric.} \tag{II.27}$$

We must assume this to be so, although its significance within the framework of the equations of motion is obscure.

By inspection of Eqs. (II.21) and (II.22) we see that

$$R_{B'B} = g_B^2\, \delta_{B'B}\, I(\omega) + C_{B'B} \tag{II.28}$$

where $C_{B'B}$ is independent of ω. It follows from

$$4\pi f = g(\omega - PR)^{-1}\, Pg \tag{II.29}$$

(Eqs. (II.26) and (II.27), suppressing indices for brevity), that

$$\mathcal{I}m\, f = 4\pi\, \mathcal{I}m\, I(\omega)\, f^* f, \tag{II.30}$$

which is the statement of one-meson unitarity (see the comment following Eq. II.15) above). Another consequence of Eq. (II.28) is that $R_{B'B}$ is not singular; thus Eq. (II.29) can be written as

$$4\pi f(\omega) = g\, R^{-1/2} \frac{1}{\omega - R^{1/2} P R^{1/2}} R^{1/2} P R^{1/2} R^{-1/2} g \tag{II.31}$$

$$= gR^{-1/2} \left(\sum_n |n\rangle \frac{\pi_n}{\omega - \pi_n} \langle n|\right) R^{-1/2}\, g \tag{II.32}$$

where we made the eigenvector expansion

$$R^{1/2} P R^{1/2} = \sum_n |n\rangle\, \pi_n \langle n|\,. \tag{II.33}$$

In the strong coupling limit $R \to \infty$, hence $\pi_n \to \infty$, except for those π_n which vanish identically, and so the scattering amplitude becomes

$$4\pi f = -g\, R^{-1/2} \left(\sum_n{}' |n\rangle\langle n|\right) R^{-1/2}\, g \tag{II.34}$$

where the primed sum denotes omission of n with $\pi_n = 0$.

One way that Eq. (II.27) can be satisfied is if

$$\sigma_A^o \neq 0 \quad \text{only if } g_A = g \tag{II.35}$$

because then

$$g_C^2\ \sigma_C^o = g^2\ \sigma_C^o \qquad (11.36)$$

[likewise $g_{C'} = g_C\ \sigma^o_{C'} = g\ \sigma^o_C$ which means that the static current transforms as a representation of a subgroup of U_N], and so

$$R_{B'B} = \delta_{B'B}(g_B^2\ I(\omega) - g^2 I(0))\ . \qquad (11.37)$$

We can make a further simplification by taking the limit of a large cutoff momentum, which makes $I(\omega)$ much larger than $I(\omega) - I(0)$, and hence $R_{BB}|_{g_B \neq g}$ much larger than $R_{BB}|_{g_B=g}$. Thus in Eq. (11.34) we can neglect components of the scattering amplitude f_{BA} in which $g_B \neq g$ or $g_A \neq g$. We finally get

$$f_{BA} = \frac{\Lambda_{BA}}{-4\pi[I(\omega) - I(0)]} \qquad (11.38)$$

where [cf. Eq. (11.33)]

$$\Lambda_{BA} = \begin{cases} \sum\limits_{n(\pi_n \neq 0)} \langle B|n\rangle\langle n|A\rangle, & g_B = g_A = g \\ 0\ , & g_B \neq g \text{ or } g_A \neq g. \end{cases} \qquad (11.39)$$

It is worth noting that $P_{AB}\ \sigma_B^o = 0$ [see Eq. 21] and $R_{AB}\sigma_B^o$ = const x σ_B^o [Eqs. (11.35) and (11.37)] imply that σ_A^o, and likewise g_A, is a zero eigenvector of $R^{1/2}PR^{1/2}$, and hence is projected to zero by Λ_{AB},

$$\Lambda_{AB}\, g_B = 0. \qquad (11.40)$$

This solution, Eqs. (11.38) and (11.39), based on Eq. (11.35), is not the most general; the most general is presumably a simple sum of such amplitudes, operating in disjoint subspaces.

For an s-wave meson interaction, the scattering amplitude is

$$f_{\alpha\beta}(\omega) = \frac{u^2(k)\ \Lambda_{\alpha\beta}}{-4\pi[I(\omega)-I(0)]} \qquad (11.41)$$

where

$$u(k) = \int d^3r\, e^{i\vec{k}\cdot\vec{r}}\, u(r),\; k^2 = \omega^2 - \mu^2 \tag{11.42}$$

and

$$I(\omega) - I(0) = \omega^2 \int d^3r\, u\, \frac{1}{\Omega^2(\Omega^2-\omega^2)}\, u = \frac{\omega^2}{\pi^2}\int_0^\infty dk\, \frac{k^2u^2(k)}{\omega_k^2(\omega_k^2-\omega^2)} \;. \tag{11.43}$$

No cutoff is needed; taking $u(k)$ = constant we have

$$f_{\alpha\beta}(\omega) = \frac{\Lambda_{\alpha\beta}}{-\mu-ik} \;. \tag{11.44}$$

As it should, this amplitude obeys unitarity by the one meson approximation,

$$\mathscr{I}m\, f_{\beta\alpha} = k\, f^*_{\beta\gamma}\, f_{\gamma\alpha} \;. \tag{11.45}$$

For a p-wave interaction,

$$f_{\alpha\beta}(\hat{k}',\hat{k};\omega) = \frac{1/3\, u^2(k)\, k'_a\, \Lambda_{a\alpha,b\beta} k_b}{-4\pi[I(\omega) - I(0)]} \tag{11.46}$$

where

$$I(\omega)-I(0) = \frac{\omega^2}{3}\int d^3r\, \vec{\nabla}u\, \frac{1}{\Omega^2(\Omega^2-\omega^2)}\cdot\vec{\nabla}u = \frac{1}{3}\,\frac{\omega^2}{2\pi^2}\int_0^\infty dk\, \frac{k^4u^2(k)}{\omega_k^2(\omega_k^2 - \omega^2)} \;. \tag{11.47}$$

For $|\omega|$ well below the cutoff, i.e. in the region where $u(k) \approx$ constant, we have

$$f_{\alpha\beta} \approx -a\, \frac{k'_a\, \Lambda_{a\alpha,b\beta}\, k_b}{\omega^2} \tag{11.48}$$

where

$$a^{-1} \equiv \frac{2}{\pi}\int_0^\infty dk\, \frac{k^4}{\omega_k^4}\, \frac{u^2(k)}{u^2(0)} \;. \tag{11.49}$$

This is as far as the classical theory (equations of motion method) goes; the scattering amplitude is given in terms of the static current g_A, Eqs. (11.38), (11.39), (11.33), (11.37), (11.21), (11.17) and (11.16). However, it is worth noting that the comparison of the small ω limit of Eq. (II.38)

$$f_{BA}(\omega) \stackrel{\sim}{\sim} -m\, \Lambda_{BA}/\omega^2 \;, \text{ where } m = \begin{cases} 2\mu, & \text{s-wave} \\ a\mu^2, & \text{p-wave} \end{cases} \tag{II.50}$$

with the scattering amplitude pole terms [Eqs. (IV.21) and (IV.22] due to isobars described by a mass operator M and coupling operator (=static current)g_A, yields the following relations between g_A and M

$$[g_B, g_A] = 0 \tag{II.51}$$

$$[g_B, [M, g_A]] = m\Lambda_{BA}, \quad \text{where } \Lambda_{CB}\,\Lambda_{BA} = \Lambda_{CA}\;. \tag{II.52}$$

The first of these two results we already knew, see Eq. (II.11), at least in leading order: The static current is a c-number. The second gives an equation for M. We quoted these equations above, Eqs. (I.5) and (I.6), and discussed there how these equations, combined with symmetry considerations, considerably restrict the possibilities for g_A and M.

But g_A is completely determined only by the condition that it minimizes the static part of the Hamiltonian, H^o

$$\begin{aligned} H^o &= \int d^3r\; 1/2\; \phi^o_\alpha\, \Omega^2\, \phi^o_\alpha - \mathcal{g}_A\, \bar{\phi}^o_A \\ &= I(0)\; [1/2\; g_A^2 - \mathcal{g}_A\, g_A] \end{aligned} \tag{II.53}$$

[we used (II.11)]. One first finds the largest eigenvalue of $\mathcal{g}_A g_A$, for a given g_A, and then varies g_A to minimize H^o. This is discussed in more detail in Section III. It should be remarked that although the satisfaction of the static parts of the equations of motion for Φ and $\mathcal{g}_A$ implies that H^o is at an extremum, this may not be an absolute (or even local) minimum.

We give some examples. An isovector s-wave meson is coupled to the i-spin of the bare scatterer, i.e.

$$\mathcal{g}_\alpha = g\, I_\alpha\;,\; \alpha = 1,2,3 \tag{II.54}$$

where the I_α are angular momentum matrices for spin i_o, the i-spin of the bare scatterer; so

$$[\mathscr{g}_\alpha, \mathscr{g}_\beta] = ig\, \varepsilon_{\alpha\beta\gamma}\, \mathscr{g}_\gamma \,. \tag{II.55}$$

To find $\Lambda_{\beta\alpha}$, it suffices to observe that $R_{\beta\alpha}$ is just a scalar, $= \delta_{\beta\alpha} \times$ const, and so is irrelevant, and that

$$P^2_{\alpha\beta} = P_{\alpha\lambda}P_{\lambda\beta} = i\varepsilon_{\alpha\lambda\mu}\, I^o_\mu\;\; i\varepsilon_{\lambda\beta\nu}\, I^o_\nu = \delta_{\alpha\beta}(I^o)^2 - I^o_\alpha\, I^o_\beta \;, \tag{II.56}$$

which is proportional to a projection operator, so

$$\Lambda_{\beta\alpha} = \delta_{\beta\alpha} - g_\beta\, g_\alpha \,. \tag{II.57}$$

[This cheap way of finding Λ worked because P had only three distinct eigenvalues $\pm C$ and 0, and so P^2 had only the eigenvalues C^2 and 0, so that $P^2 = C^2\,\Lambda$.]

By symmetry alone we knew $\mathbf{g}_\alpha$ to be a isovector, but to find its length we must minimize H^o. The largest eigenvalue of $\mathscr{g}_A \mathbf{g}_A = g\, I_\alpha \mathbf{g}_\alpha$ is $g\, i_o|\mathbf{g}_\alpha|$, where i_o is the i-spin of the bare scatterer. $H^o \sim 1/2\; \mathbf{g}_\alpha^{\;2} - g i_o|\mathbf{g}_\alpha|$ is then minimized by

$$|\mathbf{g}_\alpha| = g i_o \,. \tag{II.58}$$

Exceptionally, in the case $i_o = 1/2$ this is the unique value of $|\mathbf{g}_\alpha|$ $[\Sigma_\alpha\, (\chi,\, \tau_\alpha\, \chi)^2 = 1$ for arbitrary χ, so $\vec{\tau}^{\,o}$ is a unit vector] and minimization of H^o need not be considered.

Because M must be an isoscalar, it is a function of $I_\alpha^{\,2}$, $I_\alpha g_\alpha$, and $g_\alpha^{\,2}$. The relation (II.52) implies

$$M = \frac{m I_\alpha^2}{2 g_\alpha^2} + \text{const} = \frac{\mu I_\alpha^2}{g^2 i_o^2} + \text{const}, \tag{II.59}$$

using Eq. (II.58); this of course has angular momentum eigenstates $|i,\, i_3\rangle$ and eigenvalues

$$M_i = \frac{\mu i(i+1)}{g^2 i_o^{\,2}} + \text{const}, \qquad i \geq i_o \,, \tag{II.60}$$

where the lower limit on i comes from $I_\alpha \hat{g}_\alpha = i_o$. The diagonal i'=i matrix elements of g are particularly simple,

$$<i\ i'_3|g_\alpha|\ i\ i_3> = g\ \frac{i_o^2}{i(i+1)} < i\ i'_3|\ I_\alpha|i\ i_3>.$$

It is worthwhile noting that $\Lambda_{\beta\alpha}$ would follow from M; it does not have to be derived beforehand. This is important in more complicated models.

The charged scalar model is similar, with current $\mathcal{J}_a = gI_a$, except that a = 1,2 only; we use Roman letters to emphasize this. [Usually, the scatterer is taken to have only two charge states, so $i_o = 1/2$ and $I_\alpha = 1/2\ \tau_\alpha$]. The current commutator in this model does not close,

$$[\mathcal{J}_a, \mathcal{J}_b] = ig^2\ \varepsilon_{ab\gamma}\ I_\gamma\ = ig^2\ \varepsilon_{ab3}\ I_3\ . \qquad (11.61)$$

The static part of the equation of motion of the current, Eq. (11.12) reads

$$0 = \varepsilon_{ab3}\ g_b\ I_3^o\ . \qquad (11.62)$$

Of the two solutions of this, $I^o{}_3 = 0$ or $g_b = 0$, the former yields a lower value of H^o. In fact, the minimization of H^o yields $|g_a| = g\ i_o$ (as in the isovector model), and hence $I_3{}^o = 0$. To find Λ_{ba} it is simplest to work out that

$$(R^{1/2}P\ R^{1/2})^2_{ba} = \text{const} \times (\delta_{ba}\ g^2 - g_b g_a)$$

$$(R^{1/2}P\ R^{1/2})^2_{3a} = 0, \qquad (11.63)$$

and hence

$$\Lambda_{ba} = \delta_{ba} - \hat{g}_b \hat{g}_a\ . \qquad (11.64)$$

Eq. (11.38) is satisfied by

$$M = \frac{\mu I_3^2}{g^2 i_o^2} + \alpha\ I_3 + \text{const} \qquad (11.65)$$

with eigenstates $|i_3>$ and eigenvalues

$$M_{i_3} = \frac{\mu\, i_3^2}{g^2 i_o^2} + \alpha\, i_3 + \text{const} \qquad (11.66)$$

where α is arbitrary but not too large.

The p-wave π-N model is more complicated; here the current is

$$\mathcal{J}_{\alpha i} = g\, \tau_\alpha\, \sigma_i \qquad (11.67)$$

with the non-closing commutator

$$[\mathcal{J}_{\alpha i}, \mathcal{J}_{\beta j}] = 2ig^2(\delta_{\alpha\beta}\varepsilon_{ijk}\sigma_k + \varepsilon_{\alpha\beta\gamma}\,\tau_\gamma\,\delta_{ij}). \qquad (11.68)$$

Eq. (11.12) for $(\partial_t(\tau_\alpha\sigma_i))^o$ reads

$$0 = \varepsilon_{ijk}(\tau_\alpha\sigma_j)^o\,\sigma_k{}^o + \varepsilon_{\alpha\beta\gamma}(\tau_\beta\sigma_i)^o\,\tau_\gamma^o \ . \qquad (11.69)$$

Two obvious solutions of this are $(\tau_\alpha\sigma_i)^o = \tau_\alpha^o\,\sigma_i{}^o$ and $\sigma_k{}^o = \tau_\gamma^o = 0$. The corresponding equation for $(\partial_t\sigma_i)^o$ which reads

$$0 = [\sigma_i,\ \sigma_j\ \tau_\beta]^o(\sigma_j\ \tau_\beta)^o = i\varepsilon_{ijk}(\sigma_k\ \tau_\beta)^o(\sigma_j\tau_\beta)^o \qquad (11.70)$$

is automatically satisfied, and so adds no information. To determine the forms of $\sigma_i{}^o$, $\tau_\alpha{}^o$ and $(\sigma_i\ \tau_\alpha)^o$ one must minimize H^o. One finds [Pauli-Dancoff (1942), Wentzel (1943)]

$$\sigma_i{}^o = \tau_\alpha{}^o = 0, \quad (\tau_\alpha\ \sigma_i)^o = -\,0_{\alpha i}\ , \qquad (11.71)$$

where $0_{\alpha i}$ is an orthogonal matrix

$$0_{\alpha i}\ 0_{\beta i} = \delta_{\alpha\beta},\ \ 0_{\alpha i}\ 0_{\alpha j} = \delta_{ij}\ . \qquad (11.72)$$

Again, Λ is easy to find from the results

$$(R^{1/2}\ P\ R^{1/2})^2_{\beta j,\alpha i} = \text{const} \times (\delta_{ji}\delta_{\beta\alpha} - 0_{\beta i}\ 0_{\alpha j})$$
$$(R^{1/2}\ P\ R^{1/2})^2_{\beta j,\alpha} = 0. \qquad (11.73)$$

By test, this is itself proportional to a projection operator,

$$\Lambda_{\beta j,\alpha i} = 1/2\ (\delta_{\beta\alpha}\ \delta_{ji} - 0_{\beta i}\ 0_{\alpha j})\ . \qquad (II.74)$$

[This form of Λ would follow from Eq. (II.52) and the general form of M

$$M = a\vec{J}^2 + b\vec{I}^2$$

when the form $g_{\alpha i} = -g\ 0_{\alpha i}$ [Eqs. (II.61) and (II.71)] is used.]

III. The Field Theory (Hamiltonian) Method in Strong Coupling

In this outline of a general treatment of Hamiltonian strong coupling theory we shall emphasize generality and avoid algebraic complexity, at the expense of rigor and higher order effects.

As in Sec. II, the Hamiltonian H is supposed given in terms of noninteracting states, namely products of free meson field states and "elementary" states of the scatterer ("bare isobars"). The Hamiltonian, written as a "spinor" matrix H in the bare-isobar space, is of the form

$$H = H_\phi 1 + V(\bar{\Phi}_A) \qquad (III.1)$$

where V includes both M_o (the bare isobar mass matrix) and the interaction between field and scatterer, which is parametrized by unrenormalized coupling constants.

1. Freezing out of the spin

In the strong coupling limit, V becomes large, and so should be diagonalized first. With respect to the meson field, it is diagonal in the Φ-representation of the meson field, $\Phi = \phi$; the term $1/2 \int d^3r\ \Phi_\alpha \Omega^2 \Phi_\alpha$, which may also be large, is simultaneously diagonal, As for the spinor index, let $\chi_a(\bar{\phi})$ be the spinor eigenstates of $V(\bar{\phi})$,

$$V(\bar{\phi})\ \chi_a(\bar{\phi}) = V_a(\bar{\phi})\ \chi_a(\bar{\phi})\ . \tag{III.2}$$

Transforming from the bare-isobar spinor space to the V-eigenstate space, the Hamiltonian becomes the matrix

$$H_{ba} = \int d^3r\ 1/2\ \{\ \pi_\alpha^2\ \}_{ba} + \left[\int d^3r\ 1/2\ \Phi_\alpha \Omega^2 \Phi_\alpha + V_a(\bar{\Phi})\right] \delta_{ba} \tag{III.3}$$

where

$$\{\pi_\alpha^2\}_{ba} \equiv \chi_b^*(-i \frac{\partial}{\partial\Phi_\alpha})^2 \chi_a = \delta_{ba}\ \pi_\alpha^2 - 2i\,(\chi_b^* \frac{\partial}{\partial\Phi_\alpha}\ \chi_a)\pi_\alpha - (\chi_b^* \frac{\partial}{\partial\Phi_\alpha^2} \chi_a) . \tag{III.4}$$

Making a perturbation expansion in the off-diagonal part, the spinor eigenvalues of H are

$$\begin{aligned} H_a &= V_a(\bar{\phi}) + \frac{1}{2}\int d^3r\ \Phi_\alpha\ \Omega^2\ \Phi_\alpha \\ &+ \frac{1}{2}\ \int\ \{\pi^2\}_{aa} \\ &+ \sum_{b\neq a} \left|\ \frac{1}{2}\int\ \{\pi^2\}_{ba}\right|^2/(V_b - V_a) + \ldots \end{aligned} \tag{III.5}$$

In the strong coupling limit, $V_a - V_o \to \infty\ (a \neq o)$, the low-lying, "physical" states will all belong to the spinor ground state, a=o; states belonging to a > 0 will be very high lying in energy, and need not be considered. [From the static model point of view, they are certainly unphysical, since they lie above the cut-off energy.] We have assumed here for simplicity that the ground state of $V(\phi)$ is nondegenerate; if it is degenerate, the effective H, Eq. (III.10) below, is a matrix in the components of the spinor ground state. This causes no essential complication. So for "physical" states the Hamiltonian is effectively

$$H = V(\bar{\phi}) + 1/2 \int d^3r\ [\ \Phi_\alpha\ \Omega^2 \Phi_\alpha + \ \{\pi^2\}\ ] \tag{III.6}$$

where we have dropped the subscripts a=0, writing

$$V_o(\bar{\phi}) \equiv V(\bar{\Phi}), \text{ and } \{\pi^2\}_{oo} = \pi^2 - 2i(\chi_o^* \frac{\partial}{\partial\phi_\alpha}\chi_o)\pi_\alpha - (\chi_o^* \frac{\partial}{\partial\phi^2}\chi_o) \equiv \{\pi^2\}. \tag{III.7}$$

The situation can be described as one in which a Born-Oppenheimer approximation holds. In the strong coupling limit the frequencies $V_b - V_a$ of the "spin system" (the bare isobar states) are large compared to the frequencies of the meson field [that is, the coupled part of the field; the cut off prevents high frequency parts of the field from interacting]. Thus the spinor state follows the field and is not an independent degree of freedom.

2. The quadratic (pair theory) approximation

H, Eq. (III.6) would be trivially solvable, i.e. we could find its eigenstates, if it were quadratic in Φ and π [= pair theory Hamiltonian]. In fact, for solvability it is sufficient that H be nearly quadratic where its wave functions $\langle\Phi(r)|E\rangle$ are large, i.e., in the vicinity of the minimum (or minima) $\Phi_\alpha = \phi^o_\alpha(r)$ of the potential part of H,

$$U(\Phi) = V(\bar{\Phi}) + 1/2 \int d^3r\, \Phi_\alpha\, \Omega^2\, \Phi_\alpha \quad . \tag{III.8}$$

The extremum condition

$$0 = \left[\frac{\delta U}{\delta\, \Phi_\alpha(r)}\right]_{\Phi=\phi^o} = \left[\frac{\partial}{\partial\, \bar{\Phi}_{a\alpha}} V(\bar{\phi})\, u_a(r) + \Omega^2 \Phi_\alpha(r)\right]_{\Phi=\phi^o} \tag{III.9}$$

implies

$$\phi^o_\alpha(r) = -\frac{1}{\Omega^2} u_a(r) \left[\frac{\partial}{\partial\bar{\Phi}} V(\bar{\phi})\right]_{\Phi=\phi^o} \equiv \frac{1}{\Omega^2} u_a(r)\, g_{a\alpha} \,. \tag{III.10}$$

The r-dependence of $\phi^o(r)$ is of course as expected for a static field; using it, $U(\Phi)$ can be expressed in terms of $\bar{\Phi}$,

$$U(\Phi) = H(\Phi) = V(\bar{\Phi}) + \frac{1}{2I(0)} \bar{\Phi}_A^2 \text{ , for } \Phi \text{ static,} \tag{III.11}$$

where as above, (II.7),

$$\int d^3r\, u_a \frac{1}{\Omega^2 - \omega^2} u_b \equiv \delta_{ab}\, I(\omega). \tag{III.12}$$

If $U(\phi)$ has only one minimum, $\bar{\phi}^o$ is just a number; if U has several minima, $\bar{\phi}^o$ can take on several values, but the states belonging to different values have no overlap, on the assumption that H is effectively quadratic around each minimum. Most interesting is the degenerate case in which U has a continuum of minima [due to an internal symmetry, so that U depends only on the invariants of $\bar{\phi}^o$ (scalars) under that symmetry, not on its "angles"]; then $\bar{\phi}^o$ takes on a continuum of values. We shall first treat the non-degenerate case, taking $\bar{\phi}^o$ to be a number.

If we approximate $\{\pi^2\}$ by π^2, the equation of motion of the meson field is

$$(\Omega^2 + \partial_t^2) \quad \Phi_\alpha(r,t) = \mathcal{g}_{\alpha a}\, u_a(r) \tag{III.13}$$

where

$$\mathcal{g}_A = - \frac{\partial}{\partial \bar{\Phi}_A} V(\bar{\Phi}) \quad . \tag{III.14}$$

Writing $\Phi = \phi^o + \Phi'$, and expanding the current $\mathcal{g}_A$ around ϕ^o, keeping only constant and linear terms, we have

$$\mathcal{g}_A \approx g_A + C_{AB}\, \bar{\Phi}'_B \tag{III.15}$$

where g_A was defined above [Eq. (III.10)] and

$$C_{AB} = \left[\frac{\partial \mathcal{g}_A}{\partial \Phi_B}\right]_{\bar{\Phi}=\bar{\phi}^o} = -\left[\frac{\partial}{\partial \bar{\Phi}_A\, \partial \bar{\Phi}_B} V(\bar{\Phi})\right]_{\bar{\Phi}=\bar{\phi}^o} \quad . \tag{III.16}$$

Using (III.15) in (III.13), the static part of (III.13) is just (III.10) again; the non-static part is linear in Φ'. The frequency-ω part gives

$$\Phi_\alpha^\omega(r) = \Phi_\alpha^{in,\omega}(r) + \frac{1}{\Omega^2 - \omega^2} u_a(r)\, C_{a\alpha,b\beta}\, \bar{\Phi}_{b\beta}^\omega \quad . \tag{III.17}$$

Multiplying by u_c and integrating,

$$\bar{\Phi}^{\omega}_{A} = \bar{\Phi}^{in,\omega}_{A} + I(\omega)\ C_{AB}\ \bar{\Phi}^{\omega}_{B}\ ; \tag{III.18}$$

solving for $\bar{\Phi}_A$,

$$\bar{\Phi}^{\omega}_{A} = [\delta_{AB} - I(\omega)\ C_{AB}]^{-1}\ \bar{\Phi}^{in,\omega}_{B} = \sum_{\nu} \frac{0^{\nu}_{A}\ 0^{\nu}_{B}}{1 - I(\omega) C_{\nu}}\ \Phi^{in,\omega}_{B} \tag{III.19}$$

where C_ν are the eigenvalues, 0^{ν}_{A} the eigenfunctions [principal axis transformation] of C_{AB}

$$C_{AB} = \sum_{\nu} C_{\nu}\ 0^{\nu}_{A}\ 0^{\nu}_{B}\ . \tag{III.20}$$

The one-meson reduced scattering amplitude is thus

$$f_{AB} = \sum_{\nu} f_{\nu}\ 0^{\nu}_{A}\ 0^{\nu}_{B}\ ,\ \text{where}\ f_{\nu} = \frac{C_{\nu}}{4\pi(1 - I(\omega)\ C_{\nu})}\ . \tag{III.21}$$

The eigenamplitude f_ν has poles at $\omega = \pm\Delta_\gamma$, where

$$I(\Delta_\nu)\ C_\nu = 1. \tag{III.22}$$

If C_ν is sufficiently large and positive, $\frac{1}{I(\mu)} < C_\nu < \frac{1}{I(0)}$, the poles $\omega = \pm\Delta_\nu$ are on the first sheet, and there are harmonic oscillator bands of real isobars coupled to $0^{\nu}_{A}\ \Phi'_{A}$.

The isobars, labeled by their excitation in each mode, n_ν, have masses

$$M_{\{n_\nu\}} = \sum_{\nu} (n_\nu + 1/2)\ \Delta_\nu + M_o\ . \tag{III.23}$$

From the residues of the poles of $f_{BA}(\omega)$ one may deduce the coupling constants

$$\langle n_1 \ldots, n_\nu + 1, \ldots\ |g_A|\ n_1 \ldots, n_\nu, \ldots\rangle = (n_\nu + 1)\ \frac{0^{\nu}_{A}}{4\pi I'(\Delta_\nu)}\ . \tag{III.24}$$

For s-waves

$$1/4\pi\ \ I'(\Delta) = \sqrt{\mu^2 - \Delta^2}/\Delta\ . \tag{III.25}$$

The trivial solvability of a quadratic Hamiltonian may also

be exhibited by the decomposition of Φ and π into different frequency parts (normal mode decomposition), in terms of which H, and also the field commutation relations, decompose into independent parts. Here, we shall only split off the frequency $-\Delta_\nu$ parts. Multiplication of the equation of motion for Φ', Eq. (III.13), by $0^\nu_A\, u/(\Omega^2 - \Delta^2_\nu)$, summing and integrating over A and r respectively, and using Eqs. (III.15), (III.20), and (III.22), yields

$$\sum_{a,\alpha} 0^\nu_{a\alpha} \int d^3r\, u_a \frac{1}{\Omega^2-\Delta^2_\nu} (\Delta^2_\nu + \partial^2_t)\, \Phi'_\alpha = 0; \tag{III.26}$$

this states the orthogonality of the frequency $-\Delta_\nu$ part of Φ'_A to the rest of Φ'_A [taking a time derivative, one sees the same holds for π_A].

The resulting decomposition is

$$\Phi'_\alpha = \sum_\nu \frac{1}{\Omega^2 - \Delta^2_\nu} u_a(r)\, 0^\nu_{a\alpha}\, q_\nu + \Phi''_\alpha \tag{III.27}$$

$$\pi_\alpha = \sum_\nu \frac{1}{\Omega^2 - \Delta^2_\nu} u_a(r) \frac{0^\nu_{a\alpha}\, p_\nu}{I_2(\Delta_\nu)} + \pi''_\alpha \tag{III.28}$$

where

$$\delta_{ba}\, I_2(\Delta) = \int d^3r\, u_b \frac{1}{(\Omega^2 - \Delta^2)^2} u_a \tag{III.29}$$

and

$$[q_\nu, p_{\nu'}] = i\delta_{\nu\nu'} \tag{III.30}$$

$$[q_\nu, \pi''_\alpha] = [p_\nu, \Phi''_\alpha] = 0 \tag{III.31}$$

$$[\Phi''_\alpha(r), \pi''_\beta(r')] = i[\delta_{\alpha\beta}\delta(r-r') - \sum_\nu \frac{1}{\Omega^2-\Delta^2_\nu} u_a(r) \cdot \frac{1}{\Omega^2 - \Delta^2_\nu} u_b(r') \frac{0^\nu_{a\alpha}\, 0^\nu_{b\beta}}{I_2(\Delta_\nu)}] \tag{III.32}$$

$$H = \sum_\nu \frac{1}{2I_2(\Delta_\nu)} (p^2_\nu + \{I_2(\Delta_\nu)\Delta_\nu\}^2 q^2_\nu) + H_\phi(\Phi'', \pi''). \tag{III.33}$$

The conjugate pairs of operators, p_ν, q_ν describe the harmonic oscillator bands of isobars; matrix elements of q_ν, the frequency-Δ_ν part of the current, are coupling constants (agreeing with (III.24) above). Since $H_\phi(\Phi'',\pi'')$ is the free meson Hamiltonian, scattering occurs only because of the extra terms in the commutator (III.32), which express the orthogonality (III.26); thus the scattering is purely orthogonality scattering.

In order to assess the validity of the quadratic approximation to H, used in (III.15), we calculate expectation values of $\Phi'^2 = (\bar{\Phi}-\bar{\phi}^o)^2$. To simplify the notation, we shall do the calculation for s-waves and for a single principal axis of $V(\bar{\phi})$ [Eqs. (III.16) and (III.20)], and drop the suffix ν from Δ_ν and C_ν. In the n-th excited state

$$<n|(\bar{\Phi}-\bar{\phi}^o)^2|n> = \sum_{\zeta\neq n} <n|\bar{\Phi}|\zeta><\zeta|\bar{\Phi}|n>$$

$$= \sum_{\pm} |<n\pm 1|\Phi^\Delta|n>|^2 + \int d^3k|<n,k|\bar{\Phi}^{\omega_k}|n>|^2 . \qquad \text{(III.34)}$$

The first term is the contribution of intermediate isobar states; since $q = C\bar{\phi}$,

$$|<n\pm 1|\bar{\phi}^\Delta|n>|^2 = |C^{-1}<n\pm 1|q|n>|^2 = C^{-2}(n+\frac{1}{2}\pm\frac{1}{2})\sqrt{\mu^2-\Delta^2}/\Delta \qquad \text{(III.35)}$$

using the result (III.24), and (III.25) for the coupling constants.

The second term of (III.34) is the contribution of meson-containing states, where only the one meson plus isobar "n" state occurs because $\bar{\phi} \propto \bar{\phi}^{in}$. Using (III.19) we have

$$\int d^3k|<n,k|\bar{\phi}^{\omega_k}|n>|^2 = \int d^3k \left| \frac{\sqrt{4\pi}\, u_k}{\sqrt{2\omega_k}(2\pi)^3} / [1-CI(\omega_k)] \right|^2 =$$

$$\frac{1}{\pi}\int_\mu^\infty d\omega_k\, k\, u_k^2 |1-CI(\omega_k)|^2 . \qquad \text{(III.36)}$$

We make a rough evaluation of the integral: If C << a (no binding) $1-C\,I(\omega) \sim 1$, and so (III.36) becomes

$$\underset{\sim}{\sim} \frac{1}{\pi} \int_{\mu}^{\infty} d\omega \; k \; u_k^2 = c_1 \; a^{-2} \tag{III.37}$$

where c_1 is a constant of order 1. If $C \lesssim a$ (binding, or near binding) $1\text{-}C\mathcal{I}(\omega) \underset{\sim}{\sim} (\sqrt{\mu^2-\Delta^2} + ik)\, a$, for $\omega << a^{-1}$, and so (III.36) becomes

$$\frac{a^{-2}}{\pi} \int_{\mu}^{\infty} d\omega \; \frac{k\, u_k^2}{\omega^2-\Delta^2} \underset{\sim}{\sim} \frac{a^{-2}}{\pi} \; \ell n \left(\frac{c_2}{a\mu} \right) \tag{III.38}$$

where c_2 is a constant of order 1 [it depends on Δ, but only weakly.] For intermediate values of C, the results are intermediate between (III.37) and (III.38) [as always, we assume $a\mu << 1$].

As a curiosity of unknown significance, we remark that the integral in (III.36) requires for convergence a cutoff u_k which falls more rapidly than $O(\omega^{-1})$ for large ω, whereas all perturbation terms are finite for weaker cutoffs, namely those for which $\int d\omega\, \omega^{-1} \ell n^n \omega\, u_k{}^2$ converges, i.e. $u_k = O(\omega^{-\sigma})$, $\sigma > 0$. The reason is probably the failure of the strong coupling approximation at sufficiently high energy.

Putting (III.35), and (III.37) or (III.38) together, we have

$$\langle 0 | (\bar{\Phi}-\bar{\phi}^o)^2 | 0 \rangle = c_1 a^{-2}, \quad \text{for } C << a \tag{III.39}$$

and

$$\langle n | (\bar{\Phi}-\bar{\phi}^o)^2 | n \rangle = a^{-2} \left[\frac{(2n+1)\sqrt{\mu^2-\Delta^2}}{\Delta} + \frac{1}{\pi}\, \ell n \, \frac{c_2}{a\mu} \right] , \quad \text{for } C \lesssim a \tag{III.40}$$

These results give a measure of the region of $\bar{\Phi}$ space within which $V(\bar{\Phi})$ must be well approximated by a quadratic form, and $\{\pi^2\}$ must be well approximated by π^2, in order that the quadratic approximation be valid.

3. The degenerate case

Here the minima of $U(\bar{\Phi})$ lie not at discrete points in $\bar{\Phi}_A$ space, but on a surface, say n dimensional. As remarked above, this is generally a consequence of symmetry, O_A^ν and Δ_ν are now <u>continuous</u> functions of $\bar{\Phi}_A$, which ranges over the minimum surface. We shall use the static current g_A as coordinate, instead of $\bar{\Phi}_A$, where $\bar{\Phi}_A = I(0)g_A$ [Cf. (III.10) and (III.12)]; this is convenient because g_A has a direct physical significance.

The excitation energies Δ_ν corresponding to motion <u>on</u> the minimum-surface are vanishing small (in the strong coupling limit); thus the scattering amplitude Eq. (III.21) is of the form (for s-waves)

$$f_{BA}(g) = \frac{\Lambda_{BA}(g)}{-\mu-ik} + \sum_\nu{}' \frac{O_B^\nu(g)\ O_A^\nu(g)}{4\pi[C_\nu^{-1} - I(\omega)]} \tag{III.41}$$

where the primed sum is over the nondegenerate $[\Delta_\nu > 0]$ principal directions; in the sum over the degenerate directions $[\Delta_\nu \approx 0]$ we have set $\sqrt{\mu^2 - \Delta_\nu^2} \approx \mu$, and have defined

$$\Lambda_{BA}(g) = \sum_{\Delta_\nu \approx 0} O_B^\nu(g)\ O_A^\nu(g)\ . \tag{III.42}$$

Note that Λ_{BA} is a projection operator, $\Lambda_{CB}\Lambda_{BA} = \Lambda_{CA}$, projecting into the minimum surface of $H(\Phi)$. The assumption that (III.21) continues to hold locally in the degenerate case can be justified by considering scattering from states which are wave packets on the minimum surface. If the curvature of the surface is small enough, such a wave packet can be sufficiently localized so that O_A^ν is well defined, but at the same time has a small spread of energies, $\ll\mu$, so that it is static.

More formally, one may utilize the decomposition Eqs. (III.27), (III.28), with the modification that the quasi-static eigenmodes $\Delta_\nu \approx 0$ are absorbed into the static field. The conjugate field, the quasi-static part of π, π^o is small (but not identically zero, as was the static π), of order $1/\phi^o$. Note that $\{\pi^{o2}\}$ cannot be approximated by π^{o2} because the difference of these is itself of the order $(1/\phi^o)^2$. Finally, it is convenient <u>not</u> to separate out the nonquasi-static part $[\Delta_\nu > 0]$ of Φ', because usually these isobars are virtual [in fact very virtual, with $C_\nu \approx 0$]. The resulting decomposition is

$$\Phi(r) = \frac{1}{\Omega^2}\, u_a(r)\, g_{a\alpha} + \Phi'_\alpha(r) \tag{III.43}$$

$$\pi_\alpha(r) = \frac{1}{\Omega^2} u_a(r)\, \Lambda_{a\alpha,B}(g)\, p_B/I_2(0) + \pi'_\alpha(r) \qquad (III.44)$$

with $I_2(\omega)$ defined above, Eq. (III.29), and

$$[g_A, p_B] = i\delta_{AB} \qquad (III.45)$$

$$[g_A, g_B] = [g_A, \pi_\alpha'(r)] = [\Phi'_\alpha(r), p_A] = 0, \text{ etc.} \qquad (III.46)$$

$$[\Phi'_\alpha(r), \pi'_\beta(r')] = i[\delta_{\alpha\beta}\delta(r-r') - \frac{1}{\Omega^2} u_a(r)\cdot \frac{1}{\Omega^2} u_b(r') \frac{\Lambda_{a\alpha,b\beta}(g)}{I_2(0)}] \qquad (III.47)$$

$$H = \frac{\{p_A\, \Lambda_{AB}(g)\, p_B\}}{2I_2(0)} + H_\phi(\Phi', \pi') - \frac{1}{2} C'_{AB}\, \bar{\Phi}'_A\, \bar{\Phi}'_B \qquad (III.48)$$

where

$$C'_{AB} = (\delta_{AA'} - \Lambda_{AA'})\, C_{A'B'} (\delta_{B'B} - \Lambda_{B'B}) \qquad (III.49)$$

[If the degeneracy is not exact, the term

$$(1 + \frac{\bar{\Phi}'_B}{I} \frac{\partial}{\partial g_B})\ [\ \frac{1}{2} I\ g_A^2 + V(I\, g)\ ] \qquad (III.50)$$

must be added to Eq. (III.48); it is merely a constant if the degeneracy is exact.]

From Eq. (III.48), the isobar part of the Hamiltonian, i.e., the mass operator M, is given by

$$M = \frac{m}{2}\ \{p_A \Lambda_{AB}(g)\ p_B\} = \frac{m}{2}\ \chi_o^*\ p_A \Lambda_{AB}(g)\ p_B \chi_o \qquad (III.51)$$

where

$$m = I_2^{-1}(0) = \begin{cases} 2\mu & : \text{ s-wave} \\ a\mu^2 & : \text{ p-wave} \end{cases} \qquad (III.52)$$

and p_A is the canonical conjugate to g_A, Eq. (III.45). From this follows the same result as found above, Eq. (III.52),

$$[g_B, [M, g_A]] = m\ \Lambda_{BA}, \quad \text{where } \Lambda_{CB}\Lambda_{BA} = \Lambda_{CA}\ . \qquad (III.53)$$

As remarked above, this equation may suffice to determine M, as well as Λ_{BA}, at a great saving of effort over finding eigenvalues of (III.51) directly.

IV. The S-Matrix (Chew-Low) Method in Strong Coupling

We have already abstracted from field theory the "strong coupling algebra" Eqs. (II.51), (II.52) or (III.46), (III.53) which involve only observable quantities; together with the form (II.38) or (III.41) for the meson scattering amplitudes, they thus describe a class of solutions of the Chew-Low equations, as may be verified. However, there are two reasons for considering ab initio the strong coupling theory within the Chew-Low formalism. One is, if you will, a tour de force, to show that it is possible to work entirely with S-matrix quantities, never introducing fields or bare couplings (nor a cutoff, in the renormalizable s-wave case). In fact the Chew-Low theory is seen to be no more complex, and in some respects simpler, than the field theory. And, the Chew-Low techniques themselves may be of some interest, illustrating a situation where "bootstrapping" yields the exact solution. The second reason is to see whether there are any non-field theory solutions [i.e., "exceptional" Chew-Low solutions, with fewer degrees of freedom than the field theory solution; "bootstrap solutions" in the strict sense.] The answer, however, is pretty clearly no (although no rigorous proof is given); there are no strong coupling Chew-Low solutions which are not field theory solutions. [Of course, this still leaves open the question for intermediate coupling.]

In Chew-Low terms, a strong coupling solution is a set of one-meson scattering amplitudes which satisfy one-meson unitarity, and which imply vanishing meson production.

1. The no-production condition

It is well known that scattering generally implies production. For instance, the scattering process $\theta A\to\theta A$ provides the two meson production process $\theta A\to\theta^2\bar{\theta} A$ with pole terms

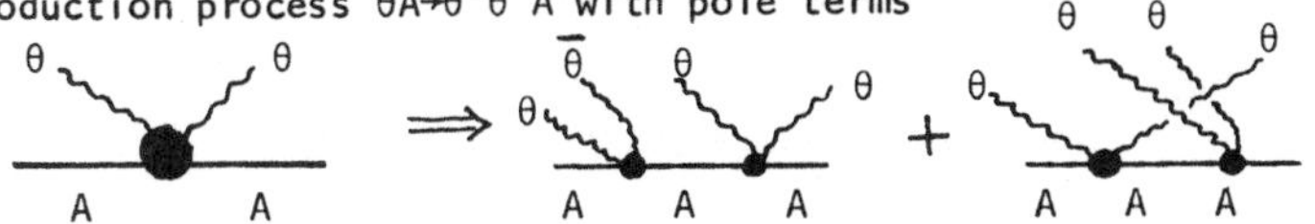

Further, if the scattering process itself has a pole term, say an s-channel pole representing $\theta A\to B\to\theta A$, then the production process $\theta A\to\theta\bar{\theta}B$

has pole terms

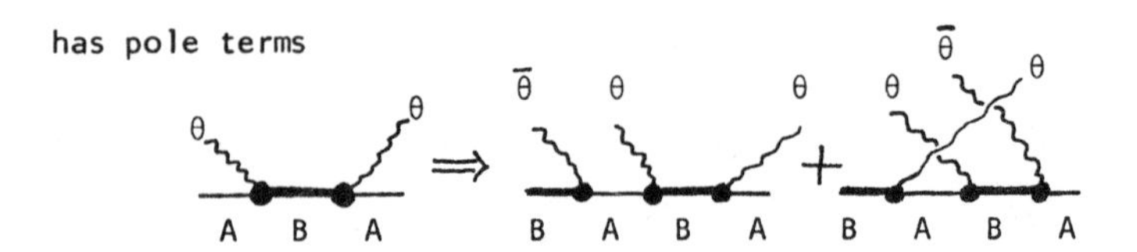

This last depends on "factorization" of the $\theta A \to B \to \theta A$ pole term into two $\theta A\bar{B}$ vertices and a B propagator, and holds no matter where the B pole is, whether on the physical sheet or not.

There are just two ways in which the scattering-implies-production theorem can fail to hold. One is the weak coupling limit, in which the production amplitudes are vanishingly small compared to the scattering amplitudes. The other is the static model strong coupling limit, in which the production amplitude pole terms vanish by cancellation. The restriction of the possibility of cancellation to the static model is clear: The cancellation must occur between poles in different channels, but these are kinematically inequivalent (i.e. cannot coincide on the Mandelstam plane), except in the static model. Here s and u are equivalent $[s-m_A^2 \approx 2m_A\omega,\ \ u-m_A^2 \approx -2m_A\omega, \text{as } m_A \to \infty]$, and t-channel poles do not exist, by hypothesis.

The calculation of the sum of the pole terms of an amplitude is best approached through an effective Hamiltonian

$$H^{eff} = H_\phi + M - g_A \bar{\Phi}_A \tag{IV.1}$$

whose noninteracting states contain free mesons and exact isobars. The Born approximation of an amplitude, calculated from this Hamiltonian, is just the desired sum of pole terms. But if the Born approximation is calculated in the usual way through diagrams, we get a sum over n! terms, corresponding to the number of different orders in which the n external mesons can be emitted or absorbed [and as it happens the cancellation, when it occurs, does not do so in any subset of terms, but only when all are taken together.]

For us, a convenient calculation of the Born approximation is through the equation of motion of the current, i.e. through repeated commutators of H^{eff} with g_A. The matrix element of g_A between exact states is given by

$$\langle f|g_A|i\rangle = \omega_{fi}^{-1} \langle f|[H^{eff}, g_A]\, i\rangle = \omega_{fi}^{-2} \langle f|[H^{eff}, H^{eff}, g_A]|i\rangle = \ldots \tag{IV.2}$$

where $\omega_{fi} = E_f - E_i$; we write the multiple commutators $[a,[b,c]]=[a,b,c]$, etc., for short. The Born approximation is obtained by requiring the field $\overline{\Phi}_A$ of the interaction term in H^{eff} to act only directly upon the mesons in the initial and final states. Note that H_ϕ can be eliminated from the multiple commutator by using its commutators $[H_\phi,M] = 0$ and $[H_\phi, g_A \overline{\Phi}_A] = i g_A \dot{\overline{\Phi}}_A$. The result is that the Born approximation to the amplitude with N external mesons (indices $A_1, A_2, \ldots, A_N$) vanishes if all multiple commutators of $g_{A_1}, g_{A_2}, \ldots, g_{A_N}$ together with M (M occurring an arbitrary number of times) vanish. That is,

$$[M, \ldots g_{A_N}, \ldots, g_{A_2}, M, \ldots, g_{A_1}] = 0, \quad N \geq 3, \tag{IV.3}$$

is the condition for meson production to vanish.

Those conditions (IV.3) which do not contain M are satisfied if and only if

$$[g_C, [g_B, g_A]\ ] = 0 \ . \tag{IV.4}$$

This in turn is satisfied if

$$[g_B, g_A] = 0 \ ; \tag{IV.5}$$

we shall see later that this stronger condition is required by the one-meson amplitude, in order that it not have poles corresponding to additional isobars, not contained in M.

The conditions (IV.3) which are first degree in M are satisfied if and only if

$$[g_C, [g_B, [M, g_A]\ ]\ ] = 0 \ . \tag{IV.6}$$

We must assume

$$[g_B, [M, g_A]\ ] \neq 0 \tag{IV.7}$$

because the vanishing of this quantity, for all A and B, would imply the vanishing of the one-meson amplitudes f_{BA}, which would mean a trivial model. The two conditions (IV.6) and (IV.7) are simultaneously possible only for infinite matrices, which means that all no-production solutions have an infinite number of isobars. As seen later, the isobars must be real bound states in s-wave models, or narrow resonances in p-wave models, except in the "harmonic oscillator" solutions where they may be arbitrarily distant on the second sheet.

Introducing p_A, the canonical conjugate of g_A, $[g_A, p_B] = i\delta_{AB}$, Eqs. (IV.6) and (IV.7) imply that M is quadratic in p_A,

$$M = A(g) + B_A(g)\, p_A + 1/2\, C_{AB}(g)\, p_A p_B \tag{IV.8}$$

where $C_{AB}(g) = [g_A, [M, g_B]]$. We shall see later that unitarity of the one-meson amplitude requires $C_{AB} = m\,\Lambda_{AB}$ where m is a constant and Λ_{AB} is a projection operator, $\Lambda_{AB}\Lambda_{BC} = \Lambda_{AC}$.

The conditions (IV.3) which are of second degree in M are satisfied if and only if

$$[g_C, M, g_B, M, g_A] = 0 \tag{IV.9}$$

$$[g_C, g_B, M, M, g_A] = 0\ . \tag{IV.10}$$

These conditions, in view of (IV.8), are equivalent to

$$C_{CD} \frac{\partial}{\partial g_D} C_{BA} = 0 \tag{IV.11}$$

which in turn imply

$$\frac{\partial}{\partial g_C}\ C_{BA} = 0. \tag{IV.12}$$

It is easily found that the conditions (IV.3) which are of third and fourth degree in M are then satisfied if and only if

$$\frac{\partial^2}{\partial g_C \partial g_B} B_A = 0 \tag{IV.13}$$

and

$$\frac{\partial^3}{\partial g_A\, \partial g_B\, \partial g_C} A = 0 \tag{IV.14}$$

respectively; all other conditions (IV.3) are then satisfied.

Equations (IV.12), (IV.13), and (IV.14) together simply require that M be a quadratic form in g_A and p_A. It is trivial to verify that this is sufficient to satisfy (IV.3).

We have written the condition (IV.3), and drawn the above conclusions, as operator equations, but in fact only their matrix elements between physical states need hold. This is the same situation discussed in Sec. III: A Hamiltonian $H(\Phi,\pi)$ is solvable if it is "effectively" quadratic in Φ and π. For instance, in the important rotational case

$$g_A \, C_{AB} = m g_A \, \Lambda_{AB} = 0 \qquad \text{(IV.15)}$$

holds, so that g_A^2 is a constant of the motion. Supposing for simplicity that it is the only constant, then

$$C_{AB} = m \, (\delta_{AB} - g^{-2} g_A \, g_B), \; g^2 = g_A^2 \; , \qquad \text{(IV.16)}$$

so that $\frac{\partial}{\partial g_A} C_{BC} = O(g^{-1})$. Hence condition (IV.3) is satisfied in, but only in, the strong coupling limit, $g \rightarrow \infty$. It should be noted that when the no-production conditions are satisfied in this way, M is no longer restricted to be quadratic in P_A; it may have terms of the form $g^{-2}p^4$, $g^{-4}p^6$ etc.

Stated more generally, there is negligible meson production if, and only if, M has the essentially quadratic form

$$M = \frac{1}{2} A_{AB}(g,p) g_A g_B + B_{AB}(g,p) g_A p_B + \frac{1}{2} C_{AB}(g,p) p_A p_B + E_A(g,p) g_A + \text{const} \qquad \text{(IV.17)}$$

where the coefficients $A_{AB}(g,p)$, etc., are sufficiently slowly varying functions for physical values of g_A and p_A, or stated differently, the matrix elements of their derivatives with respect to g_A and p_A between physical states are sufficiently small.

2. One-meson scattering amplitudes

We now turn to the problem of finding one-meson scattering amplitudes $f_{Bs',As}(E)$ [E is the total energy; A and s label the initial meson and isobar, respectively; the amplitude is actually the "reduced" amplitude in which the centrifugal and cutoff factors, which depend on the momenta of the external mesons, have been omitted.] These amplitudes are to be analytic functions (Chew-Low solutions) which satisfy

a) crossing

$$f_{Bs',As}(E) = f_{As',Bs}(M_{s'} + M_s - E), \tag{IV.18}$$

b) one-meson unitarity

$$\mathscr{I}m\, f^{-1}_{Bs',As}(E+i\varepsilon) = -\delta_{BA}\delta_{s's}u_A^2(k_s)k_s^{2\ell_A+1/}\mu^{2\ell_A}, \quad E>M_s+\mu \tag{IV.19}$$

where $k_s = \sqrt{(E-M_s)^2 - \mu^2}$, ℓ_A is the orbital angular momentum of the meson wave A, and $u_A(k)$ is the cutoff function,

c) "no-meson unitarity" : The totality of the pole terms of $f_{Bs',As}$ is to be

$$f^P_{Bs',As}(E) = \sum_{s''} \frac{\langle s'|g_B|s''\rangle\langle s''|g_A|s\rangle}{M_{s''} - E} + \frac{\langle s'|g_A|s''\rangle\langle s''|g_B|s\rangle}{M_{s''} - M_{s'} - M_s + E} \tag{IV.20}$$

where the M_s are the eigenvalues and $|s\rangle$ are the eigenvectors of an M of the form (IV.17).

It should be possible to solve the one meson scattering problem for the general form of M, Eq. (IV.17), but we here only present solutions in three special cases: A) The isobars are nearly degenerate in mass. This requires that g, the order of magnitude of the couplings $\langle s'|g_A|s\rangle$, be large; this case is strong coupling in the usual sense. B) $C_{BA}(g,p) \equiv c_{BA}$ is a constant, and $B_{BA} = 0$; this is the harmonic oscillator or pair theory case. C) C_{BA} is not constant and g is large, but B_{BA} is not small; this is the "stretched rotator" case.

A. Models in which the isobar mass differences are small (g is large).

The one-meson (reduced) amplitudes $f_{Bs',As}$ are conveniently written as the matrix elements between isobar states, $\langle s'|f_{BA}|s\rangle$, of an operator f_{BA}. From Eq. (IV.20), the pole terms of f_{BA} are

$$f^P_{BA}(E) = g_B \frac{1}{M-E} g_A + g_A \frac{1}{M-\overleftarrow{M}-\overrightarrow{M}+E} g_B \tag{IV.21}$$

where $\underset{\rightarrow}{M}$ is defined by $f(\underset{\rightarrow}{M})g_B|s\rangle = f(M_s)g_B|s\rangle$. The expansion of $\underset{\sim}{f}^P$ in powers of (isobar mass differences)/ω reads

$$f^P_{BA}(E) = -[g_B,g_A]\,\omega^{-1} - [g_B,M,g_A]\,\omega^{-2} - [g_B,M,M,g_A]\omega^{-3} - \dots \qquad \text{(IV.22)}$$

where $\omega \equiv E - \underset{\rightarrow}{M}$. Taking first the case of s-waves (ℓ=0), we can write a dispersion representation for $\underset{\sim}{f}^{-1}$ [we now suppress even the meson indices, $f_{Bs',As} = \langle Bs'|\underset{\sim}{f}|As\rangle$]; using (IV.19) but omitting the unnecessary cutoff,

$$\begin{aligned} \underset{\sim}{f}^{-1} &= -\underset{\sim}{\gamma}^{-1}\,\omega - \frac{\omega^2}{\pi}\int \frac{d\omega'\,k'}{\omega'^2(\omega'-\omega)} \\ &= -\underset{\sim}{\gamma}^{-1}\omega - (\mu+ik) \end{aligned} \qquad \text{(IV.23)}$$

where $[g_B,g_A] \equiv \gamma_{BA} = \langle B|\underset{\sim}{\gamma}|A\rangle$ has been assumed to be nonsingular. In the integrand we have neglected the isobar mass differences, approximating k_s by $k = \sqrt{(E-M_o)^2 - \mu^2}$ where M_o is a typical isobar mass; likewise we approximate $\omega = E-M \approx E-M_o$, hence $\omega^2 = k^2+\mu^2$. According to (IV.23), $\underset{\sim}{f}^{-1}$ has roots in addition to the "imput root" at ω=0, contrary to the requirement (IV.20) and the postulate that the isobar masses are nearly equal. To avoid this, all eigenvalues of $\underset{\sim}{\gamma}$ must vanish, i.e.

$$[g_B, g_A] = 0. \qquad \text{(IV.24)}$$

This conclusion is not changed by complicating (IV.23) with CDD poles or subtractions; these will only produce <u>more</u> roots of $\underset{\sim}{f}^{-1}$.

So now assuming that the first term of (IV.22) vanishes, but the next does not, the dispersion representation of $\underset{\sim}{f}^{-1}$ reads

$$\begin{aligned} \underset{\sim}{f}^{-1} &= -\underset{\sim}{C}^{-1}\,\omega^2 - \frac{\omega^3}{\pi}\int \frac{d\omega'\,k'}{\omega'^3(\omega'-\omega)} \\ &\approx -\underset{\sim}{C}^{-1}\omega^2 + \frac{\omega^2}{2\mu} - (\mu+ik) \\ &= -(\mu+ik)\,[(\underset{\sim}{C}^{-1} - \frac{1}{2\mu})(\mu-ik) + 1\,]. \end{aligned} \qquad \text{(IV.25)}$$

where $\langle B|\underset{\sim}{C}|A\rangle = C_{BA} = [g_B,[M,g_A]]$. This again has roots in addition to the (double) root at ω=0, <u>unless</u> the eigenvalues of $\underset{\sim}{C}$ equal 2μ or 0. This means $\underset{\sim}{C} = 2\mu\underset{\sim}{\Lambda}$ where $\underset{\sim}{\Lambda}$ is a projection operator, $\underset{\sim}{\Lambda}^2 = \underset{\sim}{\Lambda}$; written more explicitly, this condition is

$$\Lambda_{CB}\Lambda_{BA} = \Lambda_{CA}, \text{ where } \Lambda_{BA} = [g_B,[M/2\mu, g_A]]. \qquad (IV.26)$$

The scattering amplitude, from (IV.25), is

$$f = -\Lambda/(\mu+ik) . \qquad (IV.27)$$

[If one assumed that both the first and second terms of (IV.22) vanished, the preceding arguments would lead to the conclusion that the third term must also vanish, and f would have the form

$$f = -\tilde{\Lambda}/(\mu+ik-\omega^2/2\mu), \text{ where } \tilde{\Lambda}=[g_B,M,M,M, g_A]/8\mu^3. \qquad (IV.27a)$$

But the condition $\Lambda=0$, together with the no-production condition (IV.17 would seem to imply that $\tilde{\Lambda}=0$. If this is so, then (IV.27) is the only nonvanishing form for f which has no poles except at $\omega=0$, and which sati fies $\mathrm{Im}\, f = kf^{+}f$.]

The mass condition (IV.26) implies that the isobar mass difference ΔM are of the order

$$\Delta M \sim \mu g^{-2} \qquad (IV.28)$$

where g is the order of the coupling constants, the matrix elements of g_A. Thus the approximation made in (IV.23) and (IV.25), that ΔM could be neglected against μ, is valid if g is large compared to unity

$$g^2 >> 1 . \qquad (IV.29)$$

This was also the condition for the vanishing of meson production (except in the harmonic oscillator cases, where production vanishes even for finite values of g). Explicitly, the sum of the pole terms of a one-meson production amplitude is of order

$$\frac{g^3}{\omega^2}\left(\frac{\Delta M}{\omega}\right)^2 = g^{-1}\mu^2\omega^{-4} ; \qquad (IV.30)$$

the amplitude becomes $g^{-1}(\mu+ik)^{-2}$, schematically, when rescattering is taken into account. Thus the one-meson production cross section σ_1 is of order $g^{-2}\ell n(\frac{\omega}{\mu})/\omega^2$. Consequently $g^2 >>1$ assures the smallness of the ratio of σ_1 to the one-meson scattering cross section σ_o, $\sigma_1/\sigma_o \sim g^{-2}\ell n \frac{\omega}{\mu}$ [However, this ratio is not small for $\omega > O(\mu e^{g^2})$; the strong coupling approximation must fail there.]

For p-wave, a cutoff must be used; using $u(k)^2 = (1 + a^2k^2)^{-1}$ in (IV.19), (IV.23) becomes

$$\underset{\sim}{f}^{-1} = -\underset{\sim}{\gamma}^{-1}\,\omega - \frac{(\mu(1+a\mu)^{-1} - ik)\,a^{-1}\,\omega^2}{\mu^2(\mu-ik)(1-iak)} \tag{IV.31}$$

$$\approx -\underset{\sim}{\gamma}^{-1}\,\omega - \omega^2/\mu^2\; a(1-iak)\;, \tag{IV.32}$$

where we have assumed that the cutoff is high, $a\mu \ll 1$. We have the same result as before, that this has additional roots besides $\omega=0$, unless $\underset{\sim}{\gamma}=0$. Similarly, (IV.25) for p-wave is

$$\underset{\sim}{f}^{-1} \approx -\underset{\sim}{C}^{-1}\,\omega^2 + \frac{\omega^2}{\mu^2 a} - \omega^2/\mu^2\; a(1-iak)\;; \tag{IV.33}$$

this has no additional roots if

$$\underset{\sim}{C} = \mu^2\, a\underset{\sim}{\Lambda} \quad \text{where } \underset{\sim}{\Lambda}^2 = \underset{\sim}{\Lambda}. \tag{IV.34}$$

This is just the same as for s-wave, Eq. (IV.26), but with $\mu^2 a$ replacing 2μ as the scale of the isobar mass differences. The (reduced) scattering amplitude is

$$\underset{\sim}{f} = -\frac{\underset{\sim}{\Lambda}\; a(1-iak)}{\omega^2/\mu^2} \tag{IV.35}$$

so the complete amplitude is

$$f_{\beta\alpha} = f_{\beta j,\alpha i}\,\frac{k_j\, k_i\, u^2(k)}{\mu^2} = -\frac{k_j\,\Lambda_{\beta j,\alpha i}\,k_i}{\omega^2}\cdot\frac{a}{(1-iak)}\;, \tag{IV.36}$$

$$\approx -\frac{k_j\,\Lambda_{\beta j,\alpha i}\,k_i}{\omega^2}\;a, \quad \text{for } \omega \ll a^{-1}\;. \tag{IV.37}$$

The low energy results (IV.32) and (IV.37) hold for <u>any</u> adequate cutoff, the parameter a having the meaning

$$a^{-1} \equiv \frac{2}{\pi}\int_0^\infty dk\;\frac{k^4}{\omega^4}\;\frac{u^2(k)}{u^2(0)}\;. \tag{IV.38}$$

Compared to the s-wave case, the additional factor of $k'^2u^2(k')$ inserted into the integrands of (IV.23) and (IV.25) means that their major contributions come from $\omega' \stackrel{\sim}{\sim} a^{-1}$, and so the approximation of neglecting ΔM is valid if $\Delta M << a^{-1}$, which is a weaker condition than in the s-wave case, $\Delta M << \mu$ (assuming as always a high cutoff, $a\mu << 1$). The p-wave mass condition (IV.35) implies

$$\Delta M \sim \mu^2 a g^{-2} \tag{IV.39}$$

so that the resulting condition on g is

$$g^2 << \mu^2 a^2 . \tag{IV.40}$$

Physically speaking, the factor $k'^2u^2(k')$ provides a centrifugal barrier, so that even when an isobar is unstable, $\Delta M > \mu$, it still may be a narrow resonance which can be treated as a particle. Explicitly, the ratio of width Γ to excitation M of an isobar is of the order

$$\frac{\Gamma}{\Delta M} \sim \frac{g^2k^3/\mu^2}{\Delta M} = g^2(\Delta M)^2/\mu^2 = \mu^2a^2g^{-2} \tag{IV.41}$$

which indeed is small under the condition $\Delta M << a^{-1}$, i.e. (IV.40). Finally, checking explicitly the magnitude of the meson production, its amplitude has the order

$$g^3/\omega^2 \left(\frac{k}{\mu} \right)^3 \left(\frac{\Delta M}{\omega} \right)^2 \sim \mu a^2/g\omega \tag{IV.42}$$

resulting in

$$\sigma_1/\sigma_o \sim \left(\frac{\mu a^2}{g\omega} \right)^2 \omega^2 \, \ell n \, \frac{\omega}{\mu} = \frac{\mu^2 a^2}{g^2} \, \ell n \, \frac{\omega}{\mu} \tag{IV.43}$$

which is small under the same condition (IV.40).

B. Harmonic oscillator or pair theory models

In this case, the coefficients of M, $A_{AB}(g,p)$ etc., are constants $[C_{AB}(g,p) \equiv c_{AB}]$, and we assume $B_{BA} = E_A = 0$ for simplicity. No-production holds for any g, and so the isobar mass differences are not necessarily small. Defining Δ^2_{AB} by

$$[M, [M, g_A]] = \Delta^2_{AB}\, g_B \tag{IV.44}$$

(IV.22) can be summed to

$$f^P = - c/(\omega^2 - \Delta^2) \tag{IV.45}$$

so

$$f^{-1} = -c^{-1}(\omega^2 - \Delta^2) - \frac{(\omega^2 - \Delta^2)^2}{\pi} \int \frac{d\omega'\, k'}{(\omega'^2 - \Delta^2)^2(\omega' - \omega)}$$

$$= -c^{-1}(\omega^2 - \Delta^2) + \frac{\omega^2 - \Delta^2}{2\tilde{\mu}} - (\tilde{\mu} + ik) \tag{IV.46}$$

where $\tilde{\mu} \equiv \sqrt{\mu^2 - \Delta^2}$. So f has no additional poles only if $c = 2\tilde{\mu}$, giving the solution

$$f = -1/(\tilde{\mu} + ik) \ . \tag{IV.47}$$

We can choose the basis of the coupling operators g_A so that c and Δ^2 are both diagonal, $c_{AB}=c_A\delta_{AB}, \Delta^2_{AB}=\Delta^2_A\delta_{AB}$; then $c = 2\tilde{\mu}$ becomes

$$[g_B, [M, g_A]] = 2\tilde{\mu}_A\, \delta_{BA} \ , \quad \tilde{\mu}_A \equiv \sqrt{\mu^2 - \Delta^2_A} \ . \tag{IV.48}$$

From this and $[M, [M, g_A]] = \Delta^2_A\, g_A$ it follows that the isobars are harmonic oscillator states $|n_1 \ldots n_A \ldots>$ with masses

$$M_{n_1 \ldots n_A \ldots} = M_o + n_1\Delta_1 + \ldots + n_A\, \Delta_A + \ldots \tag{IV.49}$$

and couplings

$$<n_1 \ldots n_A \ldots |g_A| n_1' \ldots n_A' \ldots> = \sqrt{n_A}\, g_A\, \delta_{n_1,n_1'} \cdots \delta_{n_A,n_A'+1} \cdots + (n_A \leftrightarrow n_A') \tag{IV.50}$$

where Δ_A and g_A are related by

$$\Delta_A g_A^2 = \tilde{\mu}_A = \sqrt{\mu^2 - \Delta^2_A} \ , \text{ i.e. } \Delta_A = \mu/\sqrt{g^4+1} \ . \tag{IV.51}$$

In this solution the poles at $\omega^2 = \Delta^2$ can be continued through threshol onto the second sheet, that is, to negative values of μ. The isobars are then "virtual" states, rather than bound states. The solution con- tinues to be exact, and is equivalent to a pair-theory Hamiltonian model,

$$H = \sum H_A, \quad H_A = \frac{1}{2}\int d^3r \, [\pi_A^2 + \Phi_A(\mu^2 - \nabla^2)\Phi_A] - \frac{1}{2}\lambda_A \, \Phi_A^2(0). \qquad \text{[no sum]} \tag{IV.52}$$

Another Hamiltonian which has the same solution is one in which the elementary isobar states consist of a ground state and excited states, coupled together through the meson fields, i.e. a Hamiltonian of the form

$$H = \sum H_A, \quad H_A = \frac{1}{2}\int [\pi_A^2 + \Phi_A \, (\mu^2 - \nabla^2) \, \Phi_A] + \Delta_A^o(\frac{1+\tau_3}{2}) - g_A^o \, \tau_1 \, \Phi_A(0), \qquad \text{[no sum]} \tag{IV.53}$$

when the limits $\Delta_A^o \to \infty$, $g_A^o \to \infty$ are taken.

C. "Stretched rotator" models

Here M has a term $B_{AB} g_A p_B$. The simplest model is charged scalar, whose internal symmetry allows a term of M of the form

$$\Delta\varepsilon_{AB} \, g_A p_B = \Delta Q, \quad \varepsilon_{12} = -\varepsilon_{21} = 1, \; \varepsilon_{11} = \varepsilon_{22} = 0. \tag{IV.54}$$

It is convenient to choose the basis of $\mathfrak{g}_A$ so that $B_{AB} = \Delta\varepsilon_{AB}$ is diagon [at the expense that the $\mathfrak{g}_A$ are not Hermitian, contrary to our usual practice]. The components of $\mathfrak{g}$ are then $\mathfrak{g}_\varepsilon$, $\varepsilon = \pm 1$, with

$$\mathfrak{g}_\varepsilon^+ = \mathfrak{g}_{-\varepsilon} \tag{IV.55}$$

$$[M, \mathfrak{g}_\varepsilon] = \varepsilon\Delta g_\varepsilon + 0(\frac{1}{g}) \; ; \tag{IV.56}$$

$\mathfrak{g}_\varepsilon$ is the vertex operator for absorbing a meson of charge ε (in units of 1), $[Q, g_\varepsilon] = -\varepsilon g_\varepsilon$. The isobars can be labeled by their charge Q. So

$$M = M(Q) = M_o + \Delta Q + \frac{c}{2} Q^2 \tag{IV.57}$$

and it follows that

$$C_{\eta\varepsilon} = [g_\eta^+ , [M, g_\varepsilon]] = c\eta\varepsilon\, g_\eta^+ g_\varepsilon \, . \tag{IV.58}$$

Since $g_+^+ g_+ = g_- g_+$ commutes with M, it is a number, $\equiv g^2$.

The one-meson scattering matrix $\langle Q_2 | f_{\eta\varepsilon} | Q_1 \rangle$ can be labeled by η, ε and Q, the total charge, i.e. $\langle Q-\eta | f_{\eta\varepsilon} | Q-\varepsilon \rangle \equiv f_{\eta\varepsilon}^Q$. The poles of $f_{\eta\varepsilon}^Q$ occur near $0=E-M_Q \equiv \omega$, and

$$k_{\eta\varepsilon}^Q = \delta_{\eta\varepsilon} \sqrt{(E-M_{Q-\varepsilon})^2 - \mu^2} \approx \delta_{\eta\varepsilon} \sqrt{(\omega + \varepsilon\Delta)^2 - \mu^2} \equiv \delta_{\eta\varepsilon}\, k_\varepsilon \tag{IV.59}$$

where the approximation is valid as $g \to \infty$, but does not assume Δ to be small.

Figure: Singularities of $f_{++}^Q(\omega)$

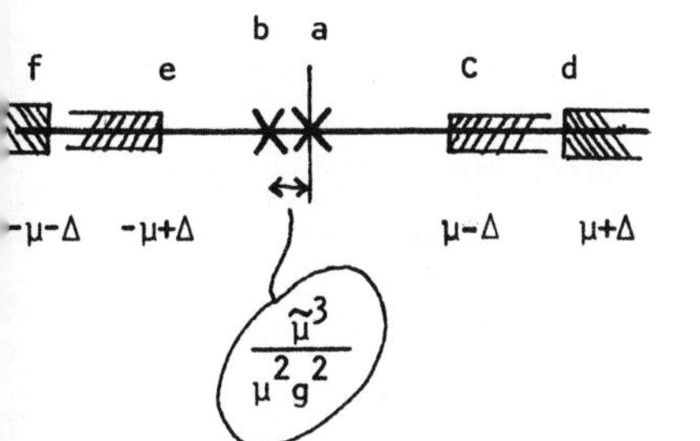

a: direct pole, isobar Q

b: crossed pole, isobar Q-2

c: direct threshold branchpoint, isobar, meson Q-1, +

d: direct threshold branchpoint, isobar, meson Q+1, -

e: crossed threshold branchpoint, isobar, meson Q-3, +

f: crossed threshold branchpoint, isobar, meson Q-1, -

For $\underset{\sim}{f}^{-1}$ [cf. Eq. (IV.25)] we have then

$$(f_{\eta\varepsilon}^Q)^{-1} = - C_{\eta\varepsilon}^{-1}\, \omega^2 + \frac{\mu^2}{2\tilde{\mu}^3}\, \omega^2\, \delta_{\eta\varepsilon} - d_\varepsilon\, \delta_{\eta\varepsilon} \tag{IV.60}$$

where

$$d_\varepsilon = \frac{\tilde{\mu}^2 - \varepsilon\omega\Delta}{\tilde{\mu}} + ik_\varepsilon \; , \quad \tilde{\mu} = \sqrt{\mu^2 - \Delta^2} \, . \tag{IV.61}$$

As above, the condition for $\underset{\sim}{f}^{-1}$ to have roots only at $\omega = 0$ is

$$C_{\eta\varepsilon} = \frac{2\tilde{\mu}^3}{\mu^2} \Lambda_{\eta\varepsilon}, \text{ where } \underset{\sim}{\Lambda}^2 = \underset{\sim}{\Lambda} . \tag{IV.62}$$

Comparing with (IV.58), and observing that $(\eta\rho\, \mathcal{g}_\eta^+ \mathcal{g}_\rho)(\rho\varepsilon\, \mathcal{g}_\rho^+ \mathcal{g}_\varepsilon) =$ $2\, g^2 (\eta\varepsilon\, \mathcal{g}_\eta^+ \mathcal{g}_\varepsilon)$, we see that

$$\Lambda_{\eta\varepsilon} = \frac{\eta\varepsilon}{2g^2} \mathcal{g}_\eta^+ \mathcal{g}_\varepsilon \quad , \text{ and } \quad c = \frac{\tilde{\mu}^3}{\mu^2 g^2} \tag{IV.63}$$

so

$$M = M_o + Q\Delta + \frac{Q^2 \tilde{\mu}^3}{2g^2 \mu^2} . \tag{IV.64}$$

Note that because $c>0$, M [Eq. (IV.57)] has a lower bound and there wil be a ground state (perhaps a doublet); of course this isobar is no mor "elementary" than any other. To get $f^Q_{\eta\varepsilon}$, invert (IV.60) by expanding in $C_{\eta\varepsilon}$; <u>then</u> use (IV.62) and

$$\Lambda_{\eta\rho}\, d_\rho\, \Lambda_{\rho\varepsilon} = \frac{1}{2} (d_+ + d_-)\, \Lambda_{\eta\varepsilon} = \frac{1}{2} [2\tilde{\mu} + i(k_+ + k_-)]\, \Lambda_{\eta\varepsilon} . \tag{IV.65}$$

The result is

$$f^Q_{\eta\varepsilon} = -2\Lambda_{\eta\varepsilon}/[2\tilde{\mu} + i(k_+ + k_-)] . \tag{IV.67}$$

If g is large but fixed, the last term of M, Eq. (IV.57) or (IV.64), is no longer small when Q is sufficiently large. But this is trivially avoided by varying Δ with Q; instead of (IV.64), we shoul write

$$\frac{dM}{dQ} = \Delta, \quad \frac{\partial^2 M}{\partial Q^2} = \frac{d\Delta}{dQ} = \tilde{\mu}^3/\mu^2 g^2 \; , \quad \text{where } \tilde{\mu} \equiv \sqrt{\mu^2 - \Delta^2} . \tag{IV.68}$$

This is valid if Δ changes slowly with Q, i.e. if $g^2 >> 1$. Integrati

$$\Delta = \frac{Q - Q_o}{\sqrt{g^4 + (Q - Q_o)^2}}\, \mu \tag{IV.69}$$

$$M = M_o + \sqrt{g^4 + (Q - Q_o)^2}\; \mu \cdot \tag{IV.70}$$

The results (IV.67) and (IV.70) were first found by Pais and Serber (1957) by Hamiltonian methods. This solution exhibits centrifugal stretching: Near $Q=Q_o$, $M \approx \frac{\mu}{2g^2}(Q-Q_o)^2 + M_o$, just the energy of a rigid rotator in charge space, with moment of inertia g^2/μ; Δ increases linearly with Q, $\Delta \approx \frac{\mu}{g^2}(Q-Q_o)$. But for larger $|Q-Q_o|$, M increases less rapidly than quadratically, corresponding to an increased moment of inertia. Asymptotically, M increases linearly, and Δ approaches μ; but since $\Delta < \mu$, all isobars are stable against meson emission.

It would appear that charged scalar is the only model with this property that the strong coupling approximation remains valid for highly excited isobars whose spacing is not small compared to μ. But in fact other models may approach the structure of charged scalar just in this case. Take for instance the symmetric scalar model, in which

$$< I', I_3' |g_\varepsilon| I, I_3 > = g \begin{pmatrix} I' & I & 1 \\ -I_o & I_o & 0 \end{pmatrix} \begin{pmatrix} I' & I & 1 \\ -I'_3 & I_3 & \varepsilon \end{pmatrix}. \qquad \text{(IV.71)}$$

Consider the elastic amplitude of a charge $\varepsilon = +$ meson incident on an isobar $| I, I_3 = I >$. It has three direct pole terms, at $\omega \approx \Delta$, 0, and $-\Delta$ $(\Delta \approx \mu I/g^2)$ corresponding to intermediate isobars of i-spin $I=I+1$, I, and $I-1$ respectively; but for large I [and $I_o=0(1)$] the residues of last two are smaller than the first by a factor I^{-2}. Hence (if $g^2 >> 1$) when $\Delta \approx \mu I/g^2$ becomes appreciable, the amplitude has essentially only the one pole, resembling charged scalar and so having a solution of the same form. [This result is also given by the classical calculation.]

It could be conjectured that this is generally true for all s-wave models (one would expect in general the solution to resemble the direct product of several charges).

Whether centrifugal stretching holds in p-wave models is doubtful. The difficulty is that since a^{-1} plays roughly the role of a scale that μ does in s-wave models, the stretching would occur when Δ was appreciable compared to a^{-1}. But then the isobars are rather broad, so that the approximation of treating them as particles is bad.

APPENDIX A

Coupling Renormalization

The physical, or renormalized, coupling constants are the matrix elements of the current between exact isobar states. The bare, or un-renormalized, coupling constants are the matrix elements of the current between bare isobar states. A ratio of the two couplings, physical/bare , is a renormalization factor, denoted by Z. Of course, it is defined only for those isobars which have bare isobar counterparts. [We shall not discuss the isobar "wave function" renormalization, i.e. the amplitude of the no-meson component of an exact isobar state vector expanded into bare states. This is uninteresting; it vanishes both in the limit of no cutoff and in the strong coupling limit.]

To avoid algebraic complexity, we shall treat the specific model $(SU_2, 3)$, with $i_o = 1/2$. In terms of bare states, the current of this model is

$$\vec{\mathcal{J}} = g_o \, \vec{\tau} \tag{A.1}$$

where , in contrast to Secs. II and III, we write a subscript on the bare coupling constant, which occurs in the Hamiltonian. [Also note the factor of 2 difference from (II.54) in the normalization of the coupling constant.] There is only one Z, because there is only one bare coupling constant. We can express Z as

$$Z = \frac{\langle \tfrac{1}{2} | \mathcal{J}_3 | \tfrac{1}{2} \rangle}{\langle \tfrac{1}{2} | \mathcal{J}_3 | \tfrac{1}{2} \rangle_o} = \frac{\langle \tfrac{1}{2} | \tau_3 | \tfrac{1}{2} \rangle}{\langle \tfrac{1}{2} | \tau_3 | \tfrac{1}{2} \rangle_o} = \langle \tfrac{1}{2} | \tau_3 | \tfrac{1}{2} \rangle \tag{A.2}$$

where "$|\tfrac{1}{2}\rangle$" is the "proton", i.e., the $i=\tfrac{1}{2}$, $i_3 = +\tfrac{1}{2}$ isobar. Note that $-1 < Z < 1$.

Put into words, Z is the expectation value of τ_3 in the clothed proton state, or

$$Z = P_{+\frac{1}{2}} - P_{-\frac{1}{2}} \tag{A.3}$$

where $P_{+\frac{1}{2}[-\frac{1}{2}]}$ is the probability that the core (i.e., the spinor state) of the proton is a proton ($i_3 = +\tfrac{1}{2}$) [neutron ($i_3 = -\tfrac{1}{2}$)]; $P_{+\frac{1}{2}} + P_{-\frac{1}{2}} =$

In the limit of no cutoff $P_{+\frac{1}{2}} \to P_{-\frac{1}{2}}$, so

$$Z \to 0, \text{ when } a \to 0 \,, \tag{A.4}$$

where a is the cutoff distance. On the other hand, in the strong coupling limit we have [see Eqs. (10) and (23) below]

$$Z \to 1/3 \,, \quad \text{when} \quad g_o \to \infty \,. \tag{A.5}$$

In a sense, the clothed proton state becomes chaotic in the limit $a \to 0$, and it is equally probable to find the core a proton or a neutron. But in the limit $g_o \to \infty$, the clothed state becomes ordered. If both $a \to 0$ and $g_o \to \infty$, the value of Z depends on the order in which the limits are taken.

An important question is the region of validity of the strong coupling approximation. We shall show that it is valid for large <u>physical</u> coupling

$$g = Z\, g_o >> 1. \tag{A.6}$$

Hence even if $g_o >> 1$, as $a \to 0$ (hence $Z \to 0$) the strong coupling approximation fails.

For weak coupling, Z can be calculated by perturbation series

$$\langle|\not{q}_\alpha|\rangle = {}_0\langle|[\, \not{q}_\alpha + [\not{q}_\beta, \not{q}_\alpha]\not{q}_\beta \int d^3k \; \frac{4\pi u_k^2}{8\pi^3 2\omega_k^3} + \ldots]\, |\rangle_0 \tag{A.7}$$

In $(SU_2, 3)$ with $i_o = \frac{1}{2}$, this yields

$$Z = 1 - g_o^2 \frac{4}{\pi} \int_0^\infty dk \, \frac{k^2}{\omega_k^3} \; u_k^2 + 0(g_o^4)$$

$$= 1 - g_o^2 \; \frac{4}{\pi} \, \ell n \, \frac{c}{a\mu} \; + 0(g_o^4) \,. \tag{A.8}$$

The quantity $g_o^2 \; \frac{2}{\pi} \, \ell n \, \frac{c}{a\mu}$ is the probability, to first order in g_o^2, of finding the proton as a neutron plus a positive meson; it diverges $\to \infty$, as $a \to 0$. In the Landau approximation, which consists of keeping the highest power of $\ell n(1/a\mu)$ in each order of g_o^2, one finds

$$Z \approx [1 + g_o^2 \, \frac{4}{\pi} \; \ell n(1/a\mu)]^{-1} \,. \tag{A.9}$$

This is better behaved than (8) as $a \to 0$, but it is not clear that it is trustworthy.

In the strong coupling limit, we have from Eq. (II.60a),

$$Z = \left. \frac{i_o^2}{i(i+1)} \right|_{i_o = i = \frac{1}{2}} = 1/3 \;. \qquad (A.10)$$

Similarly in general in other models, Z is finite in the strong coupling limit. [In terms of the structure of the physical "proton" state, the 1/3 comes about as follows: The probabilities of the meson cloud having i-spin L=0 and L=1 are both equal to 1/2; in the former state $<\tau> = 1$, and in the latter $<\tau> = -1/3$; so $Z = \frac{1}{2}(1 - 1/3) = 1/$

We now look at the correction to Z of order g_o^{-2}. This is simply found as follows: The low-lying states belong to the eigenvalue $+g_o$ of $\vec{\mathcal{G}} \cdot \hat{g} = \vec{\tau} \cdot \vec{g}$; hence for them

$$\vec{\mathcal{G}}^2 = g_o^2 \;. \qquad (A.11)$$

Take the matrix element of this in say the proton state $|\frac{1}{2}>$, and insert a complete set of low-lying states

$$g_o^2 = <\tfrac{1}{2}|\vec{\mathcal{G}}^2|\tfrac{1}{2}> = <\tfrac{1}{2}|\mathcal{G}_j|n><n|\mathcal{G}_j|\tfrac{1}{2}>$$

$$= \sum_s <\tfrac{1}{2}|g_j|s><s|g_j|\tfrac{1}{2}> + \prod_{(n)} \left(\frac{u_i^2 k_i d\omega_i}{\pi}\right) \cdot \left|f_{n,\frac{1}{2}j}\right|^2$$

$$= g^2 + \frac{1}{4\pi^2} \sum_j \int dE \, \frac{k\,\sigma_{\frac{1}{2}j}(E)}{u_k^2} \qquad (A.12)$$

where we have separated the no-meson and meson-containing intermediate state contributions. In the last line $\sigma_{\frac{1}{2}j}(E)$ is the total cross secti for meson j, with energy $\omega = E - M_{\frac{1}{2}}$, incident on isobar "½", i.e., proton

The strong coupling sum rule (13) for g_o^2 is similar to the Chew-Low sum rule, but not identical. The difference is that in the Chew-Low sum rule no distinction is made between low lying and high lying states (this distinction can be made of course only in strong coupling From $\vec{\mathcal{G}} = g_o \vec{\tau}$ follows directly

$$\vec{\mathcal{G}}^2 = 3g_o^2 \qquad (A.13)$$

instead of (A.11) and the Chew-Low sum rule is

$$3g_o^2 = \langle \tfrac{1}{2} | \vec{\mathscr{g}}^2 | \tfrac{1}{2} \rangle = \langle \tfrac{1}{2} | \mathscr{g}_i \; | n \rangle \langle n | \mathscr{g}_i \; | \tfrac{1}{2} \rangle \quad . \tag{A.14}$$

[Chew-Low actually used " $\mathscr{g}_3 = g_o \tau_3$, hence $g_o^2 = \mathscr{g}_3^2$ ". For our purposes, it is a little more convenient to sum over the three components]. This differs from (12) by a) the factor 3 on the left hand side, b) the extension of the summation on the right hand side over all states, not just low lying. In the strong coupling limit the Chew-Low sum rule (14) is inconvenient, in fact useless, because $\sigma_{\frac{1}{2}i}(E)$ is not known for large energy. [There are sum rules, based on the commutators of $\mathscr{g}_\alpha$, which are similarly useless in the strong coupling limit, because $[\mathscr{g}_\alpha, \mathscr{g}_\beta] = 0$ if only low-lying states are kept.]

Before turning to the evaluation of (12) let us digress a bit more on the subject of the very high energy behavior of the scattering amplitude. The strong coupling approximation certainly fails at an energy of the order of $g^2 a^{-1}$, the excitation energy of the higher spinor states. Of course we know one thing about the high energy cross section: By subtracting (12) from (14) we have the sum rule

$$g_o^2 = \frac{1}{8\pi^2} \sum_j \int_{E_c}^{\infty} dE \; \frac{k\, \sigma_{\frac{1}{2}j}(E)}{u_k^2} \tag{A.15}$$

where E_c is somewhat larger than $M_{\frac{1}{2}} + a^{-1}$. If it were not for the cutoff function u_k in the denominator of the integral, it would be impossible for this sum rule to hold for large g_o, because the right hand side would be of order $(g_o)^o = 1$. To see how the sum rule might be satisfied, suppose that there is a "resonance" of width Γ at the upper spinor state, i.e.

$$k^2 \sigma_{\frac{1}{2}j}(E) = O(1) \text{ for } |\omega - g_o^2 \; a^{-1}| \lesssim \Gamma \quad ; \tag{A.16}$$

then assuming $u_k^2 \approx (ak)^{-2n}$ for large k [n must be > 1 in order that the contribution of the low-lying states to the sum rule (12) converges] we find that (15) is satisfied by a width

$$\Gamma \approx g_o^{-4(n-1)} \; a^{-1}, \; n-1 > 0. \tag{A.17}$$

Thus the width of this "resonance" is small compared to its energy

$g_o^2 a^{-1}$.

This "resonance" at the upper spinor state may not exist for weak enough cutoff, $a^{-1} >> \mu e^{-g^2}$, because the strong coupling approximation already fails at the energy

$$\omega = c_1 \mu\, e^{c_2 g^2} \qquad c_1, c_2 = O(1) \tag{A.18}$$

due to the violation of the unitarity limit by the strong coupling approximation to the one-meson production cross section [see the estimate following Eq. (IV.30). One estimate for the two-meson production cross section reaches the unitarity limit at a much lower energy, namely, $g^2\mu$; it is probably wrong.]

Another evidence that the strong coupling amplitude, e.g., Eq. (IV.27), must fail at high energies is that it does not obey Levinson's theorem, that is, the asymptotic behavior of its eigenphases is not correct; they are too low. Thus at a sufficiently high energy the eigenphases must rise to the correct values; this may be related to the "resonance" behavior discussed in the preceding paragraph.

A final comment on the high energy behavior of the strong coupling solutions is the reminder that this is completely unphysical. As stated in the Introduction, the cutoff is only an artifice to simulate features of relativistic field theory which are omitted in the static model. One should not attach any physical significance to the behavior of amplitudes at energies near to or above the cutoff energy a^{-1}. A corrolary is that the sum rules (14) and (15) are unphysical.

We now return to the evaluation of (12). When the states $|n\rangle$ are low-lying, only the one-meson states contribute (no meson production) and $f_{sk,\frac{1}{2}j}$ is given by (II.41), so (12) becomes

$$g_o^2 = g^2 + \frac{\mathrm{Tr}\Lambda}{\pi} \int_\mu^\infty \frac{d\omega\, k\, u_k^2}{|I(\omega) - I(0)|^2}\ , \qquad \mathrm{Tr}\Lambda = 2 \text{ in}(SU_2, 3). \tag{A.19}$$

where we have used

$$\langle \tfrac{1}{2}|\Lambda_{ik}|s\rangle\langle s|\Lambda_{ki}|\tfrac{1}{2}\rangle = \langle \tfrac{1}{2}|\Lambda_{ik}\Lambda_{ki}|\tfrac{1}{2}\rangle = \langle \tfrac{1}{2}|\mathrm{Tr}\Lambda|\tfrac{1}{2}\rangle = \mathrm{Tr}\ \Lambda \tag{A.20}$$

which is an invariant of the rotational band (representation).

This calculation is similar to that performed above for $<\bar{\phi}^2>$, Eqs. (III.34) - (III.40), and the integral which occurs is the same

$$\int_{\mu}^{\infty} \frac{d\omega\ k\ u_k^2}{|I(\omega)-I(0)|^2} \approx \int_{\mu}^{a^{-1}} \frac{d\omega\ k}{\omega^2} + [\ \int dk\ u_k^2]^{-1} \int_{a^{-1}}^{\infty} dk\ k\ u_k^2 \qquad (A.21)$$

$$\approx \ln(1/a\mu),\ \text{if}\ a\mu << 1.$$

Using (21) in (18), the result is

$$g_o^2 = g^2 + \frac{2}{\pi} \ln(1/a\mu),\ g_o\ \text{large}, \qquad (A.22)$$

or

$$Z = \frac{1}{3} [1 - \frac{1}{g_o^2} \frac{1}{\pi} \ln(1/a\mu) + 0(g_o^{-4})]\ ,\quad a\mu << 1. \qquad (A.23)$$

This gives Z to order g_o^{-2}.

APPENDIX B

Dynamic Binding from the S-Matrix Point of View

If a scattering amplitude has poles with large residues, these poles are not independent, but instead strongly correlated both in position and residue. Roughly speaking, this results from the following consideration. Schematically, the dispersion representation of an amplitude f is

$$f = f^P + \widetilde{\mathcal{I}m\ f} \qquad (B.1)$$

which exhibits f as the sum of its pole terms f^P and of the Hilbert transform $\widetilde{\mathcal{I}m\ f}$ of its imaginary part. In the physical region of energy, f is limited in magnitude because of unitarity, hence, so is $\mathcal{I}m\ f$ and $\widetilde{\mathcal{I}m\ f}$. Consequently, f^P must be limited in magnitude in the physical region. Thus pole terms with large residues must cancel one another sufficiently, to avoid violating the unitarity limit. We can put this more quantitatively by using the N/D method. We treat explicitly the two simplest models, in which there is just one meson which is either self conjugate (neutral scalar) or not (charged scalar).

1. Charged scalar

The model is specified as follows: There is a state of the scatterer, isobar B, which is stable but can virtually decay into isobar A and an s-wave meson θ, with coupling g_{BA}; likewise A can virtually dec into B and $\bar{\theta}$, with the same coupling.

We take θ to be distinct from its antiparticle θ; one can imagine θ to carry one unit of a "charge", $\bar{\theta}$ minus-one unit, and isobar B to have one more unit than A.

We consider the elastic scattering amplitude $f_B(\omega)$ for the proces $\theta B \to \theta B$. [$f_B(-\omega)$ is the amplitude for $\bar{\theta}B \to \bar{\theta}B$ at energy ω]. $f_B(\omega)$ has the "crossed pole" term

$$f_B^P(\omega) = g_{BA}^2/(\omega + A - B) \tag{B.2}$$

where A, B stand for the masses of the isobars. From the viewpoint of the θB system [i.e. for $\omega > \mu$] this amplitude is positive and represents an attractive interaction; from the viewpoint of the $\bar{\theta}B$ system [i.e. for $\omega < -\mu$] it is a repulsion. Finally, we suppose that we are given the ratio $\sigma/\sigma_{el} \equiv \mathcal{R}_B(\omega)$, where $\sigma(\sigma_{el})$ is the total (elastic) θB cross section. Notice that because $\sigma > \sigma_{el}$, we have $\mathcal{R}_B(\omega) > \theta(|\omega| - \mu)$.

Let us write

$$f_B(\omega) = f_B^P(\omega)/D_B(\omega) \tag{B.3}$$

where $D_B(B-A) = 1$, so that $f_B(\omega) \to f_B^P(\omega)$ as $\omega \to B-A$.

We have

$$\mathcal{I}m\, D_B(\omega) = -k\, f_B^P(\omega)\, \mathcal{R}_B(\omega) \quad . \tag{B.4}$$

Thus we can write a dispersion representation [D.R.] for D_B

$$D_B(\omega) = 1 - g_{BA}^2 \; \frac{\omega+A-B}{\pi} \int \frac{d\omega' \; k' \; \mathcal{R}(\omega')}{(\omega' + A-B)^2 \; (\omega'-\omega)} \quad . \tag{B.5}$$

[This once subtracted D.R. converges if $\mathcal{R}(\omega) < 0(|\omega|)$ for large $|\omega|$; we shall assume this to be so. The unsubtracted D.R. for D would also have converged if $|\mathcal{R}(\omega) - \mathcal{R}(-\omega)| < 0(1/\ell n\; \omega)$; the normalization condition $D_B(B-A) = 1$ would then require

$$1 = - \frac{g_{BA}^2}{\pi} \int d\omega \; \frac{k\, \mathcal{R}_B(\omega)}{(\omega+A-B)^2} \quad . \tag{B.6}$$

This is unlikely to be fulfilled in general. In any case, we can continue to use (5), whether or not (6) is satisfied.]

In (5) we have not written any pole terms (CDD poles) and so have not increased the number of free parameters in the solution beyond the original two parameters, g_{BA} and B-A.

Since $\mathcal{R}_B(\omega) \geq 0$, we see from

$$\partial_\omega \; D_B(\omega) = - \frac{g_{BA}^2}{\pi} \int \frac{d\omega' \; k' \; \mathcal{R}_B(\omega')}{(\omega'-\omega)^2 (\omega'+A-B)} \tag{B.7}$$

that $D_B(\omega)$ decreases monotonically in the "gap" $-\mu<\omega<\mu$. Hence we have the satisfactory behavior that higher order effects make the $\bar{\theta}B$ scattering length, $f_B(-\mu)$, which is repulsive, become smaller, and the θB scattering length, $f_B(\mu)$, become larger (for not too large g_{BA}^2). For a sufficiently large value of g_{AB}^2, $D_B(\omega)$ will pass through zero at some ω_B, $B-A < \omega_B < \mu$. Thus $f_B(\omega)$ will have a pole at $\omega = \omega_B$, representing a bound state of the θB system; let us call it C. The mass of C is given by $C = \omega_B + B$.

The larger that $\mathcal{R}_B(\omega)$ is, the more deeply bound C is. Because $\mathcal{R}_B(\omega)$ has the lower limit $\mathcal{R}_B > \theta(|\omega| - \mu)$, we have an upper limit on the mass of C. For example, if $g_{BA}^2 > \sqrt{1-(B-A/\mu)^2}$, then $D_B(\mu) < 0$, and hence C must be bound. This type of bound on g^2, which puts an upper limit on

g^2 under the assumption that there is no bound state, was first found by Gasiorowicz and Ruderman (1958). It holds only because of the assumed short range nature of the meson-scatterer interaction.

The θBC coupling constant is given by the residue of the pole of $f_B(\omega)$ at $\omega = \omega_B$,

$$g_{CB}^2 = \frac{g_{BA}^2}{(\omega_B + A-B)[-\partial_\omega \; D_B(\omega)]_{\omega=\omega_B}} = g_{BA}^2 - \left(\frac{I_{2,2}}{I_{1,2}I_{2,1}}\right)_{\omega=\omega_B} \qquad (B.8)$$

where

$$I_{m,n} = \frac{1}{\pi}\int d\omega \; \frac{k\,\mathcal{R}(\omega)}{(\omega+A-B)^m(\omega-\omega_B)^n} \; . \qquad (B.9)$$

The larger g_{BA}^2 is, the closer ω_B is to B-A. For sufficiently large g_{BA}^2 we can neglect the difference between ω_B and B-A in the integrand of D_B in Eq. (5), and so write

$$\omega_B + A-B \approx \frac{\nu_B}{g_{BA}^2} \; , \qquad g_{BA}^2 >> 1 \qquad (B.10)$$

where

$$\nu_B = \left[\frac{1}{\pi}\int d\omega \frac{k\,\mathcal{R}_B(\omega)}{(\omega+A-B)^3}\right]^{-1} \leqslant 2\mu\left[1 - \left(\frac{B-A}{\mu}\right)^2\right]^{3/2} \lesssim 2\mu. \qquad (B.11)$$

Thus

$$C-B = B-A + \nu_B/g_{BA}^2 + O(g_{BA}^{-4}) \qquad (B.12)$$

i.e., the excitation energy of C above B is equal to the excitation energy of B above A, plus the amount ν_B/g_{BA}^2, which is smaller than $2\mu/g_{BA}^2$.

Likewise neglecting the difference between ω_B and B-A in the formula for g_{BC}^2, Eq. (B.8), we have

$$g_{CB}^2 \approx g_{BA}^2 - X, \text{ where } X \lesssim \tfrac{1}{2}\left[1 + \frac{|B-A|}{\mu}\right]^2\left[1 - \left(\frac{B-A}{\mu}\right)^2\right]^{\frac{1}{2}} \leqslant 1.035 \qquad (B.13)$$

that is,

$$g_{CB}^2/g_{BA}^2 = 1 + O(g_{BA}^{-2}) \; . \qquad (B.14)$$

If g_{BA}^2 is sufficiently large, so that C is bound, we can consider the scattering process θC→θC. The preceding argument repeats itself, with the replacements A→B, B→C. We deduce a bound state of θC, call it D, with mass given by

$$D-C = C-B+ \nu_C/g_{CB}^2 + O(g_{CB}^{-4}) \qquad (B.12')$$

where ν_C is given by Eq. (11) upon the replacements A→B, B→C. The coupling g_{DC} is given by a similar modification of Eq. (13). Then we can consider the scattering process θD→θD, and deduce a bound state of θD, call it E, and so on; we deduce a chain of isobars. Suppose for definitiveness that B was heavier than A, B-A>0 [we can always make this so, by making the redefinitions A↔B and $\theta \leftrightarrow \bar{\theta}$]. Then the masses of the chain of isobars A,B,C,D,..., increase, and the mass differences increase, in steps of ν_i/g_{ij}^2. If $\mathcal{R}_i(\omega)$ and $\mathcal{R}_i(-\omega)$ are not too different, the couplings g_{AB}^2, g_{BC}^2, g_{CD}^2,... decrease at most by steps of order 1. In Sec. IV, Eqs. (IV.54) - (IV.70), we have solved the present model completely and found that in fact the couplings are all equal, from which it follows [Eqs. (B.11) and (B.12), or Eq. (IV.68)], that the chain of isobars extends upward indefinitely.

By considering the scattering $\bar{\theta}A \to \bar{\theta}A$ one finds a $\bar{\theta}A$ bound state, call it Z. Similarly, there is a $\bar{\theta}Z$ bound state, Y, and so on, so that the chain also extends downward. The masses, however, do not decrease indefinitely, because as we proceed down the chain the mass differences decrease in steps of ν_i/g_{ij}^2 and eventually become negative. Thus there is a lowest mass isobar, the ground state of the scatterer. Clearly, our treatment we have made here does not qualitatively distinguish the ground state from the other isobars; all isobars of the chain are equally describable as bound states of θ and the isobar below in the chain, or of $\bar{\theta}$ and the isobar above. The choice of the two isobars A and B with which to start our considerations was entirely arbitrary.

Figure: Part of the isobar spectrum

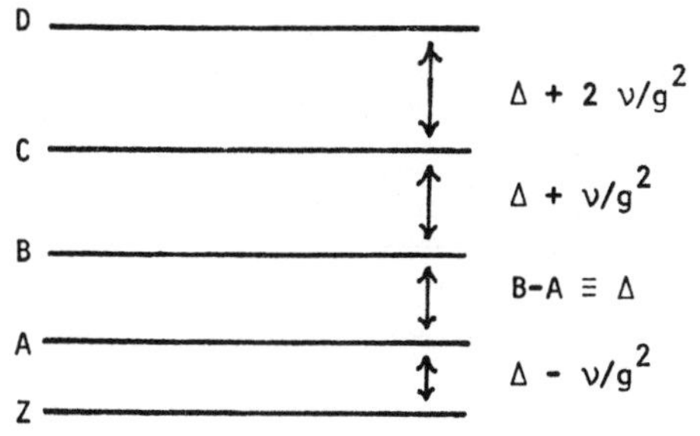

D

$\Delta + 2\,\nu/g^2$

C

$\Delta + \nu/g^2$

B

$B-A \equiv \Delta$

A

$\Delta - \nu/g^2$

Z

The nature of the mass spectrum of this chain of isobars will become clearer if we make the specializing assumption that the ν_i are all the same. This will be true if a) we restrict ourselves to the part of the chain where the mass differences are small compared to μ, and b) the $\mathcal{R}_i(\omega)$ are the same [this is actually true in the strong coupling limit of the scalar model, where $\mathcal{R}_i(\omega) = 2\theta(|\omega|-\mu)$; see Goebel (1958) and above, section IV]. Instead of calling the isobars Y,Z,A,B,C,...., let us label them by their charge index Q, which runs over integrally spaced values. Then the relation between masses, Eq. (12), reads

$$M_{Q+1} - M_Q = M_Q - M_{Q-1} + \nu/g^2 \tag{B.15}$$

where

$$\nu \approx [\frac{1}{\pi} \int \frac{d\omega\ k\,\mathcal{R}(\omega)}{\omega^3}]^{-1} \lesssim 2\mu \; ; \tag{B.16}$$

hence

$$M_Q = \frac{\nu}{g^2} (\frac{1}{2} Q^2 + \alpha Q) + \beta \tag{B.17}$$

where α and β are constants, which depend on the choice of origin of Q and of energy, respectively. M_Q has the form of the energy levels of a two dimensional rigid rotator, with angular momentum quantum number Q; the chain of isobars is a rotational band.

2. Neutral scalar ("odd parity")

The simplest model with an s-wave <u>self-conjugate</u> meson, [$\bar{\theta}=\theta$; all scattering amplitudes are crossing symmetric $f(-\omega) = f(\omega)$] would have one isobar "A" to which the meson was coupled. Since the direct and crossed pole terms [,] of the θA scattering amplitude $f(\omega)$ cancel, the model is trivial unless $f(\omega)$ is made nonvanishing by specifying, say, $f(0) = \lambda$. [This is equivalent to a pair theory Hamiltonian model.] Then the dispersion representation for $f^{-1}(\omega)$ is

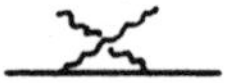

$$f_A^{-1}(\omega) = \lambda^{-1} - \frac{\omega^2}{\pi} \int \frac{d\omega'\ k'\,\mathcal{R}(\omega')}{\omega'^2(\omega' - \omega)} = \lambda^{-1} - \omega^2 \frac{2}{\pi} \int_0^\infty \frac{dk'\ k'^2\mathcal{R}(\omega')}{\omega'^2(\omega'^2 - \omega^2)} \tag{B.18}$$

where as before, $\mathcal{R}(\omega) \equiv \sigma/\sigma_{e\ell}$. It is not possible for λ to be negative, for then $f_A(\omega)$ would have a "ghost" pole on the negative ω^2 axis; this is an example of the theorem that a zero range force cannot produce a nonvanishing repulsive scattering amplitude. For small positive λ, $f_A(\omega)$ has a pole on the second sheet (= a virtual state); for λ larger than the critical value λ_c

$$\lambda_c = [\ \frac{2}{\pi}\int_0^\infty dk\ \frac{\mu^2}{\omega^2}\ \mathcal{R}_A(\omega)\ ]^{-1} \leqslant \mu^{-1} \tag{B.19}$$

the pole is on the first sheet (= bound state). The inequality $\lambda_c \leqslant \mu^{-1}$ follows from $\mathcal{R}(\omega) \geqslant 1$, and is a Gasiorowicz-Ruderman type of theorem. Calling the θA bound state "B", with excitation energy $M_B - M_A = \Delta$, the θAB coupling constant g_{BA} is

$$g_{BA}^2 = [-\partial_\omega\ f_A^{-1}(\omega)]^{-1}_{\omega=\Delta} = g^2(\Delta,\mathcal{R}) \tag{B.20}$$

where

$$g^2(\Delta,\mathcal{R}) \equiv [\ \frac{4\Delta}{\pi}\int_0^\infty dk\ \frac{k^2}{(\omega^2-\Delta^2)^2}\ \mathcal{R}_A(\omega)\ ]^{-1} \leqslant \frac{\sqrt{\mu^2-\Delta^2}}{\Delta}\ . \tag{B.21}$$

Notice that we can ascribe an odd "parity" to the meson θ (if we set to zero the θAA coupling, which had no effect in the one-scatterer system) and opposite parities to isobars A and B. The model can be described as neutral "pseudoscalar" s-wave.

One may assume the two isobars A and B from the start, with $M_B - M_A \equiv \Delta$ and $g_{BA} \equiv g$ as given parameters [this is equivalent to a Hamiltonian model with bare scatter states B_o and A_o and bare parameters Δ_o and g_o]. The A scattering amplitude then has the pole terms

$$f_A^P = 2\Delta g^2/(\Delta^2 - \omega^2), \tag{B.22}$$

so

$$f_A(\omega) = f_A^P(\omega)/D_A(\omega)\ , \tag{B.23}$$

where

$$D_A(\omega) = 1 - 2\Delta g^2(\Delta^2 - \omega^2)\ \frac{2}{\pi}\int_0^\infty \frac{dk'\ k'^2\ \mathcal{R}_A(\omega')}{(\omega'^2 - \Delta^2)^2(\omega'^2 - \omega^2)}\ . \tag{B.24}$$

If $f(\omega)$ is to have no ghost pole at negative ω^2, g^2 must not exceed the value $g^2(\Delta,\mathcal{R})$ given in Eq. (21); if g^2 equals $g^2(\Delta,\mathcal{R})$ the present two parameter [Δ and g^2] model reduces to the one parameter [λ or Δ or g^2] model of the previous paragraph.

The θB scattering amplitude $f_B(\omega)$ has just the form of Eqs. (22), (23) and (24), but with the sign of $2\Delta g^2$ reversed and $\mathcal{R}_A$ replaced by $\mathcal{R}_B$. There is a pole corresponding to a bound state, isobar "C", if

$$2\Delta g^2 \frac{2}{\pi} \int^{\infty} dk \quad \frac{(\mu^2-\Delta^2)}{(\omega^2-\Delta^2)^2} \mathcal{R}_B(\omega) > 1 . \qquad \text{(B.25)}$$

The inequality $g^2 \leqslant g^2(\Delta,\mathcal{R})$ (equality in the case of the one-paramete model) means that the quantity on the l.h.s. of (25) has the upper boun

$$\int_0^{\infty} dk \quad \frac{(\mu^2-\Delta^2)}{(\omega^2-\Delta^2)^2} \mathcal{R}_B(\omega) \;/\; \int_0^{\infty} dk \quad \frac{k^2}{(\omega^2-\Delta^2)^2} \mathcal{R}_A(\omega). \qquad \text{(B.26)}$$

If the ratio $\mathcal{R}_B(\omega)/\mathcal{R}_A(\omega)$ were constant, this would equal $\mathcal{R}_B/\mathcal{R}_A$. Thu it would appear that C would be bound only if the inelasticity of θB scattering exceeded that of θA.

But this amplitude f_B can be seen to be unsatisfactory, by the following somewhat indirect argument. If we take the strong coupling limit, i.e. g^2_{BA} to its absolute upper bound $g^2(\Delta,1) = \sqrt{\mu^2-\Delta^2}/\Delta$, we must have $\mathcal{R}_A(\omega) = 1$. Thus, the vanishing of meson production in θA scattering, can occur only if the pole terms of each production amplitude vanish. In particular, the pole terms of $\theta A \rightarrow \theta\theta B$ can vanish only by cancellation, which requires an isobar C, with $M_C - M_B = M_B - M_A$ ($=\Delta$) and $g^2_{CB} = 2g^2_{BA}$ ($=2g^2$). The coupling of the bound state "C" described in the preceding paragraph cannot, however, exceed g_{BA}.

The way out of this impass is to insert a CDD zero into the θB scattering amplitude $f_B(\omega)$. [It is easy to see from the requirement $g^2_{CB} \rightarrow 2g^2_{BA}$ as $g^2_{BA} \rightarrow g^2(\Delta,1)$ that in this limit the CDD zero must approach $\omega^2 = \Delta^2$ from below; presumably, as g^2_{BA} is increased past a critical value, the CDD zero first appears at $\omega^2 = -\infty$, and then moves to the right along the ω^2 axis.] From the point of view of the elasti amplitude $f_B(\omega)$, this represents an additional pair of parameters, wher as a complete solution of all amplitudes would fix the CDD zero, as wel as the entire function $\mathcal{R}(\omega)$, in terms of Δ and g^2_{BA}. Although in general we do not know the complete solution, we do know that in the strong coupling limit, as we have said, we must have $\mathcal{R}_A(\omega) = 1$. This in turn requires $M_C - M_B = \Delta$, and $g_{CB}^2 = 2g_{BA}^2$, which fixes the CDD zero of $f_B(\omega)$.

The argument can be repeated, and it shows that $g_{CB}^2 = 2g_{BA}^2 = 2g^2(\Delta,1)$ is possible only if $\mathcal{R}_B(\omega) = 1$, and this in turn requires an isobar "D", with $M_D - M_C = \Delta$ and $g_{DC}^2 = 3g_{BA}^2$. Thus, the entire solution in the strong coupling limit is determined through the requirement that all scattering is purely elastic. There is a sequence of equally spaced isobars A,B,C,... with $\Delta = M_B - M_A = M_C - M_B = \ldots$ and couplings $\sqrt{\mu^2 - \Delta^2}/\Delta = g_{BA}^2 = g_{CB}^2 - g_{BA}^2 = g_{DC}^2 - g_{CB}^2 = \ldots.$ There is only elastic scattering, and all elastic scattering $\theta X \rightarrow \theta X$ amplitudes are the same,

$$f_X(\omega) = \frac{1}{-\sqrt{\mu^2 - \Delta^2} - ik} . \tag{B.27}$$

This is the simplest (one-dimensional) "harmonic oscillator" strong coupling solution.

3. General Models

The above simple model, in which each amplitude had just two poles, had Abelian internal symmetry, a "charge conservation" gauge group. The amplitudes of a model with a larger, non-Abelian, internal symmetry will have more poles, with constraints among them; the isobars will form a net rather than a chain. In weak coupling, in particular in the one-meson approximation, it is best to first take account of the symmetry constraints among the amplitudes, and express crossed singularities with the aid of crossing matrices. In strong coupling it is instead simplier to first find the dynamical constraints between poles, their positions and residues. This can be done with a many-channel N/D formulation [Goebel, 1966; Sec. D], and the resulting constraints are the "bootstrap equations" (II.51) and (II.52) or (III.46) and (III.53). Instead of repeating this approach here, in Sec. IV we first find the conditions for no-meson production; the one meson scattering amplitude is then easy to treat.

APPENDIX C

"Generalized" Static Models

When many meson waves, in particular different partial waves, are strongly coupled, the mass condition (III.53) becomes

$$\sum_B C_{CB}\, \rho_B\, C_{BA} = C_{CA} \tag{C.1}$$

where

$$C_{BA} = [g_B, [M, g_A]] \tag{C.2}$$

and

$$\rho_B = \frac{1}{2\pi^2} \int d^3k\, k^{2\ell_B}\, U^2(k)/\omega_k^4 \ . \tag{C.3}$$

Sakita (1968) pointed out a great simplification. One begins with the observation that whatever meson waves are coupled, M must still be a scalar and at most quadratic in the J_i, i.e. of the form

$$M = \tfrac{1}{2}\ J_i\ C_{ij}\ J_j + B_i\ J_i + A. \tag{C.4}$$

Carrying out (2),

$$C_{BA} = [g_B, J_i]\ C_{ij}[J_j, g_A] \tag{C.5}$$

so that (1) reads

$$[g_C, J_i]\ C_{ij}\ R_{jk}\ C_{k\ell}\ [J_\ell, g_A] = [g_C, J_i]\ C_i\ [J_\ell, g_A] \tag{C.6}$$

where

$$R_{jk} \equiv \sum_B\ [J_j, g_B]\ \rho_B[g_B, J_k]\ . \tag{C.7}$$

It can be shown that one may "factor off" $[g_C, J_i]$ and $[J_\ell, g_A]$ from (C.6); that is, not only is

$$C_{ij}\ R_{jk}\ C_{k\ell} = C_{i\ell} \tag{C.8}$$

sufficient for (6) to hold, but it puts no real additional condition on C_{ij}. However, one must also satisfy

$$\mathrm{Tr}(CR) = C_{ij} R_{ji} = N_c - N_{L.G.} \tag{C.9}$$

where N_c is the dimension of the internal symmetry group C and $N_{L.G.}$ is the dimension of the L.G. of $\mathcal{g}_A$.

The reason why Eq. (8) may be a great simplification over Eq. (1) is that the range of the indices in latter may be arbitrarily large, depending on how many meson waves are strongly coupled, whereas the range of the indices in the former is fixed by the symmetry group. As a simple example we may take $(SU_2 \times SU_2, 3.3 + 3.1 + 1.3)$, in which there are three meson wave multiplets: J^P, $I = 1^+$, 1 (p-wave π), 1^+, 0 (p-wave η), and 0^+,1 (s-wave δ). In the representations in which i_o and j_o are invariants, the coupling operators have the forms

$$\mathcal{g}^{(\pi)}_{\alpha_i} = g_\pi m_\alpha n_i, \quad \mathcal{g}^{(\eta)}_i = g_\eta n_i, \quad \mathcal{g}^{(\delta)}_\alpha = g_\delta m_\alpha \tag{C.10}$$

respectively, where m_α and n_i are unit vectors; the body-axis i-spin and spin projections are

$$m_\alpha I_\alpha = i_o, \quad n_i J_i = j_o \tag{C.11}$$

respectively. Using $[m_\alpha, I_\beta] = i\varepsilon_{\alpha\beta\gamma} m_\gamma$, $[m_\alpha, J_i] = 0$, etc., the matrix R_{jk}, Eq. (7) is calculated to be

$$\begin{aligned} R_{jk} &= (g_\pi^2 + g_\eta^2)\, \rho_P\, \Lambda_{jk} \\ R_{\beta k} &= R_{j\gamma} = 0 \\ R_{\beta\gamma} &= (g_\pi^2\, \rho_P + g_\delta^2\, \rho_S)\, \Lambda_{\beta\gamma} \end{aligned} \tag{C.12}$$

where ρ and ρ_P are the s-wave and p-wave forms of (3), i.e., $\ell_B = 0$ and 1 respectively, and

$$\Lambda_{jk} = \delta_{jk} - n_j n_k, \quad \Lambda_{\beta\gamma} = \delta_{\beta\gamma} - m_\beta m_\gamma \,. \tag{C.13}$$

The solution of (8) is

$$C_{ij} = \xi\, \Lambda_{jk} / (g_\pi^2\, \rho_P + g_\eta^2\, \rho_P)\,, \qquad \xi = 0 \text{ or } 1$$

$$C_{\beta k} = C_{j\gamma} = 0 \tag{C.14}$$

$$C_{\beta\gamma} = \zeta\, \Lambda_{\beta\gamma}/(g_\pi^2\, \rho_P + g_\delta^2\, \rho_S)\ ,\quad \zeta = 0 \text{ or } 1\ .$$

Since $C=SU_2 \times SU_2$ and the L.G. is $U_1 x U_1$, we have $N_C - N_{L.G.} = 6-2=4$, and so (9) requires $2\xi + 2\zeta = 4$, i.e. $\xi = \zeta = 1$. So finally, (4) reads

$$M = \frac{\vec{j}^2}{2(g_\pi^2 + g_\eta^2)\rho_P} + \frac{\vec{\imath}^2}{2(g_\pi^2 \rho_P + g_\delta^2\, \rho_S)} \tag{C.15}$$

thus determining the isobar masses in terms of the coupling parameters g_π, g_η and g_δ [cf Ramachandran (1965)].

As another example we may take representations of $SU_2 x SU_2$ in which Y_3 is invariant. The reduced coupling constants are [Goebel (1966b)]

$$(i'j'y'\,|\underline{g}^{(\iota,\lambda)}|\,ij\ y) = \sum_\upsilon g_\upsilon^{(\iota,\lambda)} \sqrt{N_\upsilon N_{i'} N_{j'} N_{y'} N_i N_j N_y}$$

$$\begin{Bmatrix} i' & \iota & i \\ j' & \lambda & j \\ y' & \upsilon & y \end{Bmatrix} \begin{pmatrix} y' & \upsilon & y \\ -y_3 & 0 & y_3 \end{pmatrix} \tag{C.16}$$

where ι,λ are the i-spin and spin respectively of the meson wave, i,j,y label the i-spin, spin, and $|\vec{Y}|$ respectively of isobars; y_3 and the g_υ are invariants of the representation. [It might be noted that if $y_3 = 0$, the well known selection rule on the 3-j symbol in (C.16), namely $(-)^{y'+\upsilon+y} = +1$, implies that the isobar $|ijy)$ can be assigned a parity $(-)^y$ if the meson waves are assigned a parity π in the sense that $g^{(\iota,\lambda;\pi)}$ vanishes unless $(-)^\upsilon = \pi$. If $y_3 \neq 0$, parity can only be introduced inorganically, resulting in parity doubling.] The matrix R_{jk} is most simply calculated by the same method as in the previous paragraph. The invariance of Y_3 means that $\underline{g}_A$ is axially symmetric in the space reached by transforming i-spin (spin) indices by $A_{\alpha r}$ (B_{ir}) [cf. Eq.(1.15)] and hence is expressible in terms of a unit vector n_s in that space. We shall not do explicit calculations here [cf. Sakita (1968); note that he did not impose the condition Tr(CR) = 5]. The most important result is independent of the explicit calculations, namely that M is of a simple form, no matter how many meson waves are coupled. In the present case, $SU_2 \times SU_2$ representations in which Y_3 is invariant, M has the form

$$M = a\,\vec{J}^2 + b\,\vec{I}^2 + c\vec{Y}^2 + d\,I_3^2 \qquad (C.17)$$

where I_3 is the body axis projection of i-spin, i.e. $I_\alpha A_{\alpha s} n_s$. The first three terms of (17) are diagonal in $|ijy\rangle$ states. Similarly, in $(SU_2 \times SU_3, 3.8)$, in representations in which $(I_o + J)_3 = Y_3$ is invariant, M has the form

$$M = a\,\vec{J}^2 + b\,F_a^2 + c\,I_o^2 + dY^2 + eI_3^2\,. \qquad (C.18)$$

The isobar mass spectra predicted by "generalized" static models contain few parameters, and so the confrontation with experimental baryon masses is easy. The result is not encouraging and leads to the statement at the beginning of Sec. I.2.

APPENDIX D

$O(g^{-2})$ Corrections in Chew-Low Theory

In Chew-Low theory, the correction of order μg^{-4} to isobar masses comes from $O(g^{-2})$ terms of $\mathcal{R} = \sigma/\sigma_{e\ell}$. These have two sources. The one is the shifts of the thresholds for isobar excitation relative to the elastic threshold, of order ΔM; this implies a term $(\delta\mathcal{R})_{\text{one-meson}} = \frac{\mu}{k^2}\,\delta$, where $\delta = O(\Delta M)$. The other g^{-2} correction to $\mathcal{R}$ comes from single meson production, $(\delta\mathcal{R})_{prod} = \sigma_{prod}/\sigma_{el}$. These two corrections to $\mathcal{R}$ are negative and positive respectively, and so raise and lower respectively the isobar excitation energies. In the charged scalar model, at least, their net effect is to raise the isobar excitation energy above the lowest order $[O(\mu g^{-2})]$ value [cf. the field theory calculation of Nickle and Serber (1960)].

APPENDIX E

Here we show that in the quadratic approximation, the scattering is purely orthogonality scattering, as stated following Eq. (III.33). Inserting the general form [cf. Eq. (II.4)]

$$\Phi_\alpha^\omega = \Phi_\alpha^{\omega,in} + \frac{1}{\Omega^2-\omega^2} u_a \mathcal{J}_{a\alpha}^\omega$$

into the orthogonality statement Eq. (III.26), we get an equation for the current $\mathcal{J}_A^\omega$

$$[I(\omega) - I(\Delta_\nu)]\, O_A^\nu \mathcal{J}_A^\omega = -O_A^\nu\, \bar{\Phi}_A^{\omega,in} . \tag{E.1}$$

This can be solved for $\mathcal{J}_A^\omega$ by using the completeness of the O_A^ν,

$$\sum_\nu O_B^\nu O_A^\nu = \delta_{BA}$$

and also

$$\sum_\nu I(\Delta_\nu)\, O_B^\nu O_A^\nu = C_{BA}^{-1} \tag{E.2}$$

which follows from (III.20) and (III.22). The result is

$$\mathcal{J}_A^\omega = -[I(\omega)\,\delta_{AB} - C_{AB}^{-1}]^{-1}\, \bar{\Phi}_B^{\omega,in} \tag{E.3}$$

and hence

$$4\pi f_{AB}(\omega) = -[I(\omega)\,\delta_{AB} - C_{AB}^{-1}]^{-1} \tag{E.4}$$

in agreement with (III.21).

It should be noted that the scattering ($\omega > \mu$) part of Φ_α is orthogonal to the bound state parts $\Phi_\alpha^\nu = [\Omega^2 - \Delta_\nu^2]^{-1} u_b O_{\beta b}^\nu g^\nu$, not to the static field $\Phi_\alpha^o = \Omega^{-2} u_b g_\alpha^o$. Hence in the degenerate case, Φ_α^ω is orthogonal to the quasistatic modes,

$$\sum_{b,\beta} \Lambda_{A,b\beta} \int u_b \frac{1}{\Omega^2} \Phi_\beta^\omega = 0 , \tag{E.5}$$

but not to the static field, i.e.

$$\sum_{b\beta} g_{b\beta} \int u_b \frac{1}{\Omega^2} \Phi_\beta^\omega \neq 0 .$$

APPENDIX F

"y-Excitations" in $SU_2 \times SU_2$

Landovitz and Margolis (1958, 1959a) presented a calculation exhibiting fairly low lying isobars in the π-N static model [$SU_2 \times SU_2$, 3·3] which were not members of the y=0 (∴ I=J) ground state band, but were excited in y. From similar considerations they wrote (1959b) a mass spectrum for the π-Σ static model, which had the Λ as ground state. We shall describe their calculation in our terms here. In our opinion, it is an open question whether their results are correct. In particular their π-Σ model spectrum disagrees with the one derived by Wentzel (1963).

We begin by defining [cf.(1.15)] the orthogonal matrices $A_{\alpha r}$, B_{ir} which transform $g_{\alpha i}$ to diagonal form

$$g_{\alpha i} = \sum_r A_{\alpha r} B_{ir} g_r \tag{F.1}$$

but we do not now make the assumption that the g_r are equal, and so the Y_r

$$Y_r \equiv \tilde{I}_r + \tilde{J}_r = I_\alpha A_{\alpha r} + J_i\ B_{ir} \qquad \text{(F.2)}$$

are not L. G. generators and the eigenvalues $\vec{Y}^2 = y(y+1)$, $Y_3 = y_3$ are not invariants. [We use the notation $\tilde{X}_{r\ldots s\ldots}$ for tensors in the intrinsic frame, $\tilde{X}_{r\ldots s\ldots} = X_{\alpha\ldots i\ldots}\ A_{\alpha r}\ldots B_{is}\ldots$; thus $\tilde{g}_{\alpha i} = \delta_{\alpha i} g_\alpha$]. We now find M.

M has the form

$$\frac{M}{m} \equiv \mathcal{M} = \frac{1}{2}\, C_{\alpha\beta}\ I_\alpha I_\beta + D_{\alpha i} I_\alpha J_i + \frac{1}{2}\, E_{ij}\ J_i J_j \qquad \text{(F.3)}$$

where $\underset{\sim}{C}$, $\underset{\sim}{D}$ and $\underset{\sim}{E}$ are functions of $g_{\alpha i}$. From this we calculate $\Lambda_{BA} = [g_B, [\mathcal{M}, g_A]]$, and express it in the intrinsic frame

$$\tilde{\Lambda}_{\beta j,\alpha i} = \sum_r \varepsilon_{\beta rj}\ \varepsilon_{\alpha ri} [C_r\ g_j g_i - D_r (g_j g_\alpha + g_\beta g_i) + E_r\ g_\beta g_\alpha] \qquad \text{(F.4)}$$

where we have exploited the fact that the diagonality of $\tilde{g}_{\alpha i}$ implies the diagonality of $\tilde{C}$, $\tilde{D}$, and $\tilde{E}$. Because of the $\varepsilon\ldots$ factors of (F.4), the sum over r is empty, and $\tilde{\Lambda}_{\beta j,\ \alpha i}$ vanishes for $\alpha = i$ or $\beta = j$. Because there is no continuous little group when the g_r are unequal, the number of angular degrees of freedom of $g_{\alpha i}$ is the full number of symmetry group generators, 3 + 3 = 6; this must be the dimension of the space projected out by $\tilde{\underset{\sim}{\Lambda}}$, i.e. $\mathrm{Tr}\underset{\sim}{\Lambda} = 6$. One easily finds that this plus the property that $\Lambda_{\beta j,\ \alpha i}$ vanishes for $\alpha = i$ or $\beta = j$ implies

$$\tilde{\Lambda}_{\beta j,\ \alpha i} = \begin{cases} \delta_{\beta\alpha}\ \delta_{ji} & \alpha \neq i,\ \beta \neq j \\ 0 & \alpha = i \text{ or } \beta = j. \end{cases} \qquad \text{(F.5)}$$

Setting the form (F.4) equal to this and solving for C_r, D_r and E_r, one finds

$$C_3 = E_3 = (g_1{}^2 + g_2{}^2)/(g_1{}^2 - g_2{}^2)^2$$

$$D_3 = 2g_1g_2/(g_1{}^2 - g_2{}^2)^2 \tag{F.6}$$

and cyclically. [One might observe that it would be very difficult to express directly the matrices $C_{\alpha\beta}$, etc., in terms of the matrices $g_{\alpha i}$.] Consequently (F.3) becomes

$$\mathcal{M} = \frac{1}{4} \sum_{\text{cyclic}} \left[\left(\frac{\tilde{I}_3 + \tilde{J}_3}{g_1 - g_2}\right)^2 + \left(\frac{\tilde{I}_3 - \tilde{J}_3}{g_1 + g_2}\right)^2 \right]. \tag{F.7}$$

We now consider the limiting case in which the g_r become equal. The first term of (7)

$$\frac{1}{4} \sum_{\text{cyclic}} \frac{Y_e{}^2}{(g_1 - g_2)^2} \equiv \mathcal{M}_o \tag{F.8}$$

becomes large, and so eigenstates of $\mathcal{M}$ become eigenstates of $\mathcal{M}_o$. $\mathcal{M}_o$ is just the Hamiltonian of a three-dimensional rigid rotator with moments of inertia $\mathcal{I}_3 = 2(g_1 - g_2)^2$, etc.; the Y_i are the spin projections on the principal axes. As well known, the eigenstates of $\mathcal{M}_o$ are eigenstates of $\vec{Y}^2$, with 2y + 1 states, generally non-degenerate, belonging to each value of $\vec{Y}^2 = y(y+1)$. For instance,

$$\begin{aligned}
y = 0:&\quad \text{one state, } \mathcal{M}_o = 0, \\
y = \tfrac{1}{2}:&\quad \text{two states, } \mathcal{M}_o = \frac{1}{16} \sum_{\text{cyclic}} \frac{1}{(g_1 - g_2)^2} \text{ for both,} \\
y = 1:&\quad \text{three states, } \mathcal{M}_o = \frac{1}{4} \left[\frac{1}{(g_1 - g_2)^2} + \frac{1}{(g_2 - g_3)^2} \right] \\
&\quad \text{and cyclically.}
\end{aligned} \tag{F.9}$$

We can write $\mathcal{M}$, Eq. (7), in the form

$$\mathcal{M} = \mathcal{M}_o + \frac{\vec{I}^2 + \vec{J}^2 - \frac{1}{2}\vec{Y}^2}{8g^2} + 0\left(\frac{(\Delta g)^2}{g^2}\right) \tag{F.10}$$

where Δg stands for the differences $g_i - g_j$, and g is the common value of the g_i in the limit $\Delta g \to 0$. The low lying states belong to $y = y_{min}$, and their mass dependence according to Eq. (10) is

$$\mathcal{M} = \frac{\vec{I}^2 + \vec{J}^2}{8g^2} + \text{const.} \tag{F.11}$$

We contrast this result with the consequences of assuming the g_r to be equal from the start. Here $\tilde{g}_{\alpha i} = g\delta_{\alpha i}$ is scalar [thus invariant to an O_3 L. G. generated by the Y_r] and so $\tilde{C}$, $\tilde{D}$, and $\tilde{E}$ are scalar. Hence (4) reads

$$\tilde{\Lambda}_{\beta j,\alpha i} = g^2 \, (C-2D+E) \sum_r \varepsilon_{\beta rj} \, \varepsilon_{\alpha ri} ; \tag{F.4'}$$

$\Lambda^2 = \Lambda$ and $\mathrm{Tr}\Lambda = N_C - N_{L.G.} = 6 - 3 = 3$ are satisfied by

$$\tilde{\Lambda}_{\beta j,\alpha i} = \frac{1}{2} \sum_r \varepsilon_{\beta rj} \, \varepsilon_{\alpha ri} = \frac{1}{2}(\delta_{\beta\alpha} \, \delta_{ji} - \delta_{\beta i} \, \delta_{j\alpha}) = \Lambda_{\beta j,\alpha i} \tag{F.5'}$$

[cf. Eq. (II.74)], hence

$$C - 2D + E = 1/2g^2, \tag{F.6'}$$

which results in

$$\mathcal{M} = \frac{\vec{I}^2 + \vec{J}^2}{8g^2} + \frac{C - E}{4} \, (\vec{I}^2 - \vec{J}^2) + \frac{1}{2} \, D\vec{Y}^2 \, . \tag{F.7'}$$

The last term is irrelevant, since $\vec{Y}^2$ is an invariant; the coefficient of the second term, C - E, is left undetermined. Comparing (7') with (11), we see the difference that the latter lacks the term in $\vec{I}^2 - \vec{J}^2$. The reason for this difference between the forms of $\mathcal{M}$, Eqs. (11) and (7'), in the cases $\Delta g \to 0$ and $\Delta g = 0$ respectively is of course that the $\underset{\sim}{\Lambda}$ in the two cases, Eqs. (5) and (5') are different.

In a short note, Landovitz and Margolis (1959b) stated Eq. (11) as the form for M in the π - Σ static model. However, this seems to be contradicted by a straightforward treatment of the π - Σ - Λ model [Wentzel (1963)]. The π - Σ - Λ model has two bare couplings, $g(\pi\Sigma\Lambda) = g_o$ and $g(\pi\Sigma\Sigma) = g_1$. The secular equation for the eigenvalues of the interaction term (of Eq. (III. 2) is complicated [Wentzel, loc. cit., Eq. (4)], but it can be shown that it, together with the minimization conditions

$$\frac{\partial H_o}{\partial g_A} = \frac{\partial}{\partial g_A} \left[\frac{1}{2} \Sigma g_B{}^2 - V(g) \right] = 0, \qquad \text{(F.12)}$$

can always be satisfied by a solution in which the g_r are equal. It is probable that this is <u>the</u> solution, i.e. that which gives the absolute minimum of H_o; this has been shown explicitly in the vicinity of the values $g_o = g_1$ and $g_o = 0$ [this last is the π-Σ model, in which no bare Λ is coupled.] [Exception: When $g_1 = 0$, Eq. (12) does not constrain each g_r, but yields only $\Sigma g_r{}^2$ = const; as a result $g_{\alpha i}$ has the higher symmetry 0_9, as remarked following Eq. (I.14) above]. If the g_r are equal, (F.7') holds, and the coefficient of the $\vec{I}^2 - \vec{J}^2$ term (which is an independent parameter from the point of view of representation theory) is determined in terms of g_1/g_o (Wentzel, loc. cit.). A simple way to find this, which does not require complex calculations, is to write I_α and J_i in terms of the (i-) spin of the bare scatterer and of the meson field

$$I_\alpha = t_\alpha + \varepsilon_{\alpha\beta\gamma}\, g_{\beta i}\, \dot{g}_{\gamma i} \qquad \text{(F.13)}$$

$$J_i = s_i + \varepsilon_{ijk}\, g_{\alpha j}\, \dot{g}_{\alpha k}$$

(we have set the mass scale equal to unity) where

$$\dot{g}_{\alpha i} = i\, [\mathcal{M}, g_{\alpha i}]\ .$$

Taking $\mathcal{M}$ of the form (7'), and using $g_{\alpha i} = -g O_{\alpha i}$, $O_{\alpha i}$ orthogonal, one finds that Eqs. (13) are satisfied if the ratio of the coefficients of the $\vec{I}^2$ and $\vec{J}^2$ terms of $\mathcal{M}$ is

$$\frac{s_i O_{\beta i} I_\beta}{t_\alpha I_\alpha} = \frac{s_j J_j}{t_\alpha O_{\alpha i} J_i} = \frac{s_i \tilde{Y}_i}{t_i \tilde{Y}_i} . \tag{F.14}$$

The quantities appearing in this formula are determined by the eigenspinor of the interaction term, and the result is

$$\mathcal{M} = \frac{(1-v)\vec{I}^2 + v\,\vec{J}^2}{4g^2}$$

where

$$v = \frac{2\ (\alpha + \sqrt{3 + \alpha^2})}{3\ \sqrt{3 + \alpha^2}}, \quad \alpha = g_1/g_o . \tag{F.15}$$

In the case $g_o = 0$, the π-Σ model, this yields $v = 4/3$, and M is not of the form (F.11).

The Landovitz-Margolis (1958, 1959a) theory of the y-excited states of the π-N model is as follows: The static Hamiltonian is of the form

$$H_o = C_1\ \Sigma(g_r - g)^2 + C_2\mathcal{M} \quad \begin{cases} C_1 = O(a^{-3}) \\ C_2 = O(a) \end{cases} \tag{F.16}$$

where $\mathcal{M}$ is given in (F.10) and (F.9); it is approximately diagonal in y. Since $\mathcal{M}$ does not contain momenta conjugate to the g_r, the low lying eigenvalues are obtained by simply minimizing H_o with respect to the g_r. For $y = 0$, the $C_2\mathcal{M}$ term is negligible compared to the first term, and minimization with respect to the g_r results in the expected result $g_r = g$. For $y \neq 0$, the term $C_2\mathcal{M}_o$ cannot be neglected, because $\mathcal{M}_o$ blows up for $g_r \to g$. Minimization results in $|g_r - g| = O\ [(C_2/C_1)^{1/4}] = O(a)$, and an excitation energy

$$(H_o)_{y\neq o} - (H_o)_{y=o} = 0\,[(C_1C_2)^{\frac{1}{2}}] = 0(a^{-1})\,. \qquad (F.17)$$

This excitation energy is just the cut-off energy, and so the $y \neq 0$ states cannot be taken very seriously as physical states. It is not even clear whether they are to be taken seriously within the static model; this would depend partly on their width, which has not been calculated.

BIBLIOGRAPHY

Bhabha, H. J. (1941), Phys. Rev. 59, 100.
Brueckner, K.A. (1952), Phys. Rev. 86, 106.
Chew, G.F. (1954), Phys. Rev. 95, 1669.
Chew, G.F. and Low, F.E. (1956), Phys. Rev. 101, 1570.
Chew, G.F. (1962), Phys. Rev. Letters 9, 233.
Cook, T. and Sakita, B. (1967), J. Math. Phys. 8, 708.
Dancoff, S. (1950), Phys. Rev. 78, 382.
Dullemond, C. (1965), Ann. Phys. (N.Y.) 33, 214.
Dullemond, C. and von der Linden, F.J.M., (1967), Ann. Phys. (N.Y.) 41, 372.
Dyson, Ross, Salpeter, Schweber, Sundaresan, Visscher,and Bethe, (1954), Phys. Rev. 95, 1644.
Edwards, S.F. and Matthews, P. (1957), Phil. Mag. 2, 467.
Gasiorowicz, S. and Ruderman, M. (1958), Nuovo Cimento 8, 860; also Ruderman (1962), Phys. Rev. 127, 312.
Goebel, C. (1958), Phys. Rev. 109, 1846.
(1966a), Phys. Rev. Letters 16, 1130.
(1966b), In NonCompact Groups in Particle Physics, Chow, Ed., Benjamin, N.Y.
Heisenberg, W. (1939), Zeit.f. Physik 113, 61.

Heitler, W. (1940), Nature 145, 29; (with Ma), Proc. Roy. Soc. A176, 368.
Houriet, A. (1945), Helv. Ph. Acta 18, 473.
Huang, K. and Mueller, A.H. (1965), Phys. Rev. 140, B365.
Kaufman, A.N. (1954), Phys. Rev. 92, 468.
Landovitz, L. and Margolis, B. (1958), Phys. Rev. Letters 1, 206.
(1959a), Annals of Physics (NY) 7, 52.
(1959b), Phys. Rev. Letters 2, 318.
Lee, T.D. (1954), Phys. Rev. 95, 1329.
Miyazawa, H. (1956), Phys. Rev. 101, 1564.
Nickle, H. and Serber,R. (1960), Phys. Rev. 119, 449.
Oppenheimer, R. and Schwinger, J. (1941), Phys. Rev. 60, 150.
Pais, A. and Serber, R. (1957), Phys. Rev. 105, 1636.
(1959), Phys. Rev. 113, 955.
Pauli, W. (1946), Meson Theory of Nuclear Forces, Interscience.
Pauli, W. and Dancoff, S. (1942), Phys. Rev. 62, 85.
Ramachandran, R. (1965), Phys. Rev. 139, B110 and B121.
Sakita, B. (1968), Phys. Rev. 170, 1453.
Schwartz, C. (1965), Phys. Rev. 137, B212.
Tamm, I. (1945), J. Phys. (U.S.S.R.) 9, 449.
Tomonaga, S. (1947), Prog. Theor. Phys. 2, 6.
Wentzel, G. (1940), Helv. Ph. Acta. 13, 269, also 14, 633.
(1943), Helv. Ph. Acta 16, 222 and 551.
(1947), Revs. Modern Phys. 19, 1.
(1952), Phys. Rev. 86, 437.
(1962), Phys. Rev. 125, 771.
(1963), Phys. Rev. 129, 1367.
(1965)? unpublished.
Wick, G. C. (1955), Revs. Modern Phys. 27, 339.
Wilson, K. (1964)?

Received 5/1/6

On the Charged Scalar Mesotron Field

by

J. Schwinger
Harvard University

Stimulated by Wentzel's pioneering work on strong coupling theory, this paper was written in early 1941. It was referred to in a note published jointly with J. R. Oppenheimer in July of that year* as "to be published soon." Although practically complete, it remained unfinished and unpublished. It is offered now, still incomplete and lacking references, in homage to Gregor Wentzel.

The mesotron field theory of nuclear forces has been developed in analogy with the quantum electrodynamical treatment of the Maxwell field equations. This analogy suffers by the two characteristic properties of the mesotron field: the large magnitude of the heavy particle coupling required to explain nuclear forces, and the charged nature of the field. The first circumstance effectively prevents a satisfactory discussion of these problems by the conventional perturbation treatment, and the second prohibits any attempt to develop a classical mesotron field theory along the lines of that investigated by Bhabha for the neutral mesotron field--no classical description exists for the discontinuous change in the charge state of a heavy particle occasioned by mesotron absorption or emission. These difficulties are particularly evident in the discussion of mesotron scattering by nucleons (neutron, proton). The scattering of a light quantum by a charged particle, described by the Thomson formula, involves the mass of the scatterer and, in particular, vanishes for an infinitely massive body. This, of course, is a simple consequence of the fact that the sole degree of freedom contributing to the scattering is the translational motion of the particle. Such also

---

*J. R. Oppenheimer and J. Schwinger, Phys. Rev. <u>60</u>, 150 (1941).

is the result of the classical theory developed for neutral mesotrons. Quite otherwise is the situation with respect to charged meson scattering, for the additional degree of freedom associated with the fluctuating charge state of the nucleon contributes a further scattering essentially independent of the heavy particle mass. (A second non-classical degree of freedom, the spin of the nucleon, also contributes in like fashion, but with this we shall not be concerned.) Indeed, the scattering cross sections obtained by current perturbation treatments are much too large to be reconciled with experiment.

From the viewpoint of the perturbation calculus, the immediate cause of these difficulties is the prohibition, by the principle of charge conservation, of certain intermediate states in the scattering process. This has led Bhabha and Heitler to postulate the existence of excited states of the nucleon with charges of 2, 3, ... and -1, -2, ... units. The states of 1 and 0 units of charge (proton and neutron) are conceived of as merely the ground states of an infinite sequence of levels. To find a mechanism for the description of these isobaric states one need seek no further than the mesotron field itself, for the strong nuclear coupling should serve to bind mesotrons in stationary states around the nucleon. This note is devoted to a discussion of such effects on the scalar mesotron theory. The treatment of spin phenomena, based on the pseudo-scalar theory, will be deferred to a subsequent paper.

<u>General theory</u>

The model upon which we shall base our considerations is that of a Pauli-Weisskopf scalar field interacting with an infinitely massive nuclear particle through a finite-distance coupling. The introduction of the finite-distance coupling, although primarily motivated as a mathematical artifice for achieving non-divergent results, may possess some physical significance associated with the inapplicability of spatio-temporal description. This interpretation need not be insisted upon, however, for the physical quantities in which we shall be interested behave properly in the limit of point coupling. The principal liabilities of a theory involving finite-distance interactions are its lack of relativistic and gauge invariance. The former difficulty is specious since the assumption of an infinitely massive nucleon implies a natural reference frame. The lack of gauge invariance is a fundamental difficulty necessarily present in any theory which does not guarantee point conservation of charge and current.

The usual treatment of such a model is one in which the kinetic energy of the field is diagonalized and the coupling term regarded as a perturbation. In this paper we shall exploit the opposite approximation, regarding the coupling constant g as so large that the coupling energy must be made approximately diagonal. The actual situation, of course, is intermediate between these extremes. For convenience, we shall summarize the essential results, as they appear in the limit of point-coupling. It is found that charge may indeed be bound to the heavy particle, producing excited states of higher charge. The excitation energy of an isobaric state of charge q is given by:

$$\left(\frac{g^2}{\hbar c}\right)^{-1} \mu c^2 \left(q - \frac{1}{2}\right)^2 \quad . \tag{1}$$

The energy of excitation of a doubly-charged proton is therefore small compared with the rest energy of a mesotron, μc^2, for the assumption of strong coupling finds its mathematical expression in the statement:

$$\frac{g^2}{\hbar c} \gg 1 \quad . \tag{2}$$

In its essentials, the treatment of mesotron scattering is identical with the zero-range approximation for neutron-proton scattering. The wave function of the bound mesotron is simply the Yukawa potential

$$\frac{e^{-\kappa r}}{r} \,, \quad \frac{1}{\kappa} = \frac{\hbar}{\mu c} \,, \tag{3}$$

which by the condition of orthogonality determines the phase shifts of the perturbed continuum S states, and, thereby, the scattering cross section,

$$\sigma = 4\pi \left(\frac{\hbar}{\mu c}\right)^2 \left(\frac{\mu c^2}{E}\right)^2 , \tag{4}$$

for a mesotron of total energy E. However, the detailed discussion shows that only half of the field is to be so treated; the other half remaining unperturbed; the actual scattering cross section is therefore $\frac{1}{2}\sigma$. A further result is that the scattered particle has, with equal probability, both signs of charge, irrespective of the charge of the incident particle or of the nucleon (the latter statement is only correct when the mesotron kinetic energy is large compared with the charge splitting of the nuclear states). It is of some interest that the theory gives no indication of multiple processes, despite the large magnitude of the nuclear coupling.

The Hamiltonian of the theory:

$$\begin{aligned} H = \int \{c^2\pi^*(\underset{\sim}{r})\pi(\underset{\sim}{r}) + \nabla\phi^*(\underset{\sim}{r}) \cdot \nabla\phi(\underset{\sim}{r}) + \kappa^2\phi^*(\underset{\sim}{r})\phi(\underset{\sim}{r})\} d\underset{\sim}{r} \\ - (8\pi)^{1/2} g \int \phi(\underset{\sim}{r})K(\underset{\sim}{r})d\underset{\sim}{r}\, \frac{1}{2}(\tau_x + i\tau_y) - (8\pi)^{1/2} g \\ \int \phi^*(\underset{\sim}{r})K(\underset{\sim}{r})d\underset{\sim}{r}\, \frac{1}{2}(\tau_x - i\tau_y), \end{aligned} \tag{5}$$

is composed additively of the kinetic energy of the mesotrons and of a coupling term which describes the emission and absorption of charged mesotrons accompanied by corresponding changes in the charge state of the heavy particle stationed at the origin. The description of the latter process is by the isotopic spin operators τ_x, τ_y, which act on the charge coordinate of the nucleon. The function K(r) replaces the delta-function appropriate to point interactions and represents the ability of the nucleon to create or destroy charge at a finite distance from the origin.

With the normalization:

$$\int K(\underset{\sim}{r})\, d\underset{\sim}{r} = 1 \ , \tag{6}$$

the coupling constant g assumes the dimensions of charge (the factor $(8\pi)^{1/2}$ is inserted in order that the "charge" be measured in Gaussian rather than Heaveside-Lorentz units).

The Hamiltonian (5) must be supplemented by the operator of total charge, to completely specify the system. The definition:

$$Q/e = \frac{1}{2}\,(1+\tau_z) + \frac{i}{\hbar}\int \{\phi^*(\underset{\sim}{r})\pi^*(\underset{\sim}{r}) - \phi(\underset{\sim}{r})\pi(\underset{\sim}{r})\}\, d\underset{\sim}{r} \tag{7}$$

guarantees that the total charge Q shall be preserved in time, by virtue of the commutation laws of the mesotron field quantities:

$$[\phi(\underset{\sim}{r}),\pi(\underset{\sim}{r}')] = i\hbar\delta(\underset{\sim}{r}-\underset{\sim}{r}'),\ [\phi^*(\underset{\sim}{r}),\pi^*(\underset{\sim}{r}')] = i\hbar\delta(\underset{\sim}{r}-\underset{\sim}{r}'), \tag{8}$$

and of the Pauli matrix representation of the isotopic spin τ.

We shall introduce a complete set of real ortho-normal functions $\phi_s(\underset{\sim}{r})$ in terms of which the field quantities are expanded:

$$\frac{1}{c}\,\phi(\underset{\sim}{r}) = 2^{-1/2}\sum_s (x_s - iy_s)\phi_s(\underset{\sim}{r}),\ \frac{1}{c}\,\phi^*(\underset{\sim}{r}) = 2^{-1/2}\sum_s (x_s + iy_s)\,\phi_s(\underset{\sim}{r})$$

$$c\pi(\underset{\sim}{r}) = 2^{-1/2}\sum_s (P_{x_s} + iP_{y_s})\,\phi_s(\underset{\sim}{r}),\quad c\pi^*(\underset{\sim}{r}) = 2^{-1/2}\sum_s (P_{x_s} - iP_{y_s})\,\phi_s(\underset{\sim}{r}) \tag{9}$$

The real expansion coordinates x_s, y_s, P_{x_s}, P_{y_s} satisfy the usual commutation relations of independent coordinates:

$$[x_r,\ P_{x_s}] = i\hbar\delta_{rs},\quad [y_r,\ P_{y_s}] = i\hbar\delta_{rs},\ \text{etc.} \tag{10}$$

in consequence of the field commutation laws (8) (together with the unwritten vanishing commutators of all other quantities). The functions $\phi_s(\underset{\sim}{r})$ will be chosen so that one of them, $\phi_o(\underset{\sim}{r})$, is proportional to the source function $K(\underset{\sim}{r})$

$$\phi_o(\underset{\sim}{r}) = \frac{K(\underset{\sim}{r})}{(\int K^2(\underset{\sim}{r})d\underset{\sim}{r})^{1/2}} , \qquad K(\underset{\sim}{r}) = \frac{\phi_o(\underset{\sim}{r})}{\int \phi_o(\underset{\sim}{r})d\underset{\sim}{r}} , \tag{11}$$

with all others orthogonal thereunto. No further specification of these latter functions will be necessary. Expressed in these new coordinates, the energy and total charge of the system become:

$$H = \frac{1}{2} \sum_s (P^2_{x_s} + P^2_{y_s}) + \frac{1}{2} \sum_{r,s} \omega_{rs} (x_r x_s + y_r y_s)$$

$$- K(x_o \tau_x + y_o \tau_y) \tag{12}$$

$$Q/e = \frac{1}{2} (1+\tau_z) + \frac{1}{\hbar} \sum_s (x_s P_{y_s} - y_s P_{x_s}) .$$

The elements of the symmetrical matrix ω are defined by:

$$\omega_{rs} = c^2 \int \phi_r (\underset{\sim}{r}) (\kappa^2 - \nabla^2) \phi_s(\underset{\sim}{r}) d\underset{\sim}{r} , \tag{13}$$

while the constant K represents

$$K = (4\pi)^{1/2} gc \int \phi_o(\underset{\sim}{r}) K(\underset{\sim}{r}) d\underset{\sim}{r} = (4\pi)^{1/2} gc / \int \phi_o(\underset{\sim}{r}) d\underset{\sim}{r} . \tag{14}$$

In accordance with the program of investigating strong coupling, we seek to diagonalize $x_o\tau_x + y_o\tau_y$. It is convenient to replace the coordinates x_o, y_o by polar coordinates,

$$x_o = r_o \cos \theta_o, \qquad y_o = r_o \sin \theta_o . \tag{15}$$

The problem is then to determine the state function of the system, Ψ, which simultaneously satisfies the eigenvalue equations:

$$\{\frac{1}{2}\sum_{s\neq 0}(P^2_{x_s}+P^2_{y_s})+\frac{1}{2}\sum_{r,s\neq 0}\omega_{rs}(x_r x_s+y_r y_s)$$

$$+\sum_{s\neq 0}\omega_{os}r_o\ (\cos\theta_o x_s+\sin\theta_o y_s)$$

$$+\frac{1}{2}(P^2_{r_o}+\frac{P^2_{\theta_o}-\frac{\hbar^2}{4}}{r_o^2})+\frac{1}{2}\omega_{oo}r_o^2-K\,r_o\ (\cos\theta_o\tau_x+\sin\theta_o\tau_y)$$

$$-E\}\ \Psi=0\ ,$$

$$\{\frac{1}{2}(1+\tau_z)+\frac{1}{\hbar}\sum_{s\neq 0}(x_s P_{y_s}-y_s P_{x_s})+\frac{1}{\hbar}P_{\theta_o}-q\}\ \Psi=0\ , \tag{16}$$

with E the energy eigenvalue and q the charge eigenvalue, in units of e. Here

$$P_{\theta_o}=\frac{\hbar}{i}\frac{\partial}{\partial\theta_o}\ ,\qquad P_{r_o}=\frac{\hbar}{i}r_o^{-1/2}\frac{\partial}{\partial r_o}r_o^{1/2}\ . \tag{17}$$

The charge eigenvalue equation determines the dependence of the state function Ψ upon the angular coordinate θ_o:

$$\Psi=A(\theta_o)\Phi,\ A(\theta_o)=\exp[i\{q-\frac{1}{2}(1+\tau_z)-\frac{1}{\hbar}\sum_{s\neq 0}(x_s P_{y_s}-y_s P_{x_s})\}\ \theta_o]$$

and thereby defines a canonical transformation which eliminates the coordinate θ_o:

$$\{\frac{1}{2}\sum_{s\neq 0}(P^2_{x_s}+P^2_{y_s})+\frac{1}{2}\sum_{r,s\neq 0}\omega_{rs}(x_r x_s+y_r y_s)+\sum_{s\neq 0}\omega_{os}\,r_o\,x_s$$

$$+\frac{1}{2}\omega_{oo}r_o^2+\frac{1}{2}P^2_{r_o}+(\hbar^2/2r_o^2) \tag{19}$$

$$\times[\{q-\frac{1}{2}-\frac{1}{2}\tau_z-\frac{1}{\hbar}\sum_{s\neq 0}(x_s P_{y_s}-y_s P_{x_s})\}^2-\frac{1}{4}]$$

$$-K\,r_o\tau_x-E\}\ \Phi=0\ .$$

In deriving this result we have employed the relations

$$A^{-1}\ (\cos\ \theta_o x_s + \sin\ \theta_o y_s)\ A = x_s$$
$$A^{-1}\ (\cos\ \theta_o \tau_x + \sin\ \theta_o \tau_y)\ A = \tau_x, \tag{20}$$

which are evident from the interpretation of the "angular momenta", $\frac{1}{2}\hbar\tau_z$, $x_s p_{ys} - y_s p_{xs}$, as rotation operators, or which may be shown directly from the commutation relations by observing, through differentiation, that the left-hand members are independent of θ_o.

The diagonalization of the coupling energy is achieved by choosing Φ as an eigenfunction of τ_x. Of the two eigenvalues ($\neq 1$), we must take the one with the same sign as K, for the other solution will be much higher in energy since the coupling is assumed to be strong. For definiteness we shall take g, and hence K, to be positive. Nothing in the theory is changed if the opposite choice is made. The operator τ_z has no diagonal elements and serves only to couple these fundamental solutions. The neglect of this coupling, which introduces an error of relative order $(\hbar c/g^2)^2$, corresponds to the obvious statement that, in the limit of strong coupling, the nucleon occurs in the neutron and proton states with equal probability.

A suggestive change of notation is introduced by regarding r_o as a new x-coordinate of the "zero" oscillator. We shall rewrite it as simply x_o, bearing in mind that it differs from the original x-coordinates--principally in that it may not assume negative values, i.e.,

$$x_o \equiv r_o \geq 0. \tag{21}$$

The exact Hamiltonian of Eq. (19) is thus replaced by the approximate Hamiltonian:

$$H = \frac{1}{2}\sum_s P_{x_s}^2 + \frac{1}{2}\sum_{r,s} \omega_{rs}\, x_r x_s - K\, x_o$$

$$+ \frac{1}{2}\sum_{s\neq o} P_{y_s}^2 + \frac{1}{2}\sum_{r,s\neq o} \omega_{rs}\, y_r y_s \tag{22}$$

$$+ (\hbar^2/2x_o^2)\,[q - \frac{1}{2} - \frac{1}{\hbar}\sum_{s\neq o} (x_s P_{y_s} - y_s P_{x_s})]^2 .$$

This Hamiltonian is somewhat analogous to that describing the vibration-rotation energy of a molecular system. The strong coupling limitation has the effect of ensuring that the "rotational" energy be small compared to the "vibrational" energy, and thus facilitates diagonalization. The equilibrium positions of the x-coordinates are approximately determined by:

$$\sum_s \omega_{rs}\, x_s^{(eq.)} = K\, \delta_{r,o} \tag{23}$$

and are therefore related to the elements of the inverse kinetic energy matrix,

$$(\omega^{-1})_{rs} = (1/4\pi c^2) \int \phi_r(\mathbf{r})\, (e^{-\kappa|\mathbf{r}-\mathbf{r}'|}/|\mathbf{r}-\mathbf{r}'|)\, \phi_s(\mathbf{r}')d\mathbf{r}\, d\mathbf{r}' \tag{24}$$

by

$$x_s^{(eq.)} = K(\omega^{-1})_{so} . \tag{25}$$

Hence, under the transformation

$$x_s = x_s' + \zeta_s, \qquad \zeta_s = (\omega^{-1})_{so} \tag{26}$$

$$P_{x_s} = P_{x_s'} \quad ,$$

the Hamiltonian (22) assumes the form:

$$H = -\frac{1}{2}K^2\zeta_o + \frac{1}{2}\sum_s P_{x_s'}^2 + \frac{1}{2}\sum_{r,s} \omega_{rs}\, x_r' x_s' \tag{27}$$

(equation continued on next page)

$$+ \frac{1}{2} \sum_{s\neq o} P_{y_s}^2 + \frac{1}{2} \sum_{r,s\neq o} \omega_{rs}\, y_r y_s$$

$$+ (\hbar^2/2K^2\zeta_o^2)\ [1 + x_o'/K\zeta_o]^{-2}\ [q - \frac{1}{2} - \frac{1}{\hbar} \sum_{s\neq o} (x_s'\, P_{y_s} - y_s\, P_{x_s'})$$

$$- K/\hbar \sum_{s\neq o} \zeta_s\, P_{y_s}\,]^2. \qquad (27)$$

(continue

The strong coupling approximation is introduced by imposing the inequalities:

$$K_o\zeta_o >> x_o'$$

$$(K/\hbar) \sum_{s\neq o} \zeta_s\, P_{y_s} >> q - \frac{1}{2} - (1/\hbar) \sum_{s\neq o} (x_s'\, P_{y_s} - y_s\, P_{x_s'}) \qquad (28)$$

which permit the expansion of the "rotational" energy in descending powers of K. The limitations which are thereby imposed upon g in order to maintain the validity of this procedure will be discussed in a later section. It should be noted that the first of these inequalities guarantees that the condition (21) shall not be violated.

The first few terms of the Hamiltonian (27), in expanded form,

$$H = H^{(o)} + H^{(1)} + H^{(2)} + \ldots ,$$

are (discarding the self-energy term, $-\frac{1}{2} K^2\zeta_o$):

$$H^{(o)} = \frac{1}{2} \sum_{s} P_{x_s'}^2 + \frac{1}{2} \sum_{r,s} \omega_{rs} x_r' x_s'$$

$$+ \frac{1}{2} \sum_{s\neq o} P_{y_s}^2 + \frac{1}{2} \sum_{r,s\neq o} \omega_{rs}\, y_r y_s + (1/2\zeta_o^2)\ (\sum_{s\neq o} \zeta_s\, P_{y_s})^2,$$

$$H^{(1)} = -(\hbar/2K\zeta_o^2)\ [(\sum_{s\neq o} \zeta_s\, P_{y_s})\, Q + Q\, (\sum_{s\neq o} \zeta_s\, P_{y_s})], \qquad (29)$$

$$H^{(2)} = (\hbar^2/2K^2\zeta_o^2)\ Q^2 - (x_o'/K\zeta_o)\ H^{(1)},$$

$$Q = q - \frac{1}{2} - (1/\hbar) \sum_{s\neq o} (x_s'\, P_{y_s} - y_s\, P_{x_s'}) + (x_o'/\hbar\zeta_o) \sum_{s\neq o} \zeta_s\, P_{y_s} .$$

Although a term involving the total charge q appears in $H^{(1)}$, its contribution to the energy first appears in the second order of a perturbation treatment. Added to the similar term contained in $H^{(2)}$, it represents a charge dependence of the energy proportional to the inverse square of the coupling constant. The perturbation calculation can be avoided by the introduction of the transformation:

$$y_s = y'_s$$
$$P_{y_s} = P_{y'_s} + \frac{\hbar(q - \frac{1}{2})}{K \sum_r \zeta_r^2} \zeta_s \tag{30}$$

which serves to eliminate q from $H^{(1)}$. The effect of this transformation is expressed by:

$$H^{(o)} = \frac{1}{2} \sum_s P^2_{x'_s} + \frac{1}{2} \sum_{r,s} \omega_{rs} x'_r x'_s + \frac{1}{2} \sum_{s\neq o} P^2_{y'_s} + \frac{1}{2} \sum_{r,s\neq o} \omega_{rs} y'_r y'_s + \frac{1}{2\zeta_o^2} (\sum_{s\neq o} \zeta_s P_{y'_s})^2 ,$$

$$H^{(1)} = (\hbar/2K\zeta_o^2)[(\sum_{s\neq o} \zeta_s P_{y'_s}) Q' + Q'(\sum_{s\neq o} \zeta_s P_{y'_s})] , \tag{31}$$

$$H^{(2)} = (\hbar^2/2K^2 \sum_s \zeta_s^2)(q - \frac{1}{2})^2 - (\hbar^2/K^2 \sum_s \zeta_s^2)(q - \frac{1}{2}) [Q' - 1/\hbar\zeta_o^2 \sum_{s\neq o} \zeta_s P_{y'_s} \sum_s \zeta_s x'_s] + (\hbar^2/2K^2\zeta_o^2) Q'^2 - (x'_o/K\zeta_o) H^{(1)},$$

$$Q' = (1/\hbar) \sum_{s\neq o} (x'_s P_{y'_s} - y'_s P_{x'_s}) - (x'_o/\hbar\zeta_o) \sum_{s\neq o} \zeta_s P_{y'_s} .$$

The term in $H^{(2)}$ linear in q can be simplified by observing that

$$(1/\zeta_o^2) \sum_{s\neq o} \zeta_s P_{y'_s} \sum_{s\neq o} \zeta_s x'_s = - \sum_{s\neq o} (x'_s P_{y'_s} + y'_s P_{x'_s}) + \frac{d}{dt} (\sum_{s\neq o} x'_s y'_s) , \tag{32}$$

the time derivative to be computed from $H^{(o)}$; it may be discarded to the order of approximation we are considering. This replacement is effectively produced by the canonical transformation:

$$
\begin{aligned}
x'_s &= \xi_s \\
P_{x'_s} &= P_{\xi_s} - \frac{\hbar(q - \frac{1}{2})}{K^2 \sum_s \zeta_s^2} \eta_s \qquad (\eta_o = 0) \\
y'_s &= \eta_s \\
P_{y'_s} &= P_{\eta_s} - \frac{\hbar(q - \frac{1}{2})}{K^2 \sum_s \zeta_s^2} \xi_s ,
\end{aligned}
\tag{33}
$$

which gives the final expanded form of the Hamiltonian,

$$
\begin{aligned}
H^{(o)} &= \frac{1}{2} \sum_s P_{\xi_s}{}^2 + \frac{1}{2} \sum_{r,s} \omega_{rs}\, \xi_r \xi_s \\
&\quad + \frac{1}{2} \sum_{s \neq o} P_{\eta_s}{}^2 + \frac{1}{2} \sum_{r,s \neq o} \omega_{rs}\, \eta_r \eta_s + (1/2\zeta_o^2)\Big(\sum_{s \neq o} \zeta_s P_{\eta_s} \Big)^2 , \\
H^{(1)} &= (\hbar/2K\zeta_o^2)\ \Big[\Big(\sum_{s \neq o} \zeta_s P_{\eta_s} \Big) Q' + Q'\Big(\sum_{s \neq o} \zeta_s P_{\eta_s} \Big)\Big], \\
H^{(2)} &= (\hbar^2/2K^2 \sum_s \zeta_s^2)(q - \tfrac{1}{2} - Q'')^2 + (\hbar^2/2K^2\zeta_o^2)\ Q'^2 \\
&\quad - [\hbar^2/2K^2 \sum_s \zeta_s^2]\ Q''^2 - (\xi_o/K\zeta_o)\ H^{(1)}, \\
Q' &= (1/\hbar) \sum_{s \neq o} (\xi_s P_{\eta_s} - \eta_s P_{\xi_s}) - (\xi_o/\hbar\zeta_o) \sum_{s \neq o} \zeta_s P_{\eta_s} , \\
Q'' &= (2/\hbar) \sum_{s \neq o} \xi_s P_{\eta_s} - (2\xi_o/\hbar\zeta_o) \sum_{s \neq o} \zeta_s P_{\eta_s} .
\end{aligned}
\tag{34}
$$

In these coordinates, the strong coupling restrictions become

$$
\begin{aligned}
K\zeta_o &\gg \xi_o , \\
(K/\hbar) \sum_{s \neq o} \zeta_s P_{\eta_s} &\gg (\zeta_o^2/\sum_s \zeta_s^2)(q - \tfrac{1}{2}) - Q' .
\end{aligned}
\tag{35}
$$

It has thus been shown that when the fields described by the coordinates ξ_s, η_s are unexcited, the total energy (apart from fluctuation energy) is not zero but consists of the charge dependent term

$$E_q = \frac{\hbar^2}{2K^2\sum_s \zeta_s^2} (q - \tfrac{1}{2})^2, \tag{36}$$

which must be interpreted as the energy of the system: nucleon + bound mesotrons. The existence of bound mesotron charge will be demonstrated in detail in the next section. It is easily verified that

$$\sum_s \zeta_s^2 = (1/8\pi\kappa c^4) \int \phi_o(\mathbf{r})\, e^{-\kappa|\mathbf{r}-\mathbf{r}'|}\, \phi_o(\mathbf{r}')\, d\mathbf{r}\, d\mathbf{r}', \tag{37}$$

by employing the complete nature of the function set $\phi_s(\mathbf{r})$, and the theorem:

$$\int (e^{-\kappa|\mathbf{r}-\mathbf{r}''|}/|\mathbf{r}-\mathbf{r}''|)\, (e^{-\kappa|\mathbf{r}''-\mathbf{r}'|}/|\mathbf{r}''-\mathbf{r}'|)\, d\mathbf{r}'' = (2\pi/\kappa)\, e^{-\kappa|\mathbf{r}-\mathbf{r}'|}. \tag{37'}$$

In the limit in which the linear extension of the function $\phi_o(r)$ is small compared with the mesotron Compton wave length; i.e., $\kappa a \ll 1$, the exponential in Eq. (37) may be placed by unity, and one obtains immediately:

$$E_q = \left(\frac{g^2}{\hbar c}\right)^{-1} \mu c^2 \left(q - \frac{1}{2}\right)^2, \quad a \ll \frac{\hbar}{\mu c}. \tag{1}$$

In the opposite limit, $\kappa a \gg 1$, the exponential is effectively a delta function, $e^{-x|\mathbf{r}-\mathbf{r}'|} \rightarrow \frac{8\pi}{\kappa^3}\delta(\mathbf{r}-\mathbf{r}')$, and therefore $\sum_s \zeta_s^2 = (\kappa c)^{-4}$. To examine the transition between these limits, it is sufficient to choose a simple analytic form for $\phi_o(\mathbf{r})$; the essential dependence on κa is realized with any function of range a. The function

$$\phi_o(\mathbf{r}) = (2\pi a)^{-1/2}\, e^{-r/a}/r\ , \tag{38}$$

which is particularly convenient for this purpose, gives

$$E_q = (g^2/\hbar c)^{-1} (1 + \kappa a)^3 \mu c^2 (q - \tfrac{1}{2})^2. \tag{39}$$

It is evident from the nature of the development which led to this result that the charge dependent energy must be small compared to the energy of the field, $\gtrsim \mu c^2$. We can therefore obtain necessary conditions for the validity of this procedure, viz:

$$\begin{aligned} &(g^2/\hbar c) >> 1, && \kappa a << 1 \\ &(g^2/\hbar c) >> (\kappa a)^3, && \kappa a >> 1. \end{aligned} \tag{40}$$

Mesotron scattering

The Hamiltonian $H^{(o)}$ (Eq. 34) is composed of independent Hamiltonians for the ξ and η coordinates, $H_\xi^{(o)}$ and $H_\eta^{(o)}$. The former is simply the Hamiltonian of a real, unperturbed Pauli-Weisskopf field; the associated eigenfunctions are plane waves. $H_\eta^{(o)}$ differs from the Pauli-Weisskopf Hamiltonian by the presence of the term $\frac{1}{2\zeta_o^2} (\sum_{s \neq o} \zeta_s P_{\eta_s})^2$ and by the exclusion of the state $s = 0$. The corresponding eigenfunctions cannot be plane waves and consequently describe elastic scattering of mesotrons by the nucleon. The Hamiltonian $H^{(1)}$ is of the third order in the field coordinates and therefore produces non-linear equations of motion which describe multiple scattering processes (in the sense of a change in the number of particles). The probability of such processes is of order $(g^2/\hbar c)^{-1}$ relative to that of elastic scattering, which makes them comparatively unimportant. The Hamiltonian $H^{(2)}$ contains, in addition to E_q, a quadratic term proportional to $(q - \frac{1}{2})Q''$ which describes the effect of the nuclear charge on elastic scattering, and various quartic terms which correspond to higher order multiple processes.

The introduction of the real, canonically conjugate, fields

$$(1/c)\ \Phi_1\ (\underset{\sim}{r}) = \sum_s \xi_s\ \phi_s\ (\underset{\sim}{r}),\ c\ \pi_1\ (\underset{\sim}{r}) = \sum_s p_{\xi_s}\ \phi_s\ (\underset{\sim}{r}) \tag{41}$$

restores $H_\xi^{(o)}$ to the Pauli-Weisskopf form:

$$H^{(o)} = \frac{1}{2} \int \{c^2\ \pi_1^2 + \underset{\sim}{\nabla}\ \Phi_1 \cdot \underset{\sim}{\nabla}\ \Phi_1 + \kappa^2\ \Phi_1^2\}\ d\underset{\sim}{r}\ . \tag{42}$$

The expansion in normal coordinates is well known:

$$\begin{aligned} \Phi_1(\underset{\sim}{r}) &= \sum_{\underset{\sim}{p}} \hbar\ c\ (2E_{\underset{\sim}{p}})^{-1/2}\ [\alpha_{\underset{\sim}{p}}\ V^{-1/2}\ e^{i\underset{\sim}{p}\cdot\underset{\sim}{r}/\hbar} + \alpha_p^*\ V^{-1/2}\ e^{-i\underset{\sim}{p}\cdot\underset{\sim}{r}/\hbar}] \\ \pi_1(\underset{\sim}{r}) &= \sum_{\underset{\sim}{p}} (i/c)\ (\tfrac{1}{2}\ E_{\underset{\sim}{p}})^{1/2}\ [-\alpha_{\underset{\sim}{p}}\ V^{-1/2}\ e^{i\underset{\sim}{p}\cdot\underset{\sim}{r}/\hbar} + \alpha_p^*\ V^{-1/2}\ e^{-i\underset{\sim}{p}\cdot\underset{\sim}{r}/\hbar}], \end{aligned} \tag{43}$$

and

$$H_\xi^{(o)} = \sum_{\underset{\sim}{p}} E_{\underset{\sim}{p}}\ (\alpha_{\underset{\sim}{p}}^*\ \alpha_{\underset{\sim}{p}} + \tfrac{1}{2}),\ [\alpha_{\underset{\sim}{p}},\ \alpha_{\underset{\sim}{p}'}^*] = \delta_{\underset{\sim}{p},\underset{\sim}{p}'},\ \text{etc.} \tag{44}$$

To diagonalize $H_\eta^{(o)}$, it is convenient to introduce the real fields

$$\begin{aligned} (1/c)\ \Phi_2\ (\underset{\sim}{r}) &= \sum_{s\neq o} \eta_s\ \phi_s\ (\underset{\sim}{r}) \\ c\ \pi_2\ (\underset{\sim}{r}) &= \sum_{s\neq o} P_{\eta_s}\ \phi_s\ (\underset{\sim}{r}) - (1/\zeta_o)\ (\sum_{s\neq o} \zeta_s\ P_{\eta_s})\ \phi_o\ (\underset{\sim}{r}), \end{aligned} \tag{45}$$

which are <u>not</u> canonically conjugate. Indeed,

$$[\Phi_2\ (\underset{\sim}{r}),\ \pi_2\ (\underset{\sim}{r}')] = i\ \hbar\ [\delta(\underset{\sim}{r}-\underset{\sim}{r}') - (1/\zeta_o)\ (\sum_s \zeta_s\ \phi_s\ (\underset{\sim}{r}))\ \phi_o(\underset{\sim}{r}')]. \tag{46}$$

In terms of these fields, $H_\eta^{(o)}$ is formally identical with the Pauli-Weisskopf Hamiltonian:

$$H_\eta^{(o)} = \frac{1}{2} \int \{c^2\ \pi_2^2 + \underset{\sim}{\nabla}\ \Phi_2 \cdot \underset{\sim}{\nabla}\ \Phi_2 + \kappa^2\ \Phi_2^2\}\ d\underset{\sim}{r}. \tag{47}$$

This presents a rather interesting variation of the usual dynamical problem, but one which is easily re-expressed in the conventional pattern by defining new field variables which are canonically conjugate, and refer to a perturbed Hamiltonian.

It is first convenient to define a function $\chi_o(r)$ by

$$\chi_o(\underset{\sim}{r}) = (1/\zeta_o) \sum_s \zeta_s \phi_s (\underset{\sim}{r}) = (1/4\pi\zeta_o c^2) \int \frac{e^{-\kappa|\underset{\sim}{r}-\underset{\sim}{r}'|}}{|\underset{\sim}{r}-\underset{\sim}{r}'|} \phi_o (\underset{\sim}{r}') d\underset{\sim}{r}' , \quad (48)$$

the latter statement following from the definitions (Eqs. (24) and (26))of ζ_s. It is not irrelevant to remark that $\chi_o(r)$ is proportional to the Yukawa potential of the source distribution $K(\underset{\sim}{r})$. With the aid of the formulae:

$$\int \chi_o(\underset{\sim}{r}) \phi_s (\underset{\sim}{r}) d\underset{\sim}{r} = \zeta_s/\zeta_o, \quad \int \chi_o(\underset{\sim}{r}) \phi_o(\underset{\sim}{r}) d\underset{\sim}{r} = 1, \quad (49)$$

it is at once evident that $\pi_2(\underset{\sim}{r})$ is orthogonal to $\chi_o(\underset{\sim}{r})$. Indeed, the orthogonality of π_2 to χ_o, and of Φ_2 to ϕ_o, is contained in the commutation relation (46):

$$[\Phi_2(\underset{\sim}{r}), \pi_2(\underset{\sim}{r}')] = i\hbar \{\delta(\underset{\sim}{r}-\underset{\sim}{r}') - \chi_o(\underset{\sim}{r}) \phi_o(\underset{\sim}{r}')\}. \quad (46')$$

It is thus suggested that new variables be introduced in such a manner that the orthogonality relations be automatically fulfilled. The transformation

$$\Phi_2(\underset{\sim}{r}) = \Phi_2'(\underset{\sim}{r}) - \chi_o(\underset{\sim}{r}) \int \phi_o(\underset{\sim}{r}') \Phi_2'(\underset{\sim}{r}') d\underset{\sim}{r}'$$
$$\pi_2(\underset{\sim}{r}) = \pi_2'(\underset{\sim}{r}) - (\chi_o(\underset{\sim}{r})/\int\chi_o^2 d\underset{\sim}{r}) \int \chi_o(\underset{\sim}{r}') \pi_2'(\underset{\sim}{r}') d\underset{\sim}{r}' \quad (50)$$

guarantees the orthogonality relations and produces the commutation relations (46') if Φ_2' and π_2' be assumed canonically conjugate variables:

$$[\Phi_2'(\underset{\sim}{r}), \pi_2'(\underset{\sim}{r}')] = i\hbar\, \delta(\underset{\sim}{r}-\underset{\sim}{r}'). \quad (51)$$

The Hamiltonian (47), expressed in these canonical variables, is

$$H_\eta^{(o)} = \frac{1}{2} \int \{c^2 \pi_2'^2 + \nabla \Phi_2' \cdot \nabla\Phi_2' + \kappa^2 \Phi_2'^2\} \, d\mathbf{r}$$

$$-(c^2/2\int\chi_o^2 \, d\mathbf{r}) \, [\int\chi_o \, \pi_2' \, d\mathbf{r}]^2 - (1/2\zeta_o c^2) \, [\int\phi_o \, \Phi_2' \, d\mathbf{r}]^2 . \tag{52}$$

The equations of motion are, therefore,

$$\frac{\partial}{\partial t} \Phi_2'(\mathbf{r}) = c^2 \, \pi_2'(\mathbf{r}) - (c^2/\int\chi_o^2 d\mathbf{r}) \, \chi_o(\mathbf{r}) \int \chi_o \, \pi_2' \, d\mathbf{r}$$

$$-\frac{\partial}{\partial t} \pi_2'(\mathbf{r}) = (\kappa^2 - \nabla^2) \, \Phi_2'(\mathbf{r}) - (1/\zeta_o c^2) \, \phi_o(\mathbf{r}) \int \phi_o \Phi_2' \, d\mathbf{r} . \tag{53}$$

Hence

$$(\kappa^2 - \Box^2) \, \Phi_2'(\mathbf{r}) = (1/\zeta_o c^2) \, \phi_o(\mathbf{r}) \int \phi_o \Phi_2' \, d\mathbf{r} , \tag{54}$$

for

$$\frac{\partial}{\partial t} \int \chi_o \, \pi_2' \, d\mathbf{r} = -\int \chi_o \, (\kappa^2 - \nabla^2) \, \Phi_2' \, d\mathbf{r} + (1/\zeta_o c^2) \int \phi_o \Phi_2' \, d\mathbf{r} = 0 \tag{55}$$

since

$$(\kappa^2 - \nabla^2) \, \chi_o(\mathbf{r}) = (1/\zeta_o c^2) \, \phi_o(\mathbf{r}) . \tag{56}$$

The cross sections for the elastic scattering of mesotrons are therefore determined by the solutions of the integro-differential equation (54).

The monochromatic solutions of angular frequency $E/\hbar$ are obtained from the equation:

$$(\nabla^2 + \frac{E^2}{\hbar^2 c^2} - \kappa^2) \, u(\mathbf{r}) = (1/\zeta_o c^2) \, \phi_o(\mathbf{r}) \int \phi_o(\mathbf{r}') \, u(\mathbf{r}') \, d\mathbf{r}' . \tag{57}$$

It is apparent from Eq. (56) that χ_o provides a solution of this equation associated with $E = 0$. Since (57) is essentially an eigenvalue equation with a symmetrical kernel, its solutions may be chosen to form an orthogonal set. We shall show that any solution of (57) orthogonal to χ_o corresponds to $|E| > \mu c^2$, i.e., with the exception of χ_o, all the solutions of (57) describe scattered waves. The proof of this theorem is based upon the statement that

$$\int \nabla(u^* - \chi_o \int \phi_o u^*) \cdot \nabla (u - \chi_o \int \phi_o u)\, dr$$

$$= \int \nabla u^* \cdot \nabla u\, dr - (1/\zeta_o c^2) \left|\int \phi_o u\, dr\right|^2 - \kappa^2 \int \chi_o^2\, dr \left|\int \phi_o u\, dr\right|^2$$

$$+ \kappa^2 \left(\int \phi_o u^*\, dr\right) \left(\int \chi_o u\, dr\right) + \kappa^2 \left(\int \chi_o u^*\, dr\right) \left(\int \phi_o u\, dr\right), \qquad (58)$$

which implies that if u is any function orthogonal to χ_o,

$$\int \nabla u^* \cdot \nabla u\, dr - (1/\zeta_o c^2) \left|\int \phi_o u\, dr\right|^2 > 0. \qquad (59)$$

This leads at once to the desired result, for

$$[(E^2/\hbar^2 c^2) - \kappa^2] \int u^* u\, dr = \int \nabla u^* \cdot \nabla u\, dr - (1/\zeta_o c^2) \left|\int \phi_o u\, dr\right|^2 \qquad (60)$$

Although the function $\chi_o(r)$ necessarily occurs in the Fourier expansion of the canonical variables Φ_2' and π_2', it makes no contribution to the physically significant field quantities Φ_2 and π_2.

The integro-differential equation (57) can be solved exactly, for it is essentially an inhomogeneous equation of the form:

$$(\nabla^2 + k^2)\, u(r) = -4\pi A\, \phi_o(r)$$

$$4\pi\, \zeta_o c^2 A = \int \phi_o u\, dr, \qquad k^2 = (E^2 - \mu^2 c^4)/\hbar^2 c^2 = p^2/\hbar^2. \qquad (61)$$

The solution with the asymptotic form appropriate to a scattering problem is:

$$u_p(r) = e^{ik\cdot r} + A \int \frac{e^{ik|r-r'|}}{|r-r'|}\, \phi_o(r')\, dr' \qquad (62)$$

$$p = \hbar k$$

in which the constant A is determined by

$$4\pi\,\zeta_o c^2 A = \int \phi_o\, e^{i\mathbf{k}\cdot\mathbf{r}} d\mathbf{r} + A \int \phi_o(\mathbf{r})\, \frac{e^{ik|\mathbf{r}-\mathbf{r}'|}}{|\mathbf{r}-\mathbf{r}'|}\, \phi_o(\mathbf{r}')\, d\mathbf{r}\, d\mathbf{r}',$$

or

$$A = \int \phi_o(\mathbf{r})\, e^{i\mathbf{k}\cdot\mathbf{r}} d\mathbf{r} \Big/ \!\!\Big/ \int \phi_o(\mathbf{r}) \left\{\frac{e^{-\kappa|\mathbf{r}-\mathbf{r}'|} - e^{ik|\mathbf{r}-\mathbf{r}'|}}{|\mathbf{r}-\mathbf{r}'|}\right\} \phi_o(\mathbf{r}')\, d\mathbf{r}\, d\mathbf{r}'. \tag{63}$$

The asymptotic form of the wave function (62) is

$$u_{\mathbf{p}}(\mathbf{r}) \sim e^{i\mathbf{k}\cdot\mathbf{r}} + (e^{ikr}/r)\, A \int \frac{\sin kr'}{kr'}\, \phi_o(\mathbf{r}')\, d\mathbf{r}' \tag{64}$$

and therefore the total cross section for the scattering of these spherically symmetrical waves is:

$$\sigma = 4\pi |A|^2 \left[\int \frac{\sin kr}{kr}\, \phi_o(\mathbf{r})\, d\mathbf{r}\right]^2 \tag{65}$$

or

$$\sigma = 4\pi \frac{\left[\int \frac{\sin kr}{kr}\, \phi_o(\mathbf{r})\, d\mathbf{r}\right]^4}{\left[\int \phi_o(\mathbf{r}) \frac{e^{-\kappa|\mathbf{r}-\mathbf{r}'|} - \cos k|\mathbf{r}-\mathbf{r}'|}{|\mathbf{r}-\mathbf{r}'|}\, \phi_o(\mathbf{r}') d\mathbf{r}\, d\mathbf{r}'\right]^2 + k^2 \left[\int \frac{\sin kr}{kr}\, \phi_o(\mathbf{r})\, d\mathbf{r}\right]^4} \tag{66}$$

which is always less than $4\pi/k^2$, as befits an s wave scattering cross section. In the limit in which the size of the source function is small compared with all wave lengths, the approximations:

$$\int \frac{\sin kr}{kr}\, \phi_o(\mathbf{r})\, d\mathbf{r} \doteq \int \phi_o(\mathbf{r})\, d\mathbf{r},$$

$$\int \phi_o(\mathbf{r})\, \frac{e^{-\kappa|\mathbf{r}-\mathbf{r}'|} - \cos k|\mathbf{r}-\mathbf{r}'|}{|\mathbf{r}-\mathbf{r}'|}\, \phi_o(\mathbf{r}')\, d\mathbf{r}\, d\mathbf{r}' \doteq -\kappa\left[\int \phi_o(\mathbf{r})\, d\mathbf{r}\right]^2$$

lead to the simple result:

$$\sigma = 4\pi/(k^2+\kappa^2) = 4\pi(\hbar/\mu c)^2\, (\mu c^2/E)^2. \tag{4}$$

This limiting situation is more easily treated by employing the orthogonality of the scattering wave function and $\chi_o(\underset{\sim}{r})$. Outside the region occupied by the source ($r > a$), the s wave solution of (61) is of the form $\sin(kr + \delta)/kr$, while $\chi_o(\underset{\sim}{r})$ is proportional to the Yukawa potential $e^{-\kappa r}/r$. The errors introduced by extending these forms to the origin, in the orthogonality integral:

$$\int_0^\infty \frac{\sin(kr+\delta)}{kr} \frac{e^{-\kappa r}}{r} d\underset{\sim}{r} = 0,$$

vanish in the limit of a point source. Hence $\cot\delta = -\kappa/k$ and

$$\sigma = (4\pi/k^2)\sin^2\delta = 4\pi/(k^2+\kappa^2) = 4\pi(\hbar/\mu c)^2 (\mu c^2/E)^2. \quad (4)$$

The expansion of the canonical field quantities, Φ_2' and π_2', in the normal coordinates defined by the functions $u_{\underset{\sim}{p}}$ and χ_o, is

$$\begin{aligned} \Phi_2' &= \sum_{\underset{\sim}{p}} \hbar c(2E_{\underset{\sim}{p}})^{-1/2} [\beta_{\underset{\sim}{p}} V^{-1/2} u_{\underset{\sim}{p}} + \beta_{\underset{\sim}{p}}^* V^{-1/2} u_{\underset{\sim}{p}}^*] \\ &\quad + 2^{-1/2}(\chi_o/\int\chi_o^2 d\underset{\sim}{r}) [\beta_o + \beta_o^*] \\ \pi_2' &= \sum_{\underset{\sim}{p}} (i/c) (\tfrac{1}{2} E_{\underset{\sim}{p}})^{1/2} [-\beta_{\underset{\sim}{p}} V^{-1/2} u_{\underset{\sim}{p}} + \beta_{\underset{\sim}{p}}^* V^{-1/2} u_{\underset{\sim}{p}}^*] \\ &\quad + i\hbar\, 2^{-1/2} (\chi_o/\int\chi_o^2 d\underset{\sim}{r}) [-\beta_o + \beta_o^*]. \end{aligned} \quad (67)$$

Of more interest is the expansion of the physical field quantities Φ_2 and π_2, viz:

$$\begin{aligned} \Phi_2 &= \sum_{\underset{\sim}{p}} \hbar c(2E_{\underset{\sim}{p}})^{-1/2} [\beta_{\underset{\sim}{p}} V^{-1/2} (u_{\underset{\sim}{p}} - \chi_o \int \phi_o u_{\underset{\sim}{p}} d\underset{\sim}{r}) + \beta_{\underset{\sim}{p}}^* V^{-1/2} (u_{\underset{\sim}{p}}^* - \chi_o \int \phi_o u_{\underset{\sim}{p}}^* d\underset{\sim}{r})] \\ \pi_2 &= \sum_{\underset{\sim}{p}} (i/c) (\tfrac{1}{2} E_{\underset{\sim}{p}})^{1/2} [-\beta_{\underset{\sim}{p}} V^{-1/2} u_{\underset{\sim}{p}} + \beta_{\underset{\sim}{p}}^* V^{-1/2} u_{\underset{\sim}{p}}^*], \end{aligned} \quad (68)$$

from which it is easy to obtain the form of the Hamiltonian in normal coordinates:

$$H_\eta^{(o)} = \sum_{\underset{\sim}{p}} E_{\underset{\sim}{p}} \left(\beta^*_{\underset{\sim}{p}} \beta_{\underset{\sim}{p}} + \tfrac{1}{2}\right). \tag{69}$$

We have thus diagonalized the Hamiltonian $H^{(o)}$ in terms of two sets of normal coordinates, one describing unperturbed particles, the other particles scattered with the effective cross section $\sigma(E)$ (Eq. 66). This cross section, however, does not describe the scattering of charged mesotrons, for the associated wave field Φ_2 is real. It is therefore necessary to perform one further canonical transformation which combines the two real fields Φ_1 and Φ_2 into a complex wave field representing charged mesotrons incident upon the nucleon.

To perform this final transformation it is convenient to examine an electrical and a mechanical property of the mesotron field, such as the charge density and the momentum density. The charge density will also exhibit most effectively the existence of bound charge. The operator of the mesotron charge density

$$\rho(\underset{\sim}{r}) = \frac{ie}{\hbar} [\Phi^* \pi^* - \Phi\pi] \tag{70}$$

must be subjected to the transformations (Eqs. 9, 15, 18, 26, 30, 33, 41, 45) which lead to the diagonalized form of the Hamiltonian. The result of this succession of transformations, which we shall not stop to perform in detail, is

$$\begin{aligned}\rho(\underset{\sim}{r})/e = {} & (q - \tfrac{1}{2}) \left(\chi_o^2(\underset{\sim}{r}) / \int \chi_o^2\, d\underset{\sim}{r}\right) + (\kappa\zeta_o c/\hbar)\ \chi_o(\underset{\sim}{r})\ \pi_2(\underset{\sim}{r}) \\ & + (1/\hbar)\ \{[\Phi_1\pi_2 - \Phi_2\pi_1] - \phi_o\chi_o \int [\Phi_1\pi_2 - \Phi_2\pi_1]\, d\underset{\sim}{r}\}\end{aligned} \tag{71}$$

in which terms of order K^{-1}, etc. have been discarded. The expectation value of the charge density in a state in which the fields Φ_1 and Φ_2 are unexcited is therefore

$$(q - \tfrac{1}{2})e\ \chi_o^2(\underset{\sim}{r}) / \int \chi_o^2\, d\underset{\sim}{r} \rightarrow (q - \tfrac{1}{2})e\ (\kappa/2\pi)\ (e^{-2\kappa r}/r^2), \tag{72}$$

as $a \rightarrow o$,

which thus describes the distribution of the $q - \frac{1}{2}$ units of charge bound to the nucleon. The second term in the charge density expression, linear in the coupling constant, does not contribute to the total charge ($\int \chi_o \pi_2 d\underset{\sim}{r} = 0$), but represents the large charge fluctuations which occur in the vicinity of the nucleon. Indeed it is this term (or rather the similar one in the current density) which describes the coupling of the system with the radiation field, and the consequent photo-emission of mesotrons. We are principally interested, however, in the third term of the charge density operator, for this represents the charge density of the free mesotrons:

$$\rho_f(\underset{\sim}{r}) = (e/\hbar)\ [\Phi_1 \pi_2 - \Phi_2 \pi_1] \tag{73}$$

(The fourth term of (71) is in a sense an induced charge, since it is confined to the region $r < a$, and its integral over this region equals the total free charge). The quantity of physical interest is the charge density $\rho_f(\underset{\sim}{r})$ averaged over a volume containing many wave lengths. We must therefore diagonalize that part of the averaged operator $\bar{\rho}_f(r)$ which is independent of position, for this represents the constant density of the incident mesotrons. In general, there will exist additional terms varying as $1/r^2$ which represent the charge density of the scattered mesotrons. Upon inserting in (73) the expressions (43) for the unperturbed fields Φ_1 and π_1 and the asymptotic forms of the expressions (68) for the perturbed fields Φ_2 and π_2, and performing the required averaging process, one obtains

$$\bar{\rho}_f(\underset{\sim}{r}) = (ie/V) \sum_{\underset{\sim}{p}} (\alpha_{\underset{\sim}{p}} \beta^*_{\underset{\sim}{p}} - \beta_{\underset{\sim}{p}} \alpha^*_{\underset{\sim}{p}}) . \tag{74}$$

There is no term representing scattered charge since only one of the fields contains spherically diverging waves; it is therefore equally probable that a positive or a negative mesotron emerge on scattering irrespective of the charge of the incident particle or of the nucleon. The expression (74), which thus represents the charge density of the incident mesotrons, is diagonalized by the canonical transformation

$$A_{\underset{\sim}{p}} = 2^{-1/2} [\alpha_{\underset{\sim}{p}} - i \beta_{\underset{\sim}{p}}], \quad B_{\underset{\sim}{p}} = 2^{-1/2} [\alpha_{\underset{\sim}{p}} + i \beta_{\underset{\sim}{p}}]$$

$$[A_{\underset{\sim}{p}}, A^*_{\underset{\sim}{p}'}] = \delta_{\underset{\sim}{p},\underset{\sim}{p}'}, \qquad [A_{\underset{\sim}{p}}, B_{\underset{\sim}{p}'}] = 0, \text{ etc.} \tag{75}$$

$$\bar{\rho}_f = (e/V) \sum_{\underset{\sim}{p}} (A^*_{\underset{\sim}{p}} A_{\underset{\sim}{p}} - B^*_{\underset{\sim}{p}} B_{\underset{\sim}{p}}) = (e/V) \sum_{\underset{\sim}{p}} (N^+_{\underset{\sim}{p}} - N^-_{\underset{\sim}{p}}),$$

recognizing that $A^*_{\underset{\sim}{p}} A^*_{\underset{\sim}{p}}$ and $B^*_{\underset{\sim}{p}} B^*_{\underset{\sim}{p}}$ represent the occupation number operators of a positive and a negative mesotron, respectively, in the state of momentum $\underset{\sim}{p}$.

The number density of scattered mesotrons can be obtained by calculating the charge fluctuations within the averaging volume. It is simpler, however, to consider the average value of a mechanical quantity such as the momentum density. The latter is defined by:

$$\underset{\sim}{P}(\underset{\sim}{r}) = -[\pi \underset{\sim}{\nabla} \Phi + \underset{\sim}{\nabla} \Phi^* \pi^*]. \tag{76}$$

The form which this operator assumes at large distances from the nucleon, after being subjected to the sequence of transformations, is simply

$$\underset{\sim}{P}(\underset{\sim}{r}) \sim -\frac{1}{2} [\pi_1 \underset{\sim}{\nabla} \Phi_1 + \underset{\sim}{\nabla} \Phi_1 \pi_1] - \frac{1}{2} [\pi_2 \underset{\sim}{\nabla} \Phi_2 + \underset{\sim}{\nabla} \Phi_2 \pi_2]. \tag{77}$$

Upon inserting the asymptotic forms of the two sets of field quantities, writing

$$u_{\underset{\sim}{p}}(\underset{\sim}{r}) \sim e^{i\underset{\sim}{p}\cdot\underset{\sim}{r}/\hbar} + c_{\underset{\sim}{p}} e^{ipr/\hbar}/r,$$

$$\sigma(E) = 4\pi |c_p|^2, \tag{78}$$

one obtains

$$\bar{\underset{\sim}{P}}(\underset{\sim}{r}) = (1/V) \sum_{\underset{\sim}{p}} \underset{\sim}{p}\, \alpha^*_{\underset{\sim}{p}} \alpha_{\underset{\sim}{p}} + (1/V) \sum_{\underset{\sim}{p}} (\underset{\sim}{p} + p \frac{\underset{\sim}{r}}{r} \frac{|c_{\underset{\sim}{p}}|^2}{r^2}) \beta^*_{\underset{\sim}{p}} \beta_{\underset{\sim}{p}}, \tag{79}$$

after performing the required process of averaging over a volume containing many wave lengths. This equation simply states that the Φ_1, π_1 field is unperturbed while the waves associated with the Φ_2, π_2 field are scattered radially with the cross section $\sigma(E)$. The introduction of the creation and destruction operators of the charged mesotrons replaces (79) by

$$\bar{P}(r) = (1/V) \sum_{p} \{p(A_p^* A_p + B_p^* B_p) + p \frac{r}{r} (\sigma(E)/4\pi r^2) \frac{1}{2} (A_p^* A_p + B_p^* B_p - A_p^* B_p - B_p^* A_p)\}. \tag{80}$$

The expectation value of the averaged momentum density in a state with definite numbers of mesotrons in the various momentum states is therefore

$$(1/V) \sum_{p} \{p + p(r/r) \frac{1}{2} (\sigma(E)/4\pi r^2)\} (N_p^+ + N_p^-). \tag{81}$$

This clearly means that the cross section for the scattering of a charged mesotron is $\frac{1}{2}\sigma(E)$. It may also be shown, by demonstrating the absence of correlation in charge fluctuations at different points, that only one particle emerges upon the scattering of a mesotron. This, however, is a trivial observation in view of the linear character of the theory in the strong coupling limit.

Our last task is to evaluate the limitations imposed upon g by the strong coupling conditions (35). We have already obtained the inequalities (40) by the requirement that the excitation energy of the doubly charged proton will be small compared with μc^2. We shall show that the equations (35) justify this criterion. The first condition of (35) reads:

$$K\zeta_o c >> \int \phi_o \Phi_1 dr, \tag{82}$$

or

$$(K\zeta_o/\hbar) >> \sum_{p} (2E_p)^{-1/2} [\alpha_p V^{-1/2} \int \phi_o e^{ip\cdot r/\hbar} dr + \alpha_p^* V^{-1/2} \int \phi_o e^{-ip\cdot r/\hbar} d \tag{83}$$

This equation states that the excitation of the state ϕ_o in the free mesotron field--principally by zero-point fluctuations--must be small compared with the excitation of this state in the bound mesotron field. Therefore,

$$(2K^2\zeta_o{}^2/\hbar^2) >> \sum_{\underset{\sim}{p}} (1/E_{\underset{\sim}{p}}) \frac{1}{V} \left|\int \phi_o \, e^{i\underset{\sim}{p}\cdot\underset{\sim}{r}/\hbar} d\underset{\sim}{r}\right|^2 . \tag{84}$$

We shall re-express this sum by introducing an average energy denominator, viz.:

$$\sum_{\underset{\sim}{p}} \frac{1}{E_{\underset{\sim}{p}}} \frac{1}{V} \left|\int \phi_o \, e^{i\underset{\sim}{p}\cdot\underset{\sim}{r}/\hbar} d\underset{\sim}{r}\right|^2 = \frac{1}{\bar{E}} \sum_{\underset{\sim}{p}} \left|\int \phi_o \, V^{-1/2} e^{i\underset{\sim}{p}\cdot\underset{\sim}{r}/\hbar} d\underset{\sim}{r}\right|^2 = \frac{1}{\bar{E}} \tag{85}$$

by virtue of the completeness of the plane wave eigenfunctions, and of the normalization of ϕ_o. An approximate value for $\bar{E}$ may be obtained by considering the sum:

$$\sum_{\underset{\sim}{p}} \frac{1}{E^2_{\underset{\sim}{p}}} \frac{1}{V} \left|\int \phi_o \, e^{i\underset{\sim}{p}\cdot\underset{\sim}{r}/\hbar} d\underset{\sim}{r}\right|^2 = \int \phi_o(r) \left[\sum_{\underset{\sim}{p}} \frac{\frac{1}{V} e^{i\underset{\sim}{p}\cdot(\underset{\sim}{r}-\underset{\sim}{r}')/\hbar}}{E^2_{\underset{\sim}{p}}}\right] \phi_o(r') \, d\underset{\sim}{r} \, d\underset{\sim}{r}', \tag{86}$$

for

$$(\kappa^2 - \nabla^2) \left[\sum_{\underset{\sim}{p}} \frac{\frac{1}{V} e^{i\underset{\sim}{p}\cdot(\underset{\sim}{r}-\underset{\sim}{r}')/\hbar}}{E^2_{\underset{\sim}{p}}}\right] = 1/h^2c^2 \sum_{\underset{\sim}{p}} \frac{1}{V} e^{i\underset{\sim}{p}\cdot(\underset{\sim}{r}-\underset{\sim}{r}')/\hbar} = 1/\hbar^2c^2 \, \delta(\underset{\sim}{r}-\underset{\sim}{r}'),$$

whence

$$\sum_{\underset{\sim}{p}} \frac{(1/V) \, e^{i\underset{\sim}{p}\cdot(\underset{\sim}{r}-\underset{\sim}{r}')/\hbar}}{E^2_{\underset{\sim}{p}}} = (1/4\pi\hbar^2e^2) \, (e^{-\kappa|\underset{\sim}{r}-\underset{\sim}{r}'|}/|\underset{\sim}{r}-\underset{\sim}{r}'|) . \tag{87}$$

(The above is, of course, the Green's function of the operator $\kappa^2 - \nabla^2$). Therefore,

$$\sum_{\underset{\sim}{p}} \frac{1}{E^2_{\underset{\sim}{p}}} \frac{1}{V} \left|\int \phi_o \, e^{i\underset{\sim}{p}\cdot\underset{\sim}{r}/\hbar} d\underset{\sim}{r}\right|^2 = \zeta_o/\hbar^2 , \tag{88}$$

The left side of this equation can be evaluated approximately as $\bar{E}^{-2}$, for the principal contributions to this summation and that occurring in Eq. (84) both arise from the largest momenta which lead to appreciable values for the Fourier amplitude of ϕ_o, i.e., $p \sim \hbar/a$. Hence

$$\bar{E} \sim \hbar \, \zeta_o{}^{-1/2} \qquad \text{and} \qquad \frac{2\,K^2\,\zeta_o{}^{3/2}}{\hbar} >> 1 . \tag{89}$$

We shall write this condition in terms of the quantity ε, which measures the excitation energy of isobars,

$$E_q = \varepsilon(q - \tfrac{1}{2})^2, \qquad \varepsilon = \frac{\hbar^2}{2K^2\sum_s \zeta^2}, \tag{36}$$

whence,

$$\varepsilon \ll \hbar \frac{\zeta_o^{3/2}}{\sum_s \zeta_s^2} \tag{90}$$

In the limit $\kappa a \ll 1$, $\sum_s \zeta_s^2 \sim a^3/\kappa c^4$ (Eq. 37) and $\zeta_o \sim a^2/c^2$ (Eq. 24). Therefore $\zeta_o^{3/2}/\sum_s \zeta_s^2 \sim \kappa c$ and the limitation (90) reads $\varepsilon \ll \mu c^2$. Similarly, in the opposite limit $\kappa a \gg 1$, $\sum_s \zeta_s^2 = (\kappa c)^{-4}$, and $\zeta_o = (\kappa c)^{-2}$, whence $\varepsilon \ll \mu c^2$. It is rather remarkable that, when calculated with the special choice (38) for ϕ_o; $\zeta_o^{3/2}/\sum_s \zeta_s^2 = \kappa c = \mu c^2/\hbar$ for all values of a. We have thus shown that the first of equations (35) leads to the restriction

$$\varepsilon \ll \mu c^2. \tag{91}$$

The second of equations (35), after a somewhat more complicated treatment, leads to the same result. The inequalities (40) therefore define the domain of "strong coupling."

Classical theory

Although the nuclear interaction of the charged scalar mesotron field is essentially quantum-mechanical in nature, it is possible to obtain all the preceding results by a purely classical calculation. This remarkable fact may be understood as a consequence of the strong coupling condition, which manifests itself in two ways: (a) The number of quanta in the state ϕ_o is $\sim g^2/\hbar c$ ($\kappa a \ll 1$) and is thus large compared with one in the limit of strong coupling; quantum fluctuations of the field are negligible. (b) The two states of isotopic spin are so widely separated energetically that one is inaccessible; quantum fluctuations of the source are negligible. Such classical considerations in no way supplant the correct quantum calculations, but they do facilitate an elementary discussion of the phenomena of strong coupling, in addition to providing criteria for the inapplicability of perturbation theory, since the classical calculations can be performed for all values of g.

The Hamiltonian (5) is most easily treated classically in terms of the real fields introduced by the canonical transformation:

$$\Phi(\underset{\sim}{r}) = (8\pi)^{-1/2}\left(\Phi_1(\underset{\sim}{r}) - i\,\Phi_2(\underset{\sim}{r})\right),\quad \pi(\underset{\sim}{r}) = (2\pi)^{1/2}\left(\pi_1(\underset{\sim}{r}) + i\,\pi_2(\underset{\sim}{r})\right)$$

$$[\Phi_1(\underset{\sim}{r}),\ \pi_1(\underset{\sim}{r}')] = \delta(\underset{\sim}{r}-\underset{\sim}{r}'),\quad [\Phi_1(\underset{\sim}{r}),\ \pi_2(\underset{\sim}{r}')] = 0,\ \text{etc.} \tag{92}$$

(The bracket symbols now designate classical Poisson brackets.) Hence,

$$H = (1/8\pi)\int\{16\pi^2c^2(\pi_1^2 + \pi_2^2) + \underset{\sim}{\nabla}\Phi_1\cdot\underset{\sim}{\nabla}\Phi_1 + \underset{\sim}{\nabla}\Phi_2\cdot\underset{\sim}{\nabla}\Phi_2$$

$$+ \kappa^2(\Phi_1^2 + \Phi_2^2)\}\,d\underset{\sim}{r} - g\int K(r)\,(\tau_1\Phi_1 + \tau_2\Phi_2)\,d\underset{\sim}{r}, \tag{93}$$

$$Q/e = \frac{1}{2}(1 + \tau_3) + (1/\hbar)\int\{\Phi_1\pi_2 - \Phi_2\pi_1\}\,d\underset{\sim}{r},$$

in which τ is to be considered a unit vector whose components obey the Poisson brack relations:

$$[\tau_1,\ \tau_2] = (2/\hbar)\,\tau_3,\ \text{etc.} \tag{94}$$

The equations of motion of the system are:

$$\dot{\tau}_1 = -(2g/\hbar)\,\tau_3\int K(r)\,\Phi_2\,d\underset{\sim}{r},\quad \dot{\tau}_2 = (2g/\hbar)\,\tau_3\int K(r)\,\Phi_1\,d\underset{\sim}{r},$$

$$\dot{\tau}_3 = (2g/\hbar)\int K(r)\,(\tau_1\Phi_2 - \tau_2\Phi_1)\,d\underset{\sim}{r}, \tag{95}$$

$$\dot{\Phi}_{1,2} = 4\pi c^2\,\pi_{1,2},\quad \dot{\pi}_{1,2} = (1/4\pi)\,(\nabla^2-\kappa^2)\,\Phi_{1,2} + g\,\tau_{1,2}\,K(r),$$

whence

$$(\kappa^2 - \Box^2)\ \Phi_{1,2} = 4\pi g\ \tau_{1,2}\ K(r). \tag{96}$$

We may first inquire under what conditions a static solution of these equations exists. It is evident that these conditions are, that

$$\tau_3\tau_1 = 0;\ \text{and}\ \tau_3\tau_2 = 0\ .$$

The first possibility, $\tau_3 = 0$, although admissible classically, corresponds to a system with charge $Q = \frac{1}{2}e$ and must therefore be excluded. The second type of static state, $\tau_1 = \tau_2 = 0$, $\tau_3 = \pm 1$, cannot be excluded on the same grounds since it represents either a neutron or a proton ($Q = 0, e$). We shall also disregard this possibility, however, for this state is highly unstable against the isotopic spin fluctuations which we have neglected. We seek, instead, the general time-dependent solution of the Eqs. (95) and (96) which describes an isolated, non-radiating system. The requirement that the system be non-radiating implies that in the Fourier expansion of the field quantities (or of the isotopic spin) there occurs no frequency $|\omega|$ which is greater than κc, i.e., $\hbar|\omega| < \mu c^2$. The field generated by the oscillations of the isotopic spin (Eq. 96) must be inserted into Eq. (95) to determine the reaction of the isotopic spin to the field produced by its own motion. The equations of motion thus obtained for the isotopic spin are clearly non-linear and contain infinitely high time derivates. The non-linear character of these equations strongly restricts the possible solutions which can only be of the form:

$$\tau_1 = (1 - \tau_3^2)^{1/2}\ \cos(\omega t + \delta),\ \tau_2 = (1 - \tau_3^2)^{1/2}\ \sin(\omega t + \delta),$$
$$\tau_3 = \text{const.} \tag{97}$$

for were a second frequency to be introduced, all the over-tones would be created, violating the condition $\omega < \kappa c$. The corresponding field quantities are:

$$\Phi_{1,2}(\underset{\sim}{r},t) = g\,\tau_{1,2}(t) \int \frac{e^{-(\kappa^2-\omega^2/c^2)^{1/2}|\underset{\sim}{r}-\underset{\sim}{r}'|}}{|r-r'|} K(r')\, d\underset{\sim}{r}'$$

$$\pi_1 = -(\omega/4\pi c^2)\,\Phi_2, \quad \pi_2 = (\omega/4\pi c^2)\,\Phi_1. \tag{98}$$

The equations of motion for τ are thereby satisfied if the two constants ω and τ_3 are related by:

$$\hbar\omega = 2g^2\,\tau_3 \int K(\underset{\sim}{r}) \frac{e^{-(\kappa^2-\omega^2/c^2)^{1/2}|\underset{\sim}{r}-\underset{\sim}{r}'|}}{|\underset{\sim}{r}-\underset{\sim}{r}'|} K(\underset{\sim}{r}')\, d\underset{\sim}{r}\, d\underset{\sim}{r}'. \tag{99}$$

In the limit of a point source ($\kappa a << 1$) this relation becomes

$$|\tau_3| \doteq \frac{1}{2}\hbar|\omega|/(g^2/a) < \frac{1}{2}\mu c^2/(g^2/a)$$

$$\left(= \frac{1}{2}\frac{\kappa a}{g^2/\hbar c}\right). \tag{100}$$

Therefore, when

$$g^2/\hbar c >> \kappa a, \tag{101}$$

(equivalently, when the self-energy is large compared to μc^2) the value of τ_3 is small compared to unity for all permissible frequencies: $0 < |\omega| < \kappa c$.

The basic assumption of the perturbation theory is that the probability of finding a mesotron in the field is small, and that therefore the average charge of the nucleon differs but little from the value it assumes when the coupling with the mesotron field is adiabatically removed. We have seen, however, that when the condition (101) obtains, $\tau_3 \ll 1$, and therefore the average charge of the nucleon is 1/2, which is the expression of strong coupling for the isotopic spin. It appears, therefore, that perturbation theory will be valid only when $g^2/\hbar c \ll \kappa a$, a relation that can never be satisfied by a point source.

The precession of the isotopic spin with frequency ω creates a charge-bearing field. We must select from among the continuum of classically allowed frequencies those which correspond to a half-integral number of charge units in the field, for the nucleon in the point source limit contributes but half a unit of charge. The charge in the mesotron field:

$$(Q/e) - \frac{1}{2} = (1/\hbar) \int \{\Phi_1 \pi_2 - \Phi_2 \pi_1\} \, d\underset{\sim}{r} = (\omega/4\pi\hbar c^2) \int \{\Phi_1^2 + \Phi_2^2\} \, d\underset{\sim}{r} \,. \tag{102}$$

Inserting the expression (98) for the fields and employing the theorem (37') we obtain

$$(Q/e) - \frac{1}{2} = \frac{1}{2}(g^2/\hbar c) \frac{\omega}{(\kappa^2 c^2 - \omega^2)^{1/2}} \int K(\underset{\sim}{r}) e^{-(\kappa^2 - \frac{\omega^2}{c^2})^{1/2} |\underset{\sim}{r} - \underset{\sim}{r}'|} K(\underset{\sim}{r}') \, d\underset{\sim}{r} \, d\underset{\sim}{r}' = \frac{1}{2}(g^2/\hbar c) \frac{\omega}{(\kappa^2 c^2 - \omega^2)^{1/2}} \qquad (\kappa a \ll 1), \tag{103}$$

remembering that

$$\tau_1^2 + \tau_2^2 = 1 \tag{104}$$

in the point source limit (101). Therefore, the precession frequency required to generate q-1/2 units of charge ($Q/e = q$) is:

$$\omega = (\mu c^2/\hbar) \frac{q - \frac{1}{2}}{[(q - \frac{1}{2})^2 + \frac{1}{4}(g^2/\hbar c)^2]^{1/2}} \,. \tag{105}$$

It is interesting that $|\omega|$ is always less than $\mu c^2/\hbar$, and thus all states of definite charge are stable, non-radiating systems.

To determine the excitation energy of an isobar, we must calculate the value assumed by the Hamiltonian (93) in a state of charge q. Eliminating the gradients from (93) by employing the field equations (96), one obtains

$$\begin{aligned} E &= (1/4\pi c^2)\int\{\dot{\Phi}_1^2+\dot{\Phi}_2^2\}d\underset{\sim}{r} - \frac{1}{2}g\int K(\underset{\sim}{r})(\tau_1\Phi_1+\tau_2\Phi_2)d\underset{\sim}{r} \\ &= (\omega^2/4\pi c^2)\int\{\Phi_1^2+\Phi_2^2\}d\underset{\sim}{r} + \frac{1}{2}g^2\int K(\underset{\sim}{r})\frac{e^{-\kappa|\underset{\sim}{r}-\underset{\sim}{r}'|}-e^{-(\kappa^2-\omega^2/c^2)^{1/2}|\underset{\sim}{r}-\underset{\sim}{r}'|}}{|\underset{\sim}{r}-\underset{\sim}{r}'|} \\ &\qquad\qquad \times K(\underset{\sim}{r}')\,d\underset{\sim}{r}\,d\underset{\sim}{r}' \\ &\quad - \frac{1}{2}g^2\int K(\underset{\sim}{r})\frac{e^{-\kappa|\underset{\sim}{r}-\underset{\sim}{r}'|}}{|\underset{\sim}{r}-\underset{\sim}{r}'|}K(\underset{\sim}{r}')\,d\underset{\sim}{r}\,d\underset{\sim}{r}'. \end{aligned} \tag{106}$$

The last term represents the self-energy of the system ($\sim -g^2/a$). We are interested only in the first two terms, the charge-dependent part of the energy:

$$\begin{aligned} E_q &= \hbar\omega(q-\tfrac{1}{2}) - \tfrac{1}{2}g^2\{\kappa - (\kappa^2-\tfrac{\omega^2}{c^2})^{1/2}\} \\ &= \mu c^2\{[(q-\tfrac{1}{2})^2+\tfrac{1}{4}(g^2/\hbar c)^2]^{1/2} - \tfrac{1}{2}(g^2/\hbar c)\}. \end{aligned} \tag{107}$$

In the limiting case of large coupling:

$$g^2/\hbar c >> 1:\ E_q = (g^2/\hbar c)^{-1}\mu c^2(q-\tfrac{1}{2})^2\ , \tag{1}$$

while in the limit of weak coupling:

$$g^2/\hbar c << 1:\ E_q = \mu c^2|q-\tfrac{1}{2}|. \tag{107'}$$

The ease with which Eq. (1) is reached classically is truly remarkable.

The connection between the precession frequency ω and the isobar excitation energy Eq should be noted. Indeed,

$$\omega = (1/\hbar)(\partial E_q/\partial q) \tag{108}$$

which shows that $\hbar q$ is to be interpreted as an action variable, consistent with our requirement that q shall be an integer. The theory we have developed is thus to be considered as an extension of the old Quantum Theory to field phenomena.

We shall next consider the classical theory of mesotron scattering by a nucleon. It will first be convenient to exploit to the full the point source limit, which is all that we consider in this paper. The equations of motion for τ_1 and τ_2, Eqs. (95), depend only on the value of the field in the immediate vicinity of the nucleon. It is evident from (96) that the field in this neighborhood is:

$$\Phi_{1,2}(\underset{\sim}{r},t) \sim g\,\tau_{1,2}(t) \int \frac{K(\underset{\sim}{r}')}{|\underset{\sim}{r}-\underset{\sim}{r}'|}\,d\underset{\sim}{r}'$$

provided a is small compared to all the wave lengths which occur in this otherwise arbitrary field. The equations of motion are thus:

$$\dot{\tau}_1 = -(2g^2/\hbar c)\ (c/a)\ \tau_3\tau_2,\ \dot{\tau}_2 = (2g^2/\hbar c)\ (c/a)\ \tau_3\tau_1$$

or

$$\tau_3 = \frac{1}{2}\,\frac{(a/c)}{(g^2/\hbar c)}\,\frac{d}{dt}\tan^{-1}\ (\tau_2/\tau_1).$$

If, therefore, all the frequencies which occur in the system are such that

$$(\omega/c)\ a \ll g^2/\hbar c, \quad (\hbar\omega \ll g^2/a), \tag{109}$$

then $\tau_3 \ll 1$ in an arbitrary state of motion of the field. The condition (109) of course imposes no restriction on g in the strict point source limit. For a small but finite source ($\kappa a << 1$), (109) will be satisfied for all wavelengths greater than a, provided $g^2/\hbar c >> 1$. The advantage of supposing that a is sufficiently small to permit the consistent neglect of τ_3 becomes evident on considering the equation $\dot{\tau}_3 = 0$, which now states that

$$\tau_{1,2} = \left(\int K(\underset{\sim}{r})\ \Phi_{1,2}\ d\underset{\sim}{r}\right) \Big/ \left[\left(\int K(\underset{\sim}{r})\ \Phi_1\ d\underset{\sim}{r}\right)^2 + \left(\int K(\underset{\sim}{r})\ \Phi_2\ d\underset{\sim}{r}\right)^2\right]^{1/2}. \tag{110}$$

One thereby obtains the rather bizarre field equations:

$$(\kappa^2 - \Box^2)\Phi_{1,2} = 4\pi g\ K(\underset{\sim}{r}) \int K\Phi_{1,2} d\underset{\sim}{r} \Big/ \left[\left(\int K(\underset{\sim}{r})\Phi_1 d\underset{\sim}{r}\right)^2 + \left(\int K\Phi_2 d\underset{\sim}{r}\right)^2\right]^{1/2}, \tag{111}$$

which are rigorous in the strict point source limit.

The fields which we have found to describe an isolated system in a definite charge state, viz.:

$$\Phi_{1,2}^{(o)} = \tau_{1,2}^{(o)}\ \Phi^{(o)},\quad \Phi^{(o)} = g \int \frac{e^{-(\kappa^2 - \omega^2/c^2)^{1/2}|\underset{\sim}{r}-\underset{\sim}{r}'|}}{|\underset{\sim}{r}-\underset{\sim}{r}'|} K(\underset{\sim}{r}')d\underset{\sim}{r}'$$

$$\tau_1^{(o)} = \cos \omega t, \qquad \tau_2^{(o)} = \sin \omega t \tag{112}$$

obviously satisfy these equations, which in this form show clearly the necessity for the special time dependence (97).

Let us now suppose that a mesotron wave falls upon the system in the state (112) inducing oscillations which generate the scattered wave. We seek a solution of the equations (111) in the form

$$\Phi_{1,2} = \Phi_{1,2}^{(o)} + \Phi_{1,2}^{(1)} \tag{113}$$

where $\Phi_{1,2}^{(1)}$ includes the incident plane wave and the spherically diverging scattered wave. Introducing the conventional assumption that the incident wave is extremely weak in comparison to the stationary field, or more specifically that:

$$\int K(\underset{\sim}{r})\, \Phi^{(1)}_{1,2}\, d\underset{\sim}{r} << \int K(\underset{\sim}{r})\, \Phi^{(o)}_{1,2}\, d\underset{\sim}{r},$$

we expand the field equations (111) to obtain linear equations for $\Phi^{(1)}_{1,2}$,

$$(\kappa^2-\Box^2)\Phi^{(1)}_{1,2} = 4\pi g\left[K(\underset{\sim}{r})\Big/\int K(\underset{\sim}{r})\Phi^{(o)}\, d\underset{\sim}{r}\right] \int K(\underset{\sim}{r}')\, \{\Phi^{(1)}_{1,2}(\underset{\sim}{r}') - \tau^{(o)}_{1,2}\left[\tau^{(o)}_{1}\, \Phi^{(1)}_{1}(\underset{\sim}{r}') + \tau^{(o)}_{2}\, \Phi^{(1)}_{2}(\underset{\sim}{r}')\right]\}\, d\underset{\sim}{r}'. \quad (11\ldots$$

It is convenient to introduce two independent complex fields Φ_+ and Φ_- from which the two real fields Φ_1, Φ_2 are to be derived by the operations:

$$\Phi_1 = R[\Phi_+ + \Phi_-] \qquad R[\psi] = \frac{1}{2}(\psi + \psi^*) \qquad (115)$$
$$\Phi_2 = R[i(\Phi_+ - \Phi_-)]$$

These complex fields are determined by the simultaneous equations

$$(\kappa^2-\Box^2)\Phi_+ = 4\pi g\left(K(\underset{\sim}{r})\Big/\int K(\underset{\sim}{r})\Phi^{(o)} d\underset{\sim}{r}\right) \frac{1}{2}\int K(\underset{\sim}{r})(\Phi_+ - e^{-2i\omega t}\Phi_-)\, d\underset{\sim}{r}$$
$$(\kappa^2-\Box^2)\Phi_- = 4\pi g\left(K(\underset{\sim}{r})\Big/\int K(\underset{\sim}{r})\Phi^{(o)} d\underset{\sim}{r}\right) \frac{1}{2}\int K(\underset{\sim}{r})(\Phi_- - e^{-2i\omega t}\Phi_+)\, d\underset{\sim}{r}. \qquad (116)$$

The advantage of this formulation is that Φ_+ and Φ_- describe, respectively, the positive and negative charge components of the field. Thus, an incident positive mesotron is represented by:

$$\Phi^{(in)}_+ = e^{i\underset{\sim}{k}\cdot\underset{\sim}{r}-i\nu t}, \qquad \Phi^{(in)}_- = 0 \qquad (\nu \text{ positive}). \qquad (117)$$

The solution of the Eqs. (116) corresponding to this state of affairs is of the form

$$\Phi_+ = e^{-i\nu t}\, f_+(\underset{\sim}{r}), \qquad \Phi_- = e^{-i(\nu-2\omega)t}\, f_-(\underset{\sim}{r})$$

in which f_+ and f_- obey the equations:

$$(\kappa^2 - \frac{\nu^2}{c^2} - \nabla^2) f_+ = 4\pi g \left[K(\underset{\sim}{r}) \Big/ \int K(\underset{\sim}{r}) \Phi^{(o)} d\underset{\sim}{r} \right] \frac{1}{2} \int K(\underset{\sim}{r}) (f_+ - f_-) d\underset{\sim}{r}$$

$$(\kappa^2 - \frac{(\nu - 2\omega)^2}{c^2} - \nabla^2) f_- = -4\pi g \left[K(\underset{\sim}{r}) \Big/ \int K(\underset{\sim}{r}) \Phi^{(o)} d\underset{\sim}{r} \right] \frac{1}{2} \int K(\underset{\sim}{r}) (f_+ - f_-) d\underset{\sim}{r}, \quad (118)$$

subject to the boundary conditions that f_+ becomes, asymptotically, the incident plane wave $e^{i\underset{\sim}{k}\cdot\underset{\sim}{r}}$ plus a spherically diverging wave, while f_- contains only a diverging wave, describing the absence of negative charge incident on the system. If, therefore, a positive mesotron falls upon a system in a state of positive charge (ω positive), say, the above equations indicate that both positive and negative mesotrons will emerge, the latter with the diminished frequency $\nu - 2\omega$. In accordance with the Correspondence Principle (cf. Eq. 108), we must perform the replacement

$$2\omega \rightarrow (E_{q+2} - E_q)/\hbar$$

thereby attributing to the negative mesotron the energy

$$\hbar\nu - (E_{q+2} - E_q)$$

demanded by the excitation of the system from the state q to the state q+2, which the conversion of a positive mesotron to a negative mesotron entails.

Introducing the notation

$$k^2 = (\nu^2/c^2) - \kappa^2, \qquad k'^2 = \frac{(\nu - 2\omega)^2}{c^2} - \kappa^2$$

$$c = (g/\int K(\underset{\sim}{r}) \Phi^{(o)} d\underset{\sim}{r}) \frac{1}{2} \int K(\underset{\sim}{r}) (f_+ - f_-) d\underset{\sim}{r}, \quad (119)$$

we obtain the elementary equations:

$$(\nabla^2 + k^2) f_+ = -4\pi c \, K(\underset{\sim}{r})$$

$$(\nabla^2 + k'^2) f_- = 4\pi c \, K(\underset{\sim}{r}). \quad (120)$$

The appropriate solutions are:

$$f_+ = e^{i\mathbf{k}\cdot\mathbf{r}} + c \int \frac{e^{ik|\mathbf{r}-\mathbf{r}'|}}{|\mathbf{r}-\mathbf{r}'|} K(\mathbf{r}')\, d\mathbf{r}'$$

$$f_- = -c \int \frac{e^{ik'|\mathbf{r}-\mathbf{r}'|}}{|\mathbf{r}-\mathbf{r}'|} K(\mathbf{r}')\, d\mathbf{r}', \qquad (121)$$

where the constant c is determined by (119):

$$2c \int K(\mathbf{r}) \frac{e^{-\kappa|\mathbf{r}-\mathbf{r}'|}}{|\mathbf{r}-\mathbf{r}'|} K(\mathbf{r}')\, d\mathbf{r}\, d\mathbf{r}' = \int K(\mathbf{r})\, e^{i\mathbf{k}\cdot\mathbf{r}}\, d\mathbf{r}$$

$$+ c \int K(\mathbf{r}) \frac{e^{ik|\mathbf{r}-\mathbf{r}'|} + e^{ik'|\mathbf{r}-\mathbf{r}'|}}{|\mathbf{r}-\mathbf{r}'|} K(\mathbf{r}')\, d\mathbf{r}\, d\mathbf{r}'$$

which, in the point source limit, states that

$$c = -\frac{1}{2\kappa + i(k+k')} \cdot \qquad (122)$$

The cross section for elastic scattering is simply

$$\sigma_{M^+\to M^+} = 4\pi|c|^2 = \frac{4\pi}{(k+k')^2 + 4\kappa^2} \qquad (123)$$

while the cross section for the inelastic process in which a negative mesotron is produced is

$$\sigma_{M^+\to M^-} = 4\pi(k'/k)|c|^2 = (k'/k)\,\sigma_{M^+\to M^+} \qquad (124)$$

(We have tacitly assumed that sufficient energy is available for the excitation process, i.e., that k' is real.) In the limit of strong coupling, $E_{q+2} - E_q) \ll \mu c^2$, and therefore $k' \doteqdot k$. Hence

$$\sigma_{M^+\to M^+} = \sigma_{M^+\to M^-} = \pi/(k^2+\kappa^2) = \pi(\hbar/\mu c)^2\ (\mu c^2/E)^2, \qquad (125)$$

and similarly for an incident negative mesotron,

$$\sigma_{M^-\to M^-} = \sigma_{M^-\to M^+} = \pi(\hbar/\mu c)^2\ (\mu c^2/E)^2, \qquad (126)$$

which shows most clearly that it is always equally probable that either a positive or a negative mesotron emerge, and that the total scattering cross section:

$$\sigma_{M^\pm \to M+} + \sigma_{M^\pm \to M-} = 2\pi(\hbar/\mu c)^2 \ (\mu c^2/E)^2 \tag{127}$$

which is indeed $\frac{1}{2}\sigma$ (Eq. 4). The identity of the classical results with those obtained quantum mechanically, in the strong coupling limit, can be recognized at an earlier stage than this, however. In the limit of strong coupling the time variation of $\tau_{1,2}^{(o)}$ is negligible compared to that of the field ($\omega << \nu$). The two new field quantities:

$$\begin{aligned} \Phi_\xi &= \tau_1^{(o)} \ \Phi_1^{(1)} + \tau_2^{(o)} \ \Phi_2^{(1)} \\ \Phi_\eta &= \tau_1^{(o)} \ \Phi_2^{(1)} - \tau_2^{(o)} \ \Phi_1^{(1)} \end{aligned} \tag{128}$$

are therefore determined by the equations:

$$\begin{aligned} (\kappa^2 = \Box^2) \ \Phi_\xi \ (\mathbf{r}) &= 0 \\ (\kappa^2 - \Box^2) \ \Phi_\eta \ (\mathbf{r}) &= 4\pi g \ \left(K(\mathbf{r}) \Big/ \int K\Phi^{(o)} \ d\mathbf{r}\right) \int K(\mathbf{r}') \ \Phi_\eta \ (\mathbf{r}') \ d\mathbf{r}'. \end{aligned} \tag{129}$$

The field is thus split into two components, one of which is unperturbed, the other being determined by an equation which is identical with Eq. (54).

As a final task, we shall sketch a quantum mechanical treatment of the charged scalar field that bears more resemblance to the classical development than the method of expansion in orthogonal functions, which, although very powerful, is somewhat cumbersome. We begin with the Hamiltonian and charge operators in the form (93). The operators obtained by applying the canonical transformation:

$$S_\alpha = \exp \ \{-\tfrac{1}{2} \, i \ \tau_3 \ \tan^{-1} \left(\int K\Phi_2 \ d\mathbf{r} \Big/ \int K\Phi_1 \ d\mathbf{r}\right) , \tag{130}$$

with the aid of the formulae:

$$S_\alpha^{-1}\,[\tau_1\int K(\mathbf{r})\Phi_1 d\mathbf{r} + \tau_2 \int K(\mathbf{r})\Phi_2 d\mathbf{r}]\, S_\alpha$$

$$= \tau_1\ [(\int K(\mathbf{r})\Phi_1 d\mathbf{r})^2 + (\int K(\mathbf{r})\Phi_2 d\mathbf{r})^2]^{1/2},$$

$$S_\alpha^{-1}\ \pi_1(\mathbf{r})\ S_\alpha = \pi_1(\mathbf{r}) + \frac{1}{2}\hbar\ \tau_3\ \frac{K(\mathbf{r})\ \int K(\mathbf{r}')\ \Phi_2\ d\mathbf{r}'}{(\int K(\mathbf{r}')\Phi_1 d\mathbf{r}')^2 + (\int K(\mathbf{r}')\Phi_2 d\mathbf{r}')^2}$$

$$S_\alpha^{-1}\ \pi_2(\mathbf{r})\ S_\alpha = \pi_2(\mathbf{r}) - \frac{1}{2}\hbar\ \tau_3\ \frac{K(\mathbf{r})\ \int K(\mathbf{r}')\ \Phi_1\ d\mathbf{r}'}{(\int K(\mathbf{r}')\Phi_1 d\mathbf{r}')^2 + (\int K(\mathbf{r}')\Phi_2 d\mathbf{r}')^2} \tag{131}$$

are

$$Q/e = \frac{1}{2} + (1/\hbar) \int (\Phi_1\pi_2 - \Phi_2\pi_1)\ d\mathbf{r}$$

$$H = (1/8\pi) \int \{16\pi^2c^2(\pi_1^2 + \pi_2^2) + \nabla\Phi_1 \cdot \nabla\Phi_1 + \nabla\Phi_2 \cdot \nabla\Phi_2 +$$

$$+ \kappa^2(\Phi_1^2 + \Phi_2^2)\}\ d\mathbf{r} - g\ \tau_1\ [(\int K(\mathbf{r})\Phi_1\ d\mathbf{r})^2 + (\int K(\mathbf{r})\Phi_2\ d\mathbf{r})^2]^{1/2} +$$

$$+ \pi\hbar^2c^2/2$$

(The 1941 manuscript stops with this equation left incomplete, although there are sketches of the rest of the argument.)

Received 2/1/69

Parities in Strong Coupling Models

by

C. Dullemond

University of Nijmegen

There are several mechanisms to generate mixed parity spectra in static strong coupling models.[1] It is the purpose of this article to discuss these mechanisms and some of the difficulties which may be encountered.

If the coupling between a set of meson fields and set of bare, static baryons is of Yukawa-type, then mixed parity spectra can only be obtained if not all the bare baryons have the same parity. If, moreover, the Hamiltonian, describing the interaction, is invariant under transformations of some symmetry group, the bare baryons cannot all belong to the same irreducible representation of that group and they may have different bare masses without breaking the symmetry. Except when bare mass-differences are essential, we shall assume that all bare masses are equal. Terms in the Hamiltonian corresponding to the bare baryon masses will be left out, it being understood that constant terms must be added to the resulting mass-spectra.

A simple model which leads to a nontrivial mixed parity spectrum is the following: Let $\vec{\phi}$ be a set of three bound meson fields transforming as a vector in isospin space and let $\vec{\pi}$ be the momenta canonically conjugate to $\vec{\phi}$. Furthermore, let Λ and $\vec{\Sigma}$ be the annihilation operators and $\overline{\Lambda}$ and $\vec{\overline{\Sigma}}$ the corresponding creation operators of an isosinglet and isotriplet of bare static baryons, so that (with $|>$ being the bare baryon vacuum)

$$\begin{aligned} \Lambda \overline{\Lambda} \, |> &= |> \\ \Lambda \vec{\overline{\Sigma}} \, |> &= 0 \\ \vec{\Sigma} \overline{\Lambda} \, |> &= 0 \\ \Sigma_i \overline{\Sigma}_j \, |> &= \delta_{ij} |> \end{aligned} \tag{1}$$

For $\vec{\pi}$ and $\vec{\phi}$ the following canonical commutation relations hold:

$$[\pi_i, \phi_j] = \frac{1}{i} \delta_{ij} \cdot$$ (

Now consider the static Hamiltonian

$$H = \frac{1}{2} \vec{\pi}^2 + \frac{1}{2} \mu^2 \vec{\phi}^2 + g[\overline{\Lambda}(\vec{\phi}\cdot\vec{\Sigma}) + (\vec{\overline{\Sigma}}\cdot\vec{\phi})\Lambda]$$ (

where g is a large positive number.

This model is discussed in Wentzel's paper on the Λ - Σ - interaction with scalar mesons.[2] If one now assigns a positive parity to the operators Λ and $\overline{\Lambda}$, and a negative parity to $\vec{\phi}$, $\vec{\Sigma}$ and $\vec{\overline{\Sigma}}$, (so that the parity operator changes the signs of these operators) then this parity is conserved. As usual we construct from (3) a Schrödinger equation by replacing $\vec{\pi}^2$ by $-\Delta_\phi$ where Δ_ϕ is the Laplace operator in $\vec{\phi}$ - space. The wave function for the ground state of H in the limit of large g is

$$\zeta(\phi)\ (-\phi\overline{\Lambda} + \vec{\phi}\cdot\vec{\overline{\Sigma}})\,|>$$ (4

which is to be interpreted as a physical "dressed" Λ with positive parity. Here $\phi = |\vec{\phi}|$ and $\zeta(\phi)$ is a function of ϕ which is zero except in the vicinity of $\phi_o = \frac{g}{\mu^2}$ where it is strongly peaked.

One of the eigenvalues of $H - \frac{1}{2}\vec{\pi}^2$ then reaches the lowest possible level which an eigenvalue of that operator can have, and we shall refer to this level as the "potential minimum." The set of all points at which that minimum is reached will be called a "potential valley." In the general case such a set splits up into a number o potential valleys. Each valley is defined as a connected set of such points, with the provision that different valleys are always mutually disconnected. Defined in this way, a potential valley contains one or a continuous infinity of orbits.* The first excited state has the form

$$\zeta(\phi)\ \vec{\phi}(-\phi\overline{\Lambda} + \vec{\phi}\cdot\vec{\overline{\Sigma}})\,|> .$$ (5

---

*the set of points obtained from one point by applying a continuous symmetry operation.

is is an isotriplet with negative parity to be interpreted as a ysical Σ - particle. In this way an infinite number of dressed ryons can be constructed, one irreducible multiplet for each nteger value of the isospin i. Their parities are $(-1)^i$ and their nergies (to be interpreted as physical masses)

$$E_i = E_o + \frac{\mu^4}{2g^2}\, i(i+1) \qquad (i = 0,1,2,....) \; . \tag{6}$$

pparently the multiplets have alternating parity and no repetitions ccur, i.e. the isospin is sufficient to label the multiplets.

We may not always be so lucky in finding such nontrivial mixed arity spectra. There are several unpleasant side effects. One of he reasons why we were able to obtain such a spectrum is that one annot construct a pseudoscalar from $\vec{\phi}$. The only invariant is ϕ^2 nd this is a scalar. Now, the orbit of $\vec{\phi}$ is characterized by speci- ied values of all the invariants. In our case it is a sphere with adius ϕ and since no pseudoscalars can be constructed, the point $\vec{\phi}$ lies on the orbit of $\vec{\phi}$, i.e. on the same sphere. When pseudo- calars can be constructed and when they are nonzero on the orbit, hen $\vec{\phi}$ and $-\vec{\phi}$ are points on <u>different</u> orbits. When the potential eaches a minimum only in a discrete number of distinct orbits one ould have eigenfunctions of H which are zero everywhere except in he vicinity of only one of the orbits. If that orbit is character- zed by scalars and nonzero pseudoscalars, that eigenfunction of H annot be an eigenfunction of parity. Thus there is complete parity egeneration of all eigenstates of H involving that orbit.

When a potential valley embraces a continuous infinity of orbits t may well be that two orbits which transform into each other under he parity operation belong to the same valley. Then of course there eed not be parity degeneration. This all can be demonstrated with he following example:

Let $\vec{\phi}$ and $\vec{\chi}$ be two isospin triplets of meson fields with con- ugate momenta $\vec{\pi}_\phi$ and $\vec{\pi}_\chi$ and let $\bar{B}|>$ and $\bar{C}|>$ be two baryon doublets.

Consider the following Hamiltonian

$$H = \frac{1}{2}(\vec{\pi}_\phi^{\,2} + \vec{\pi}_\chi^{\,2}) + \frac{1}{2}\mu^2(\vec{\phi}^2 + \vec{\chi}^2) + \frac{g_1}{\sqrt{2}}\bar{B}\,\vec{\phi}\cdot\vec{\tau}\,B + $$
$$+ \frac{g_2}{\sqrt{2}}(\bar{B}\,\vec{\chi}\cdot\vec{\tau}\,C + \bar{C}\,\vec{\chi}\cdot\vec{\tau}\,B) + \frac{g_3}{\sqrt{2}}\bar{C}\,\vec{\phi}\cdot\vec{\tau}\,C\,.$$

One can then assign a positive parity to B and $\vec{\phi}$ and a negative parity to C and $\vec{\chi}$. We assume that neither of the three real coupling constants, g_1, g_2 and g_3 is zero. The matrix to be diagonalized is

$$\left\| \begin{matrix} \frac{1}{2}\mu^2(\phi^2+\chi^2) + \frac{g_1}{\sqrt{2}}\vec{\phi}\cdot\vec{\tau} & \frac{g_2}{\sqrt{2}}\vec{\chi}\cdot\vec{\tau} \\ \frac{g_2}{\sqrt{2}}\vec{\chi}\cdot\vec{\tau} & \frac{1}{2}\mu^2(\phi^2+\chi^2) + \frac{g_3}{\sqrt{2}}\vec{\phi}\cdot\vec{\tau} \end{matrix} \right\|$$

The eigenvalues are functions of the scalars ϕ^2 and χ^2 and the pseudoscalar $\vec{\phi}\cdot\vec{\chi}$. A potential minimum occurs when $\vec{\phi}$ and $\vec{\chi}$ either point in the same direction or in opposite directions. We shall now consider the special case that $g_1 = g_3$. Then both $\vec{\phi}$ and $\vec{\chi}$ are always bound and $\vec{\phi}\cdot\vec{\chi}$ is unequal to zero at the potential minimum. In fact, we find that the value of ϕ at the minimum is equal to $\frac{g_1}{\mu^2\sqrt{2}}$ and the value of χ at the minimum is $\frac{g_2}{\mu^2\sqrt{2}}$.

The same value of the minimum occurs at two distinct orbits, which have the same values for ϕ^2 and χ^2 but opposite values for $\vec{\phi}\cdot\vec{\chi}$. Now suppose that we have a particular eigenstate of H with positive parity and that we multiply this with $\vec{\phi}\cdot\vec{\chi}$; then the result will be an eigenstate of H with the same eigenvalue but with negative parity. This is so, because $\vec{\phi}\cdot\vec{\chi}$ varies too slowly in the region where the wave function differs from zero, as long as one stays in one valley only. The result is apparently parity degeneration.

For the mass spectrum we obtain

$$E_i = E_o + \frac{\mu^4}{g_1^{\,2}+g_2^{\,2}}\,i(i+1) \qquad (i = \tfrac{1}{2}, \tfrac{3}{2}, \tfrac{5}{2}, \ldots)$$

again i being the isospin of the dressed baryon.

When $g_3 = 0$ the eigenvalues are independent of $\vec{\phi}\cdot\vec{\chi}$. Both the $\vec{\phi}$ and $\vec{\chi}$ fields are bound when $g_1^{\,2} < 2g_2^{\,2}$, otherwise only the $\vec{\phi}$ - field is bound. Since the Hamiltonian is not invariant under independent rotations of $\vec{\phi}$ and $\vec{\chi}$, the non-appearance of $\vec{\phi}\cdot\vec{\chi}$ in the expressions for the eigenvalues is just an accident.[3] There is a continuous infinity of orbits at which the potential reaches its absolute minimum. The potential valley is therefore not characterized by a fixed value of the quantity $\vec{\phi}\cdot\vec{\chi}$ and it is possible to change the sign of $\vec{\phi}\cdot\vec{\chi}$ by walking along the valley. In this case there need not be parity degeneration, but now every isobar state repeats itself an infinite number of times with small mass-intervals (of the same order of magnitude as the rotational mass differences) due to the degeneration of one of the vibrational degrees of freedom. When this happens the usual techniques for finding isobar spectra with strong coupling models fail.

Some other aspects of the parity problem can be demonstrated with a model involving spin and isospin. Consider an s - wave scalar field ϕ transforming as an isosinglet and consider an isovector field $\vec{\chi}$ of which only p - waves are bound, so there are 9 fields $\chi_{\rho r}$ (ρ, r=1,2,3) where the indices ρ and r refer to isospin and spin. With the help of the Pauli spin matrices we construct from this a 4 x 4 matrix of which the components are

$$\chi^{\alpha i}_{\beta j} = \frac{1}{2} \sum_{\rho,r=1}^{3} (\tau_\rho)^\alpha_\beta \, (\sigma_r)^i_j \, \chi_{\rho r} \, . \tag{10}$$

We have then $\chi^{\alpha i}_{\alpha j} = 0$ and $\chi^{\alpha i}_{\beta i} = 0$ (the Einstein summation convention is adopted.)

There are 10 conjugate momenta, which we shall call π_ϕ and $\pi_{\chi^{\alpha i}_{\beta j}}$ which satisfy

$$[\pi_\phi, \phi] = \frac{1}{i}$$

$$\left[\pi_{\chi^{\alpha i}_{\beta j}}, \chi^{\gamma k}_{\delta l}\right] = \frac{1}{i} \left(\delta^\alpha_\delta \, \delta^\gamma_\beta - \frac{1}{2} \delta^\alpha_\beta \, \delta^\gamma_\delta\right) \left(\delta^i_l \, \delta^k_j - \frac{1}{2} \delta^i_j \, \delta^k_l\right) , \tag{11}$$

the remainder of the commutators being zero. We shall define

$$\pi_\chi{}^2 = \pi_{\chi^{\alpha i}_{\beta j}} \, \pi_{\chi^{\beta j}_{\alpha i}} \text{ and } \chi^2 = \chi^{\alpha i}_{\beta j} \, \chi^{\beta j}_{\alpha i} \, .$$

In addition to these fields there are two bare baryon quadruplets $\overline{B}_{\alpha i}|>$ and $\overline{C}_{\alpha i}|>$. Besides the creation operators $\overline{B}_{\alpha i}$ and $\overline{C}_{\alpha i}$ there are the corresponding annihilation operators $B^{\alpha i}$ and $C^{\alpha i}$ satisfying

$$\left.\begin{aligned} B^{\alpha i}\,\overline{B}_{\beta j}|> &= \delta^{\alpha}_{\beta}\,\delta^{i}_{j}|> \\ C^{\alpha i}\,\overline{C}_{\beta j}|> &= \delta^{\alpha}_{\beta}\,\delta^{i}_{j}|> \\ B^{\alpha i}\,\overline{C}_{\beta j}|> &= C^{\alpha i}\,\overline{B}_{\beta j}|> = 0 \end{aligned}\right\}$$

We shall consider the following Hamiltonian

$$H = \frac{1}{2}\,(\pi_{\phi}^2 + \pi_{\chi}^2) + \frac{1}{2}\,\mu^2(\phi^2 + \chi^2) + g_1\,\overline{B}_{\alpha i}\,\phi\,B^{\alpha i} +$$

$$g_2\,[\overline{B}_{\alpha i}\,\chi^{\alpha i}_{\beta j}\,C^{\beta j} + \overline{C}_{\alpha i}\,\chi^{\alpha i}_{\beta j}\,B^{\beta j}] \quad (g_1,\, g_2 \geqslant 0)\,.$$

It is now possible to assign a positive parity to ϕ, $B^{\alpha i}$ and $\overline{B}_{\alpha i}$ and a negative parity to $\chi^{\alpha i}_{\beta j}$, $C^{\alpha i}$ and $\overline{C}_{\alpha i}$, so that this parity is conserved. When $g_1 = 0$ another definition of parity is possible, which is also conserved, but which does not commute with the parity defined before. An eigenstate of H can never be an eigenstate of both parities simultaneously, so there must be complete parity degeneration of all eigenstates. More interesting is the case that g_1 and g_2 are both unequal to zero. When $g_1^2 \geqslant \frac{3}{2}\,g_2^{\,2}$, then the χ - fields are not bound, so we consider only the case that $g_1^{\,2} < \frac{3}{2}\,g_2^{\,2}$. The potential minimum occurs then when

$$\phi = -\frac{3g_1}{2\mu^2\left(3 - \dfrac{g_1^{\,2}}{g_2^{\,2}}\right)}$$

$$\chi = \frac{3g_2\sqrt{3 - 2\,\dfrac{g_1^{\,2}}{g_2^{\,2}}}}{2\mu^2\left(3 - \dfrac{g_1^{\,2}}{g_2^{\,2}}\right)}$$

while the orbit of $\chi_{\beta j}^{\alpha i}$ at the potential valley either contains the point

$$\chi_{\beta j}^{\alpha i} = \frac{\chi}{\sqrt{12}} \sum_{m=1}^{3} (\tau_m)_\beta^\alpha (\sigma_m)_j^i$$

or its opposite. Thus there are two orbits at which the potential reaches the same minimum. These orbits transform into each other when the parity operation is applied and they represent two distinct potential valleys, characterized by two different values of the pseudoscalar

$$\det [\chi_{\rho r}] = \frac{1}{3} \chi_{\beta j}^{\alpha i} \chi_{\gamma k}^{\beta j} \chi_{\alpha i}^{\gamma k} = \pm \frac{\chi^3}{3\sqrt{3}} \neq 0 .$$

Again there is parity degeneration of all isobar states in the strong coupling limit and the energy spectrum becomes

$$E_j = E_o + \frac{\mu^4}{g_2^2} \frac{1 - \dfrac{g_1^2}{3g_2^2}}{1 - \dfrac{2g_1^2}{3g_2^2}} j(j+1) \qquad (j = \tfrac{1}{2}, \tfrac{3}{2}, \ldots .) \tag{14}$$

For all isobars the spin is equal to the isospin. The fact that not all spin-isospin combinations are realized is a desirable feature of this model and is closely connected with the fact that the 3 x 3 matrix with elements $\chi_{\rho r}$ is proportional to an orthogonal matrix of which the determinant is a nonzero pseudoscalar. Apparently, the equality of spin and isospin is bought at the expense of parity-non-degeneration and this phenomenon will have its analogue in more complicated models involving SU(2) $\otimes$ SU(3) symmetry. Experimentally, there are no indications for parity degeneration. Even if it is badly broken, there should still exist a one-to-one correspondence between isobar states with exactly the same quantum numbers but opposite parity. This is not the case. Unfortunately, the occurrence of nonzero pseudoscalars seems to be the rule rather than the exception.

Now, there is another way of obtaining mixed parity spectra, which can occur when no negative parity waves are bound, namely by considering completely independent systems. For example, take the following Hamiltonian:

$$H = \frac{1}{2}\pi_\phi^2 + \frac{1}{2}\mu^2\phi^2 + g_1 \overline{B}_{\alpha i}\, \phi^{\alpha i}_{\beta j}\, B^{\beta j} + g_2 \overline{C}_{\alpha i}\, \phi^{\alpha i}_{\beta j}\, C^{\beta j} + \delta m\, \overline{B}_{\alpha i}\, B^{\alpha i} \quad (1$$

(the bare masses of the B and C baryons are taken unequal). Here we can assign to B, $\overline{B}$ and ϕ a positive parity and to C and $\overline{C}$ a negative parity, so that parity is conserved. Now there occur two spectra:

$$E_j^{(1)} = E_o^{(1)} + \frac{\mu^4}{g_1^2}\, j(j+1) \quad \text{(positive parity spectrum)}$$

$$E_j^{(2)} = E_o^{(2)} + \frac{\mu^4}{g_2^2}\, j(j+1) \quad \text{(negative parity spectrum)} \quad (1$$

with spin $j = \frac{1}{2}, \frac{3}{2}, \ldots$ equal to the isospin. A change of δm leads to a shift of the positive and negative parity spectra with respect to each other. Both spectra occur simultaneously and we have therefore a genuine mixed parity spectrum, but it is trivial because of the lack of any relation between the two. Still, this may be the only workable solution of the parity problem in strong coupling models of any serious nature.

References

1) G. Wentzel, Helv. Phys. Acta 13, 269 (1940; ibid 14, 633 (1941), W. Pauli and S. M. Dancoff, Phys. Rev. 62, 85 (1942).
2) G. Wentzel, Phys. Rev. 125, 771 (1962).
3) A similar phenomenon occurs in an SU(3) model with d-type coupling of eight scalar mesons with eight baryons. See G. Wentzel, Prog. Theor. Phys. Suppl. Extra Number, 108 (1965).

Received 1/27/69

Meson Theory of Nuclear Forces*

by

F. Coester

Argonne National Laboratory

Elimination of the meson degrees of freedom from the nuclear Hamiltonian requires a knowledge of the meson structure of the physical one-nucleon states. Conventional relativistic one-meson exchange potentials obtain if the one-nucleon states are calculated in first-order perturbation theory. If the meson-nucleon system is treated as a relativistic particle system without fields, then the physical one-nucleon state can be obtained exactly. Some of the features of local field theories can be recaptured in such models. The static limit of the nucleon-nucleon potential is local but otherwise largely arbitrary. The velocity-dependent terms are then unambiguously determined.

1. Introduction

The complexity that dominates the nuclear-force problem was already implicit in Yukawa's[1] simple idea: Nucleons conceived as elementary point particles are the sources of a nuclear force field analogous to the electromagnetic field. As soon as the field is quantized the elementary one-nucleon states are not eigenstates of the rest energy $h = (-P^{\mu}P_{\mu})^{1/2}$. On the other hand the physical nucleon states, which are eigenstates of the rest energy, have a complex meson cloud. The forces between physical nucleons depend on the structure of that cloud just as the forces between atoms depend on the states of the electron cloud. Some solution of the one-particle problem is thus required for any understanding of the forces between two and more nucleons. That was first realized in

*Work performed under the auspices of the U.S. Atomic Energy Commission.

the strong-coupling meson theories.[2] The problem was tractable only for extended nucleons in the static limit. The ground states and the excited states of the physical nucleon were obtained by expansion in inverse powers of the coupling constant. Effective potentials for the two-nucleon system were then found in the same approximation. For the same static model but with weak coupling, both the one-nucleon states and the two-nucleon potentials may be obtained by standard perturbation theory.

These rather extreme approximations are clearly inadequate to account for the observed properties of the two-nucleon system. A wealth of new data at higher energies as well as the serious obstacles to better field theories favored an emphasis on empirically-determined two-body potentials. The purpose of such potentials must be to provide a description of multinucleon systems as well. A potential would be a redundant luxury if only a parameterization of two-body data were intended.

But nucleon data alone are in principle insufficient to determine the nuclear Hamiltonian. They must be supplemented either by conventional restrictions or by a more fundamental theory. Conventional restrictions are appealing only as long as they result in manifest simplicity. It would be nice, for instance, to require local potentials and absence of multibody forces. But the data require at least different local potentials for different partial waves, and meson theories suggest that there are velocity-dependent potentials of the form $\omega(r)p^2$ and $p^2\omega(r)$. With these features we have a large class of equivalent two-body Hamiltonians[3,4] such that their eigenfunctions have the same asymptotic behavior but may differ widely for small distances. Electromagnetic interactions probe the interior of these wave functions. But velocity-dependent potentials require at least some exchange currents[5] and the admission of more general two-body terms in the charge-current densities can void any electromagnetic argument about the short-range behavior of the nuclear wave functions.

It would be desirable that sufficiently accurate calculations for multinucleon systems should yield information about multibody potentials. That seems hopeless unless the two-body potentials are already well determined.

The point of these arguments is to emphasize the importance of more fundamental theories that derive the effective nucleon-nucleon Hamiltonian from a knowledge of the internal degrees of freedom of the nucleon. The mathematical problem is always that of constructing a reduced Hamiltonian for a subsystem from the Hamiltonian of the full system. The formal framework for such constructions will be reviewed in section 2.

Since the discovery of heavy vector bosons, the two-nucleon data have been fitted by one-meson exchange potentials that include recoil effects. Their derivation within the framework of section 2 will be discussed in section 3.

Relativistic models of particle interactions without fields[6] exist if the locality requirements of field theories are abandoned. In such models the number of internal degrees of freedom of the physical nucleon may be finite. It is then possible to assume simple vertex interactions such that the one-nucleon problem is exactly soluble. The main qualitative features of such models are examined in section 4. The static limit of the two-nucleon potential can be adjusted freely by the choice of the assumed meson-nucleon vertex, but once this is done the velocity-dependent terms can be derived unambiguously.

2. Reduced Hamiltonians

Let H be the Hamiltonian of the full system and $\mathcal{H}$ the Hilbert space of its states. A one-nucleon state ψ is given by an expression of the form

$$\psi = \sum_{\sigma=\pm 1/2} \int d\underset{\sim}{p}\ \phi\ |\underset{\sim}{p},\sigma > \chi(\underset{\sim}{p},\sigma), \tag{1}$$

where $\chi(p,\sigma)$ is any square-integrable function of its arguments and

$$H\psi = \sum_{\sigma} \int d\underset{\sim}{p}\ \phi\ |\underset{\sim}{p},\sigma > (\underset{\sim}{p}^2 + m^2)^{1/2}\ \chi(\underset{\sim}{p},\sigma). \tag{2}$$

The functions $\chi(\underset{\sim}{p},\sigma)$ are elements of a one-nucleon Hilbert space $\mathcal{H}_{1n}$. The Hamiltonian H_0 is defined as an operator on $\mathcal{H}_{1n}$ by

$$H_0\chi(\underset{\sim}{p},\sigma) = (\underset{\sim}{p}^2 + m^2)^{1/2}\ \chi(\underset{\sim}{p},\sigma). \tag{3}$$

The operator ϕ is an isometric mapping of $\mathcal{H}_{1n}$ into $\mathcal{H}$, $\phi^\dagger\phi = 1$. Equations (1) and (2) may be written more abstractly as

$$\psi = \phi\chi, \tag{4}$$

$$H\phi = \phi H_0. \tag{5}$$

The two-nucleon space $\mathcal{H}_{2n} = \mathcal{H}_{1n} \times \mathcal{H}_{1n}$ is the space of the square-integrable functions $\chi(\underset{\sim}{p}_1,\sigma_1,\underset{\sim}{p}_2,\sigma_2)$ and the operator H_0 is defined on $\mathcal{H}_{2n}$ by

$$H_0\chi(\underset{\sim}{p}_1,\sigma_1,\underset{\sim}{p}_2,\sigma_2) = \{(\underset{\sim}{p}_1^2 + m^2)^{1/2} + (\underset{\sim}{p}_2^2 + m^2)^{1/2}\}\ \chi(\underset{\sim}{p}_1,\sigma_1,\underset{\sim}{p}_2,\sigma_2). \tag{6}$$

The operator Φ defined by

$$\Phi = \phi \times \phi \tag{7}$$

maps $\mathcal{H}_{2n}$ into $\mathcal{H}$. It is ordinarily only asymptotically isometric, i.e.,

$$\lim_{t\to\pm\infty} (\Phi^\dagger\Phi - 1)\ e^{-iH_0t} = 0, \tag{8}$$

and

$$\lim_{|\underset{\sim}{a}|\to\infty} (\Phi^\dagger\Phi - 1)\ e^{i\underset{\sim}{p}\cdot\underset{\sim}{a}} = 0. \tag{9}$$

Operator limits are understood to be strong limits. The S operator is given by

$$S = \Omega_+^{\dagger}\Omega_- , \tag{10}$$

where the Møller operators $\Omega_\pm$ are

$$\Omega_\pm = \lim_{t\to\pm\infty} e^{iHt}\Phi\, e^{-iH_0 t}. \tag{11}$$

An effective two-nucleon Hamiltonian $\bar{H}$ is an operator on $\mathcal{H}_{2n}$ such that the deuteron binding energy and the phase shifts below the inelastic threshold derived from $\bar{H}$ are the same as those derived from H. The operators H and $\bar{H}$ are related by an isometric mapping U from $\mathcal{H}_{2n}$ into $\mathcal{H}$ with the properties

$$HU = U\bar{H}, \tag{12}$$

and

$$\lim_{t\to\pm\infty} (U - \Phi) e^{-iH_0 t}\, \Lambda = 0, \tag{13}$$

where the projection operator Λ projects out that part of $\mathcal{H}_{2n}$ that is below the meson production threshold. With the definitions

$$\bar{\Omega}_\pm = \lim_{t\to\pm\infty} e^{i\bar{H}t}\, e^{-iH_0 t} \tag{14}$$

and

$$\bar{S} = \bar{\Omega}_+^{\dagger}\bar{\Omega}_- , \tag{15}$$

it follows from Eq. (13) that[3]

$$\Lambda\bar{S}\Lambda = \Lambda S\Lambda. \tag{16}$$

Above the meson production threshold, $\bar{S}$ and S cannot be equal.

The procedure for constructing the operator U is well known from the perturbation theory of bound states.[7] Let

$$U_0 = \Phi(\Phi^\dagger \Phi)^{-1/2}, \tag{17}$$

and define the projection operators $\mathcal{P}$ and Q by

$$\mathcal{P} = U_0 U_0^\dagger, \tag{18}$$

$$Q = 1 - \mathcal{P}.$$

The operator

$$U = (U_0 + F)(1 + F^\dagger F)^{-1/2} \tag{20}$$

is isometric if F = QF by definition. In order to satisfy Eq. (12) we require that

$$H(U_0 + F) = (U_0 + F)\tilde{H}, \tag{21}$$

or more explicitly

$$QHU_0 = F\tilde{H} - QHQF, \tag{22}$$

where the operator $\tilde{H}$ is defined by

$$\tilde{H} = U_0^\dagger HU_0 + U_0^\dagger HF. \tag{23}$$

Equations (22) and (23) are the equations to be solved for F. The reduced Hamiltonian $\bar{H}$ is then given by

$$\bar{H} = U^\dagger HU = (1 + F^\dagger F)^{1/2} \tilde{H}(1 + F^\dagger F)^{-1/2}. \tag{24}$$

Before proceeding further with an explicit construction of the operatur U, it is convenient to eliminate the center-of-mass motion. The rest-energy operator h is defined by

$$h = (H^2 - \underset{\sim}{P}^2)^{1/2}, \tag{25}$$

where $\underset{\sim}{P}$ is the total momentum. The states $\chi \equiv \mathcal{H}_{2n}$ may be represented as functions $\chi(\underset{\sim}{P},\underset{\sim}{p},s,\sigma)$, where $\underset{\sim}{p}$ is the momentum of one of the nucleons in the rest frame of $\underset{\sim}{P}$ and s,σ are the channel-spin quantum numbers in that frame ($s = 0, 1; -s \leq \sigma \leq s$). Because of the Poincaré invariance of this theory, H may be replaced by h everywhere and the two-nucleon space $\mathcal{H}_{2n}$ may be replaced by the space $\mathfrak{h}_{2n}$ of the functions $\chi(\underset{\sim}{p},s,\sigma)$.

In order to solve Eq. (22) for the operator F, we need the eigenvectors $X|b\rangle$ and $X|\underset{\sim}{p}\rangle$ of $\tilde{h}$ given by

$$\tilde{h}X|b\rangle = X|b\rangle E_b, \tag{26}$$

and

$$\tilde{h}X|\underset{\sim}{p}\rangle = X|\underset{\sim}{p}\rangle 2(\underset{\sim}{p}^2 + m^2)^{1/2}, \tag{27}$$

where E_b is the rest energy of the deuteron. Spin indices are implied but are suppressed for convenience. Let $\bar{X}|b\rangle$ and $\bar{X}|\underset{\sim}{p}\rangle$ be a set of vectors satisfying the orthogonality relation

$$\begin{aligned} \langle b|\bar{X}^{\dagger}X|\underset{\sim}{p}\rangle &= 0, \\ \langle \underset{\sim}{p}'|\bar{X}^{\dagger}X|\underset{\sim}{p}\rangle &= \delta(\underset{\sim}{p}' - \underset{\sim}{p}), \\ \langle b|\bar{X}^{\dagger}X|b\rangle &= 1, \end{aligned} \tag{28}$$

and

$$X|b\rangle\langle b|\bar{X} + \int d\underset{\sim}{p}X|\underset{\sim}{p}\rangle\langle\underset{\sim}{p}|\bar{X}^{\dagger} = 1. \tag{29}$$

Then for the operator F we find

$$\begin{aligned} F = {} & (E_b - QhQ)^{-1}\, QhU_0X|b\rangle\langle b|\bar{X}^{\dagger} \\ & + \int dp\{2(p^2 + m^2)^{1/2} - QhQ\}^{-1}QhU_0X|\underset{\sim}{p}\rangle\langle\underset{\sim}{p}|\bar{X}^{\dagger} \end{aligned} \tag{30}$$

and hence

$$\tilde{h} = \tilde{h}(E_b)X|b\rangle\langle b|\bar{X}^\dagger + \int d\mathbf{p}\tilde{h}(2\omega)X|\mathbf{p}\rangle\langle\mathbf{p}|X^\dagger, \tag{31}$$

where $\omega = (\mathbf{p}^2 + m^2)^{1/2}$ and the operator $\tilde{h}(z)$ is defined by

$$\tilde{h}(z) = U_0{}^\dagger hU_0 + U_0{}^\dagger hQ(z - QhQ)^{-1} QhU_0. \tag{32}$$

We note in passing that the operator

$$U_0\tilde{h}(z)U_0{}^\dagger = \mathcal{P}h\mathcal{P} + \mathcal{P}hQ(z - QhQ)^{-1} Qh\mathcal{P} \tag{33}$$

has the same form as the enrgy-dependent effective Hamiltonian[8]

$$\mathcal{H}(z) = \mathcal{P}H\mathcal{P} + \mathcal{P}HQ(z - QHQ)^{-1} QH\mathcal{P} \tag{34}$$

that arises in formal derivations of optical model potentials. This comes about because of the identity

$$\mathcal{P}(z - H)^{-1} \mathcal{P} = \mathcal{P}[z - \mathcal{H}(z)]^{-1} \mathcal{P}. \tag{35}$$

In practice the reduced Hamiltonian

$$\bar{h} = U^\dagger hU = (1 + F^\dagger F)^{1/2} \tilde{h}(1 + F^\dagger F)^{-1/2} \tag{36}$$

is not likely to be useful unless the operator U may be approximated by U_0. But for any well defined model, that approximation can be justified only by the smallness of the correction terms--not by fortuitous agreement with some nuclear data. We may expect the approximation $U \simeq U_0$ to be valid for the long-range part of the potential.

3. One-meson exchange potentials

Let us consider now a simple model that yields the conventional one-meson exchange potentials. The nucleons are described in configuration space and the mesons by the field $\phi_\alpha(x)$.[9] The state vectors are functions of the nucleon momenta p_i and functionals of the meson creation operators $a^\dagger(\underset{\sim}{k},\alpha)$. The Hamiltonian is

$$H = H_n + H_\mu + H' + H'', \tag{37}$$

where the energy H_n of the bare nucleons is

$$H_n = \sum_i (\underset{\sim}{p}_i^2 + m^2)^{1/2}, \tag{38}$$

the energy H_μ of the free mesons is

$$\begin{aligned} H_\mu &= \sum_\alpha \int d\underset{\sim}{k}\; a^\dagger(\underset{\sim}{k},\alpha)\; \nu(\underset{\sim}{k})\; a(\underset{\sim}{k},\alpha), \\ \nu(\underset{\sim}{k}) &\equiv (\underset{\sim}{k}^2 + \mu^2)^{1/2}, \end{aligned} \tag{39}$$

the interaction Hamiltonian H' is

$$H' = \sum_i \sum_\alpha \int d\underset{\sim}{x}\; J_\alpha^{(i)}(\underset{\sim}{x})\phi_\alpha(\underset{\sim}{x}), \tag{40}$$

and the last term H'' is the mass renormalization counterterm.

The meson field $\phi_\alpha(x)$ is related to the creation and annihilation operators by

$$\phi_\alpha(x) = (2\pi)^{-3/2} \int dk(2\nu)^{-1/2}\, [a(\underset{\sim}{k},\alpha)e^{i\underset{\sim}{k}\cdot\underset{\sim}{x}} + a^\dagger(\underset{\sim}{k},\alpha)e^{-i\underset{\sim}{k}\cdot\underset{\sim}{x}}]. \tag{41}$$

The nucleon source operator $J_\alpha(\underset{\sim}{x})$ has transformation properties such that $\sum_\alpha J_\alpha(x)\phi_\alpha(x)$ is a scalar density. In the following we suppress the index α and all spin indices for convenience. The kernel of the source $J(\underset{\sim}{x})$ is

$$(\underset{\sim}{p}'|J(\underset{\sim}{x})|\underset{\sim}{p}) = (2\pi)^{-3/2} \exp[i(\underset{\sim}{p} - \underset{\sim}{p}') \cdot \underset{\sim}{x}]\; (\underset{\sim}{p}'|\tilde{J}|\underset{\sim}{p}). \tag{42}$$

The kernel $(\underset{\sim}{p}'|\tilde{J}|\underset{\sim}{p})$ contains an arbitrary invariant form factor

$F[(p' - p)^2]$ where $(p' - p)^2$ is the invariant four-momentum transfer

$$(p' - p)^2 = |\underset{\sim}{p}' - \underset{\sim}{p}|^2 - (\omega' - \omega)^2. \tag{43}$$

Otherwise the dependence of $\tilde{J}$ on $\underset{\sim}{p}'$ and $\underset{\sim}{p}$ is unambiguously determined by its transformation properties. The kernel of the interaction Hamiltonian H' is

$$(\underset{\sim}{p}',\underset{\sim}{k}|H'|\underset{\sim}{p}) = \delta(\underset{\sim}{p}' + \underset{\sim}{k} - \underset{\sim}{p}) \; (\underset{\sim}{p}'|\tilde{J}|\underset{\sim}{p}) \; (2\nu)^{-1/2}. \tag{44}$$

In first-order perturbation theory the nonvanishing components of the physical nucleon state are

$$(\underset{\sim}{p}'|\phi|\underset{\sim}{p}) = \delta(\underset{\sim}{p}' - \underset{\sim}{p}) \tag{45}$$

and

$$(\underset{\sim}{p}',\underset{\sim}{k}|\phi|\underset{\sim}{p}) = -(\underset{\sim}{p}',\underset{\sim}{k}|H'|\underset{\sim}{p}) \; (\omega' + \nu - \omega)^{-1}. \tag{46}$$

The operator Φ is given by the tensor product

$$\Phi = \phi \times \phi \tag{47}$$

and the evaluation of the operator

$$\bar{H} = U_0{}^{\dagger} H U_0 = (\Phi^{\dagger}\Phi)^{-1/2}\Phi^{\dagger}H\Phi(\Phi^{\dagger}\Phi)^{-1/2} \tag{48}$$

to second order in H' is straightforward. The kernel of $\Phi^{\dagger}H\Phi$ is

$$\begin{aligned}(\underset{\sim}{p}'_2,\underset{\sim}{p}'_1|\Phi^{\dagger}H\Phi|\underset{\sim}{p}_1,\underset{\sim}{p}_2) &= \tfrac{1}{2}(\underset{\sim}{p}'_2,\underset{\sim}{p}'_1|\Phi^{\dagger}\Phi H_0 + H_0\Phi^{\dagger}\Phi|\underset{\sim}{p}_1,\underset{\sim}{p}_2)\\ &+ \tfrac{1}{2}\int dk\{(\underset{\sim}{p}'_2|H'|\underset{\sim}{p}_2,\underset{\sim}{k})(\underset{\sim}{k},\underset{\sim}{p}'_1|\phi|\underset{\sim}{p}_1) + (\underset{\sim}{p}'_1|\phi|\underset{\sim}{p}_1,\underset{\sim}{k})(\underset{\sim}{k},\underset{\sim}{p}'_2|H|\underset{\sim}{p}_2)\\ &+ (\underset{\sim}{p}'_1|H'|\underset{\sim}{p}_1,\underset{\sim}{k})(\underset{\sim}{k},\underset{\sim}{p}'_2|\phi|\underset{\sim}{p}_2) + (\underset{\sim}{p}'_2|\phi^{\dagger}|\underset{\sim}{p}_2,\underset{\sim}{k})(\underset{\sim}{k},\underset{\sim}{p}'_1|H'|\underset{\sim}{p}_1)\}.\end{aligned} \tag{49}$$

The term

$$(\Phi^{\dagger}\Phi)^{1/2} H_0 (\Phi^{\dagger}\Phi)^{-1/2} + (\Phi^{\dagger}\Phi)^{1/2} H_0 (\Phi^{\dagger}\Phi)^{1/2} \tag{50}$$

does not contribute to the interaction in second-order perturbation theory. In that approximation, for $\underset{\sim}{p}_1 = \underset{\sim}{p} = -\underset{\sim}{p}_2$, $\underset{\sim}{p}'_1 = \underset{\sim}{p}' = -\underset{\sim}{p}'_2$, the second term in Eq. (49) gives

$$(\underset{\sim}{p}'|\overline{H} - H_0|\underset{\sim}{p}) = -(\underset{\sim}{p}'|\tilde{J}|\underset{\sim}{p})\ (-\underset{\sim}{p}'|\tilde{J}|-\underset{\sim}{p})\ \{(p' - p)^2 + \mu^2\}^{-1} \tag{51}$$

This potential is identical to the relativistic two-body potential derived by Schierholz.[10] All the kinematical details can be found there. In an appropriate nonrelativistic approximation, our potential reduces to that of Ueda and Green[11] provided we identify the form factors with theirs. Obviously, the same reduced Hamiltonian would result if we put

$$U_0 = \exp(-iW)$$

and determined W in perturbation theory by the requirement that it should eliminate the meson field from the Hamiltonian.[9,12]

4. Particle models without fields

At the price of weakening the locality requirement of field theories, we can buy relativistic particle models with interactions.[6] The Hilbert space $\mathcal{H}$ is a direct sum of tensor products of irreducible representation spaces of the Poincaré group. Interactions are introduced by modifying the rest-energy operators of these spaces. Relativistic invariance and cluster requirements impose only fairly mild restrictions on the interactions. The particle number need not be conserved. The arbitrariness inherent in such models is reduced if we try to recapture some of the features of local field theories.

This is accomplished by assuming only a basic vertex interaction between a bare nucleon at rest and the meson-nucleon states.

All other interactions are derived. For each vertex there is an arbitrary invariant form factor. The operator Φ can then be constructed explicitly without the use of a perturbation theory.

For a simple model we take

$$\mathcal{H} = \mathcal{H}_0 \times \mathcal{H}_n \oplus \mathcal{H}_n \times \mathcal{H}_n \times \mathcal{H}_\mu = \mathcal{H}_c \times \mathfrak{h}, \tag{52}$$

where H_0 is the space of a bare nucleon of mass m_0 and $\mathcal{H}_n$ and $\mathcal{H}_\mu$ are respectively one-nucleon and one-meson spaces. The elements of $\mathcal{H}_c$ are the functions of the total momentum $\underset{\sim}{P}$ and the elements of $\mathfrak{h}$ describe the system in the rest frame of $\underset{\sim}{P}$. That space is a direct sum

$$\mathfrak{h} = \mathfrak{h}_{0n} \oplus \mathfrak{h}_{nn\mu}. \tag{53}$$

The space $\mathfrak{h}_{0n}$ is the space of functions $\psi(\underset{\sim}{p})$, where $\underset{\sim}{p}$ is the momentum of the bare nucleon in the rest frame of P. The $\mathfrak{h}_{nn}$ is the space of functions $\psi(\underset{\sim}{p}_1,\underset{\sim}{p}_2)$, where $\underset{\sim}{p}_1$ and $\underset{\sim}{p}_2$ are the momenta of the two nucleons in the rest frame of P. The momentum $\underset{\sim}{k}$ of the meson is by definition $\underset{\sim}{k} = -(\underset{\sim}{p}_1 + \underset{\sim}{p}_2)$. The momentum of the meson in the rest frame of $\underset{\sim}{k} + \underset{\sim}{p}_i$ is called $\underset{\sim}{q}_i$. Since I want to exhibit only the main qualitative features of the model, I assume that all particles--nucleons and mesons--have spin zero. The generalization to particles with spin is straightforward. There is no spin-statistics theorem in particle theories without fields.

In the absence of interactions, the rest-energy operator h_0 is given by the kernels

$$(\underset{\sim}{p}'|h_0|\underset{\sim}{p}) = \delta(\underset{\sim}{p}' - \underset{\sim}{p})\ (\omega_0 + \omega), \tag{54}$$

and

$$(\underset{\sim}{p}'_1,\underset{\sim}{p}'_2)|h_0|\underset{\sim}{p}_1,\underset{\sim}{p}_2) = \delta(\underset{\sim}{p}'_1 - \underset{\sim}{p}_1)\ \delta(\underset{\sim}{p}'_2 - \underset{\sim}{p}_2)\ (\omega_1 + \omega_2 + \nu). \tag{55}$$

In operator form we may write

$$h_0 = (\underset{\sim}{p}_2^2 + m^2)^{1/2} + (\underset{\sim}{p}_2^2 + h_1^{02})^{1/2}, \tag{56}$$

where

$$(\underset{\sim}{p}'|\underset{\sim}{p}_2|\underset{\sim}{p}) = -\delta(\underset{\sim}{p}' - \underset{\sim}{p})\underset{\sim}{p},$$

$$(\underset{\sim}{p}'_1,\underset{\sim}{p}'_2|\underset{\sim}{p}_2|\underset{\sim}{p}_1,\underset{\sim}{p}_2) = \delta(\underset{\sim}{p}'_1 - \underset{\sim}{p}_1)\delta(\underset{\sim}{p}'_2 - \underset{\sim}{p}_2)\underset{\sim}{p}_2,$$

$$(\underset{\sim}{p}'|h_1^0|\underset{\sim}{p}) = m_0\delta(\underset{\sim}{p}' - \underset{\sim}{p}), \tag{57}$$

$$(\underset{\sim}{p}'_1,\underset{\sim}{p}'_2|h_1^0|\underset{\sim}{p}_1,\underset{\sim}{p}_2) = \delta(\underset{\sim}{p}'_1 - \underset{\sim}{p}_1)\,\delta(\underset{\sim}{p}'_2 - \underset{\sim}{p}_2)\,M(q_1),$$

and

$$M(\underset{\sim}{q}) \equiv (m^2 + \underset{\sim}{q}^2)^{1/2} + (\mu^2 + \underset{\sim}{q}^2)^{1/2}.$$

The full rest-energy operator h is

$$h = h_0 + V_1 + V_2, \tag{58}$$

where the interaction

$$V_1 = (\underset{\sim}{p}_2^2 + h_1^2)^{1/2} - (\underset{\sim}{p}_2^2 + h_1^{02})^{1/2} \tag{59}$$

is determined in terms of $h_1 = h_1^0 + h'_1$.

The kernel of the interaction energy h'_1 is

$$(\underset{\sim}{p}_1,\underset{\sim}{p}_2|h'_1|\underset{\sim}{p}) = \delta(\underset{\sim}{p}_2 + \underset{\sim}{p}) \frac{(\underset{\sim}{q}_1^2 + m^2)^{1/4}\; F(\underset{\sim}{q}_1^2)\; m^{1/2}}{(\underset{\sim}{p}_1^2 + m^2)^{1/4}\; \nu^{1/2}\; (\underset{\sim}{p}^2 + m^2)^{1/4}} \tag{60}$$

The vertex function $F(q^2)$ is arbitrary. In Eq. (60) the momentum $\underset{\sim}{q}_1$ is related to $\underset{\sim}{p}_1$ and $\underset{\sim}{p}$ by

$$\begin{aligned}(p_1 - p)^2 &= \underset{\sim}{q}_1^2 - \{(\underset{\sim}{q}_1^2 + m^2)^{1/2} - m\}^2 \\ &= 2m\,\{(\underset{\sim}{q}_1^2 + m^2)^{1/2} - m\}.\end{aligned} \tag{61}$$

In the nonrelativistic approximation, Eq. (61) reduces to

$$\mathbf{q}_1^2 = |\mathbf{p}_1 - \mathbf{p}|^2 + \frac{1}{2} m^{-2} |\mathbf{p}_1 - \mathbf{p}|^4 - \frac{1}{4} m^{-2} (\mathbf{p}_1^2 - \mathbf{p}^2)^2. \tag{62}$$

In the static limit, $m \to \infty$, we have

$$\mathbf{q} = \mathbf{k} = \mathbf{p} - \mathbf{p}_1. \tag{63}$$

A comparison with the last section suggests the identification

$$(\bar{\mathbf{p}}_1 | \tilde{J} | \mathbf{p}) = \frac{(\mathbf{q}_1^2 + m^2)^{1/4} \, F(\mathbf{q}^2) m^{1/2}}{(\mathbf{p}_1^2 + m^2)^{1/4} \, (\mathbf{p}^2 + m^2)^{1/4}} . \tag{64}$$

The operator Φ by definition must satisfy Eq. (9) and

$$\lim_{|\mathbf{a}| \to \infty} (\Phi^\dagger h \Phi - \Phi^\dagger \Phi h_0) e^{i \mathbf{p} \cdot \mathbf{a}} = 0. \tag{65}$$

These two conditions are satisfied if we put

$$(\mathbf{p}' | \Phi | \mathbf{p}) = \delta(\mathbf{p}' - \mathbf{p}) Z^{-1/2}, \tag{66}$$

and

$$\begin{aligned}(\mathbf{p}_1, \mathbf{p}_2 | \Phi | \mathbf{p}) = -\{ &(\mathbf{p}_1, \mathbf{p}_2 | h'_1 | \mathbf{p}) \; (\omega_1 + \nu - \omega)^{-1} \\ &+ (\mathbf{p}_1, \mathbf{p}_2 | h'_2 | -\mathbf{p}) \; (\omega_2 + \nu - \omega)^{-1} \} Z^{-1/2} .\end{aligned} \tag{67}$$

We have thus

$$\begin{aligned}(\mathbf{p}' | \Phi^\dagger \Phi | \mathbf{p}) = \; &\delta(\mathbf{p}' - \mathbf{p}) Z^{-1} \\ &+ \frac{1}{2} \int d\mathbf{p}_1 \int d\mathbf{p}_2 (\mathbf{p}_1, \mathbf{p}_2 | \Phi | \mathbf{p}')^* (\mathbf{p}_1, \mathbf{p}_2 | \Phi | \mathbf{p}) .\end{aligned} \tag{68}$$

From these equations it follows that

$$\delta(\underset{\sim}{p}' - \underset{\sim}{p})Z = \delta(\underset{\sim}{p}' - \underset{\sim}{p}) + \frac{1}{2}\int d\underset{\sim}{p}_1 \int d\underset{\sim}{p}_2 \sum_{i=1}^{2} (\underset{\sim}{p}_1, p_2|h'_i|p')^*(\underset{\sim}{p}_1, p_2|h'_1|\underset{\sim}{p})(\omega_1 + \nu - \omega)^{-2}. \tag{69}$$

That is,

$$Z = 1 + \int d\underset{\sim}{q}(\underset{\sim}{q}^2 + \mu^2)^{-1/2} [M(q) - m]^{-2} [F(\underset{\sim}{q}^2)]^2, \tag{70}$$

since

$$(\omega_i + \nu - \omega)^2 = -(p_i + k - p)^2 = [M(q) - m]^2. \tag{71}$$

We have therefore

$$(\underset{\sim}{p}'|\Phi^\dagger\Phi|\underset{\sim}{p}) = \delta(\underset{\sim}{p}' - \underset{\sim}{p}) + (\underset{\sim}{p}'|N|\underset{\sim}{p}), \tag{72}$$

where

$$\begin{aligned}(\underset{\sim}{p}'|N|\underset{\sim}{p}) &= \frac{1}{2} Z^{-1} \int d\underset{\sim}{p}_1 \int d\underset{\sim}{p}_2 \{(\underset{\sim}{p}_1, \underset{\sim}{p}_2|h'_1|\underset{\sim}{p}')^*(\underset{\sim}{p}_1, \underset{\sim}{p}_2|h'_2|-\underset{\sim}{p}) \\ &\quad + (\underset{\sim}{p}_1, \underset{\sim}{p}_2|h'_2|-\underset{\sim}{p}')^*(\underset{\sim}{p}_1, \underset{\sim}{p}_2|h'_1|\underset{\sim}{p})\} [(p' - p)^2 + \mu^2]^{-1} \\ &= \frac{Z^{-1}(\underset{\sim}{p}'|\tilde{J}|\underset{\sim}{p})(-\underset{\sim}{p}'|\tilde{J}|\underset{\sim}{p})}{[(\underset{\sim}{p}' - \underset{\sim}{p})^2 + \mu^2]^{1/2}\,[(p' - p)^2 + \mu^2]}.\end{aligned} \tag{73}$$

In the static limit the operator N depends only on the momentum transfer, i.e.,

$$(\underset{\sim}{p}'|N|\underset{\sim}{p}) = Z^{-1}[F(|\underset{\sim}{p}' - \underset{\sim}{p}|^2)]^2 (|\underset{\sim}{p}' - \underset{\sim}{p}|^2 + \mu^2)^{-3/2}. \tag{74}$$

The operator N is therefore local, i.e.,

$$(\underset{\sim}{x}'|N|\underset{\sim}{x}) = \delta(\underset{\sim}{x}' - \underset{\sim}{x})N(r). \tag{75}$$

Equation (65) is satisfied if m_0 and m are related by the equation

$$m_0 - m = \int d\underset{\sim}{q}(\mu^2 + \underset{\sim}{q}^2)^{-1/2} [M(\underset{\sim}{q}) - m]^{-1} [F(q^2)]^2. \tag{76}$$

It is now easy to evaluate the reduced Hamiltonian

$$\bar{h} = (\Phi^\dagger\Phi)^{-1/2}\Phi^\dagger h\Phi(\Phi^\dagger\Phi)^{-1/2} \tag{77}$$

in a nonrelativistic approximation, i.e., the static limit plus the first-order terms in an expansion in powers of p^2/m^2 and h'/m. In that approximation

$$\Phi^\dagger(h - 2m)\Phi = p^2/2m + \{(p^2/2m), (N + \mathcal{V}/m)\} + \mathcal{V}, \tag{78}$$

where

$$(\underset{\sim}{p}'|\mathcal{V}|\underset{\sim}{p}) = -Z^{-1} \frac{(\underset{\sim}{p}'|\tilde{J}|\underset{\sim}{p})(-\underset{\sim}{p}'|\tilde{J}|-\underset{\sim}{p})}{(p' - p)^2 + \mu^2} . \tag{79}$$

In the static limit this becomes

$$(\underset{\sim}{p}'|\mathcal{V}|\underset{\sim}{p}) = -Z^{-1} \frac{[F(|\underset{\sim}{p}' - \underset{\sim}{p}|^2)]^2}{|\underset{\sim}{p}' - \underset{\sim}{p}|^2 + \mu^2} . \tag{80}$$

The results of the present model are readily comparable with the conventional one-boson exchange potentials. The normalization operator N has no counterpart. That was to be expected because of the formal perturbation procedures in the conventional models.

A definite prediction of the static potential is possible only if the form factor F(q) is determined by additional requirements. The main virtue of the present model is that it incorporates some features of field theories into a particle theory and predicts the nonadiabatic terms in the nuclear force.

Appendix

Equivalent partial-wave Hamiltonians[4]

Consider the Hilbert space of the square-integrable functions $\phi_L(R)$, $a \leq R \leq \infty$, with the norm $\|\phi_L\|$ defined by

$$\|\phi_L\|^2 = \int_a^\infty dR\,|\phi_L(R)|^2. \tag{A1}$$

The partial-wave Hamiltonian is

$$H = -\frac{1}{2}\{\omega(R)\frac{d^2}{dR^2} + \frac{d^2}{dR^2}\omega(R)\} + \frac{L(L+1)}{R^2} + V_L(R). \tag{A2}$$

The twice-differentiable functions that vanish for $R = a \geq 0$ are its domain. That means we have a hard-core potential of radius a. For large R the potential $V_L(R)$ vanishes and the function $\omega(R)$ tends to 1.

Any distortion of the radial scale induces a unitary transformation on the wave functions. Let

$$R = R(r)$$

such that

$$\frac{dR}{dr} \equiv \mu^{-1/2}(r) > 0, \qquad R(b) = a, \tag{A3}$$

and $\mu(r)$ goes to 1 for large r. The transformation

$$\tilde{\phi}_L(r) = \mu^{-1/4}(r)\,\phi_L[R(r)] \tag{A4}$$

defines a unitary operator

$$\tilde{\phi}_L = U^\dagger \phi_L. \tag{A5}$$

The transformed Hamiltonian is

$$\tilde{H} = U^{\dagger}HU = -\frac{1}{2}\left\{\mu\omega\,\frac{d^2}{dr^2} + \frac{d^2}{dr^2}\,\mu\omega\right\}$$

$$+\frac{1}{4}\,\omega\left\{\frac{d^2\mu}{dr^2} + \frac{1}{4}\,\mu^{-1}\left(\frac{d\mu}{dr}\right)^2\right\} + \frac{L(L+1)}{R^2(r)} + V_L[R(r)], \tag{A6}$$

The Hamiltonians H and H have the same eigenvalues. The eigenfunctions are related by a distortion of the radial scale. The asymptotic behavior of the scattering wave functions is the same. Thus the scattering phase shifts are the same.

References

1. H. Yukawa, Proc. Phys. Math. Soc. Japan 17, 48 (1935);
   H. Yukawa and S. Sakata, Proc. Phys. Math. Soc. Japan 19, 1084 (1937
   H. Yukawa, S. Sakata and M. Taketani, Proc. Phys. Math. Soc. Japan 20, 319 (1938).
2. G. Wentzel, Helv. Phys. Acta 13, 269 (1940); 14, 633 (1941); 15, 685 (1942); 16, 233 (1943); 16, 551 (1943);
   J. R. Oppenheimer and J. Schwinger, Phys. Rev. 60, 150 (1941);
   W. Pauli and S. M. Dancoff, Phys. Rev. 62, 85 (1942);
   R. Serber and S. M. Dancoff, Phys. Rev. 63, 143 (1943);
   W. Pauli, Meson Theory of Nuclear Forces (Interscience Publishers, New York, 1946).
3. H. Ekstein, Phys. Rev. 117, 1590 (1960).
4. G. A. Baker, Phys. Rev. 128, 1485 (1962). The equivalence class generated by Baker's transformations is described in the appendix of the present paper.
5. R. K. Osborn and L. L. Foldy, Phys. Rev. 79, 795 (1950), L. L. Foldy, Phys. Rev. 92, 172 (1953).
6. F. Coester, Helv. Phys. Acta 38, 7 (1964).
7. C. Bloch, Nucl. Phys. 6, 329 (1958); C. Bloch and J. Horowitz, Nucl. Phys. 8, 91 (1958).
8. H. Feshbach, Ann. Phys. (New York) 19, 287 (1962).
9. W. Pauli, Phys. Rev. 64, 332 (1942), Sec. 4.

10. G. Schierholz, Nucl. Phys. B7, 483 (1968).
11. T. Ueda and A. E. S. Green, Phys. Rev. 174, 1304 (1968).
See also R. A. Bryan and B. L. Scott, Phys. Rev. 164, 1215 (1967); L. Ingber, Phys. Rev. 174, 1250 (1968).
12. E. C. G. Stueckelberg and J. F. C. Patry, Helv. Phys. Acta 13, 167 (1940).

Received 3/17/69

What Is "Quantum-Logic"?

by

J. M. Jauch* and C. Piron

Center for Theoretical Studies, University of Miami

and Institute for Theoretical Physics, University of Geneva

(*On leave of absence from the University of Geneva)

In this paper we present a short introduction and survey of the calculus often called "Quantum Logic", and we point out the difficulties which prevent us from considering this calculus as a logic.

It is furthermore shown that the probability calculus for a quantal system is a generalization of the classical probability calculus, which gives a more general meaning to the notion of random variable.

1. Introduction

Quantum mechanics is indispensable for modern physics. It is the basic theory for all atomic, nuclear, elementary particle and solid state physics. Yet the epistemological problems arising out of the physical interpretation of the theory are so deep and apprently so radically different from the classical mechanistic tradition that there is as yet no unified view as to the correct physical interpretation or even whether a consistent interpretation is possible.

The deepest analysis of the new epistemological situation was made by Bohr, and its essence is concisely summarized in the notion of "complementarity". With this notion Bohr set the limit to the applicability of the classical concepts in the realm of microphysics. The interpretation which takes full account of this limit is usually called, in a broad sense, the "Copenhagen Interpretation". A closer analysis reveals, however, that there are several variants even within this "orthodox" view of quantum mechanics. In addition, several heterodox interpretations have from time to time been proposed which reject complementarity as a fundamental limitation.

In view of this situation, which to many thoughtful physicists is not altogether satisfactory, it is of interest to reconstruct the theory in a new manner, by trying to remain as close as possible to the phenomena and introducing theoretical concepts only when they are directly and convincingly motivated by the experimental facts. This procedure has the advantage of removing from the discussion all those problems of quantum mechanics which are extraneous and accidental and leaving a better view of those problems which are essential. It has the added advantage of showing up possibilities for generalizations of quantum mechanics in new and uncharted directions.

One of the most promising attempts of such a phenomenological foundation of quantum mechanics is that initiated by Birkhoff and von Neumann in 1936 in a paper entitled "The Logic of Quantum Mechanics".[1] This paper, which remained for a long time unnoticed by most physicists, has in recent years been the point of departure for an axiomatic reconstruction of quantum mechanics.[2,6,7] The use of the word logic in the title, and the formal similarity of the lattice structure introduced by Birkhoff and von Neumann, with the propositional calculus, have given rise to speculations that the logic governing the propositions of microsystems is somehow different from that form of logic which is applicable to reasoning with "classical" propositions. The situation has, for instance, been compared with the role of geometry in the general theory of relativity. Claims have been made that the paradoxes in quantum mechanics disappear only if they are analyzed not with ordinary logic but with a new kind of quantum logic instead. It should perhaps be mentioned that Bohr, on several occasions, rejected this point of view and Pauli, too, has expressed similar doubts.[3] For instance, in Ref. 3 Bohr writes:

"Incidentally, it would seem that the recourse to three-valued logic, sometimes proposed as means for dealing with the paradoxical features of quantum theory, is not suited to give a clearer account of the situation, since all well-defined experimental evidence, even if it cannot be analyzed in terms of classical physics, must be expressed in ordinary language making use of common logic."

The situation is complicated by the fact that the advocates of the new "logic" have not all had the same idea of what they meant by quantum logic. For instance, Reichenbach[4] published a book on the subject in which he advocated the use of a three-valued logic where the propositions have, in addition to the usual truth values true and false, a third one called undetermined. A similar idea was formulated by C. V. Weizsäcker who goes much further in that he wants to introduce a complex-valued logic.[5] Others still maintain the usual truth values but claim a different meaning for the interpretation of the logical connectives "and", "or" and "not", so that the resulting lattice is not a Boolean lattice, that is, it does not satisfy the distributive law.

In this paper we shall indicate some of the difficulties which such interpretations have to face.

2. The empirical basis of "Quantum - Logic"

In this section we shall recapitulate some of the empirical facts which lead to the formulation of "Quantum - logic". We shall continue to use this term since it has been used by Birkhoff and von Neumann and many others who followed them. This is done without prejudice as to its meaning, but as a name referred to in quotes. We shall presently see what it designates.

One arrives at this concept if one considers a microsystem (or, in fact, any physical system) as defined by and identical to the sum of all its physical properties.

Although this procedure seems a safeguard against the pseudo-problems which often beset the foundations of physics, one must face several interconnected difficulties.

In the first place, it is difficult to limit the parts of physical realities, which we may call physical systems. If we isolate a proton in the laboratory, it seems clear enough what we mean by the physical system, "proton". Yet, even in such a simple case, one soon discovers that a proton cannot be isolated from its strongly interacting meson field, or even from its radiation field. If, for practical purposes, we draw a demarcation line, we must realize that it is to some extent arbitrary and is in fact a cause of fundamental difficulties.

Other difficulties arise from the fact that many measurements are possible and that the results of such measurements may be quite different for one and the same system. We say the measurements depend on the "state" of the system.

Certain types of experiments are characterized by the fact that they admit only one of two possible alternatives. A counter triggered by an elementary particle is a typical example. We shall call them "yes-no experiments", and we shall designate the two alternatives with "true" or "false".

A natural partial ordering is defined on the set of all yes-no experiments. For example, if the system is prepared in such a way that whenever the yes-no experiment α is true, then there may be another β which is also true with certainty. Such a relationship is a physical property of the system, which is independent of the state, and we may designate it by $\alpha < \beta$. Clearly the relationship $<$ is transitive since, from $\alpha < \beta$ and $\beta < \gamma$, it follows that $\alpha < \gamma$. It thus defines a partial order on the set of all yes-no experiments, provided that we identify equivalent yes-no experiments.

We shall define α_1 as being equivalent to α_2 if $\alpha_1 < \alpha_2$ and $\alpha_2 < \alpha_1$. The class of all equivalent yes-no experiments α_1 will be denoted by a, and we shall call it a proposition.

The partial ordering of the yes-no experiments can then be transferred to the propositions by defining

$$a \subset b \quad \text{if } \alpha < \beta \text{ for all } \alpha \in a,\ \beta \in b.$$

It follows then that

$$a \subset b \quad \text{and} \quad b \subset a \quad \text{implies } a = b.$$

This partially-ordered set of propositions has an empirically determined structure for any given physical system. These structures have been studied in detail for many actual physical microsystems and they are, in all cases, found to be complete orthocomplemented lattices. We denote them by $\mathcal{L}$.

Given any set of propositions $a_i \in \mathcal{L}$, there exist propositions $\bigcap_i a_i$ and $\bigcup_i a_i$ such that

$$x \subset \bigcap a_i \iff x \subset a_i$$
$$\bigcup a_i \subset y \iff a_i \subset y.$$

They may be called the greatest lower bound and the least upper bound. It follows from this that there exists an absurd proposition ϕ which is always false and a trivial proposition I which is always true.

The orthocomplement of the lattice is a unary operation $a \to a'$ with the properties

$$(\bigcap_i a_i)' = \bigcup_i a_i'$$

and

$$\left.\begin{array}{l} a \cap a' = \phi \\ a \cup a' = I \\ (a')' = a \end{array}\right\} \quad \forall a \in \mathcal{L}$$

If $\alpha \in a$ is a yes-no experiment in the class a, then a' contains the experiment α^{ν} (the strong negation of α), which is the same experiment as α but with its alternatives interchanged. This means that whenever α is true, then α^{ν} is false and _vice versa_.

The state of a physical system is represented by a functional p(a) on the lattice with values in [0,1] satisfying

$$p(\phi) = 0, \; p(I) = 1$$

If $a_i \subset a_k'$ $(i \neq k)$, then $p(\bigcup a_i) = \Sigma\, p(a_i)$

If $p(a_i) = 1$, then $p(\bigcap_i a_i) = 1$.

Furthermore, if $a \neq b$, then there exists a state p such that $p(a) \neq p(b)$. In other words the states separate the propositions.

The characteristic property of the quantum mechanical lattices is that they are not Boolean, that is, in general[1,2], the distributive law does not hold

$$a \cap (b \cup c) \neq (a \cap b) \cup (a \cap c).$$

This fact (and we emphasize again that, due to our way of constructing the lattice on the basis of experiments, this is an empirical fact) can be verified by discussing individual systems for which the indicated operations can be carried out*. One example of such a system, perhaps the simplest possible one, is the polarization states of a photon, that is, the system which remains if we abstract from the spatial or the energy degrees of freedom.

The states of polarization of a photon can be represented as the points on a 3-dimensional unit sphere, or equivalently as the points in a 2-dimensional plane (Poincaré parameters). Points at opposite ends of a diameter are orthocomplements of one another. The lattice of yes-no experiments has the structure depicted in Fig. 1, where we have indicated two typical points and their complements.

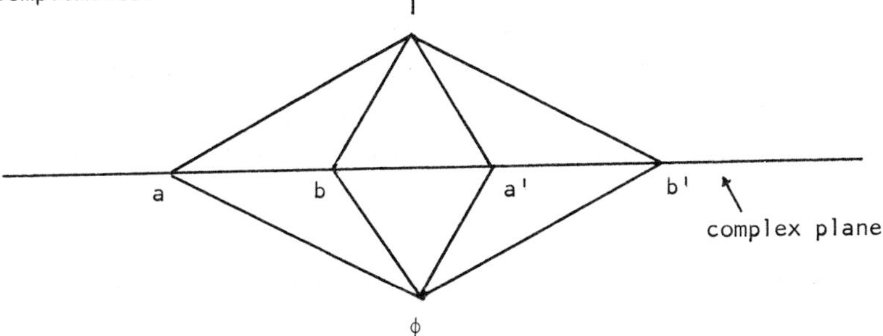

Fig. 1. The lattice for a photon

It is clear from the diagram that this lattice is not Boolean. For example, $a \cap (b \cup a') = a \cap \phi = a$, while $(a \cap b) \cup (a \cap a') = \phi \cup \phi = \phi$.

Let us now come to the main point. The operations so far introduced, the complement, the meet and the join, are (in principle

---

* In a recent publication[12], Karl R. Popper has attempted to show that such a system is mathematically inconsistent. More specifically, Popper believes he has shown that if the states separate the propositions, then the lattice is necessarily Boolean. Popper made, however, an additional assumption, viz., that for all states

$$p(a) + p(b) = p(a \cup b) + p(a \cap b),$$

which is not true in quantum mechanics. His conclusions thus prove only that his interpretation of Birkhoff and von Neumann's paper is not tenable.

if not always in practice) represented by actual physical operations, so the lattice structure is empirically given.

Let us now focus our attention on the order relation $a \subset b$. In analogy with classical logic, one might be tempted to declare $a \subset b$ a new proposition which affirms the conditional "if a then b". As such it should be an element of the lattice of yes-no experiments that we are constructing. But this interpretation of the conditional is not possible, as has been shown by Piron[10]. The difficulty is that no experimental arrangement is possible which measures the proposition $a \subset b$. This object is therefore not a new element of the lattice, but rather a relation between certain elements of the lattice, and it is therefore something entirely different from the other lattice elements.

This state of affairs is fundamentally different from a logic. The role of $a \subset b$ corresponds obviously to the conditional, and in logic the conditional is considered a proposition just as all the others. It is the proposition which affirms "if a then b". Actually in logic (that is, in a Boolean lattice) it is possible to express the proposition in terms of the other logical connectives. We shall presently see that this is not possible in "Quantum - Logic".

3. The role of the conditional in logic

For the purpose of this section we shall adopt the notation for the propositional calculus which deviates in some details from any of the standard systems in order to avoid a clash in notation with the lattice structure of quantum mechanics.

We are here concerned only with what is sometimes called primary logic, which may be described as the theory of deduction. We shall call it simply the propositional calculus. The elements of the calculus are designated by p, q, r, ... Individual propositions with a certain scope will be denoted by capitals, P, Q, R, ... The logical operations are four in number. They are listed in the table.

Logical Operation	Interpretation
$\overline{P}$	not P
$P \cup Q$	P and Q
$P \cap Q$	P or Q
$P \rightarrow Q$	P implies Q

A proposition is designated by a well-formed expression using the four logical operations.

Two operations are possible: Substitution and deduction. Substitution consists of replacing, in any well-formed expression, a letter by another well-formed expression. Deduction is the rule which permits replacing of the two propositions P and P → Q by Q. It is often written schematically by $\frac{P \quad P \rightarrow Q}{Q}$. In words it means: If P is true and if P implies Q, then Q is true too.

In primary logic there is only one rule of deduction[13,14] and without such a rule it is difficult to speak of a logic since inferences are impossible. It is therefore essential that the conditional P → Q be a proposition.

Even in classical logic the nature of the conditional has been a subject of debate since antiquity. Sextus Empiricus[8] reports on the disagreement between Diodorus of Cronus and his pupil, Philo, who defined the conditional as "the proposition which is false only when it begins with a truth and ends with a falsehood". This seems to imply that it is true for all other truth values. It is in this sense that the conditional is used in modern primary logic. Without any such convention the conditional cannot be considered a proposition. Now this interpretation of the conditional P → Q is identical to $\overline{P} \cup Q$, as one may easily verify with the truth table for the two propositions.

It is interesting to observe that one may introduce the conditional as any complex proposition involving two propositional signs and such that its conjunction with the first entails the second. If we denote this proposition by C (p,q), we must have the axiom

$$C\ ((C\ (p,q)\ p),\ q).$$

However, if this is an axiom, then one can show that, together with the other axioms which define the three other logical connectives, the conditional is uniquely determined as the Philonian conditional $\overline{p} \cup q$.

4. The conditional in "Quantum Logic"

Let us now return to the calculus of yes-no experiments in quantum theory. According to the discussion of section (2), the implication relation is not another yes-no experiment and therefore the calculus is not a logic since it does not admit any deduction. In view of the preceding section, we might be tempted to define a conditional in analogy with the Philonian conditional of classical logic. That is, we might define $a \subset b$ as identical to $a' \cup b$. However this is impossible. The easiest way to refute it is to consider the lattice of Fig. 1 representing the yes-no experiments of a photon system. For any $a' \neq b$ we have $a' \cup b = I$. This means that for any such pair of yes-no experiments the implication relation should be satisfied. A glance at Fig. 1 shows that this is false. Thus this transfer of the Philonian conditional to "Quantum - Logic" is seen to be impossible.

There is a general theorem, recently proved by Gy. Fahy[21], according to which the transitivity of the relation $a' \cup b = I$ implies distributivity of the lattice. More precisely, the following theorem holds:

If $a' \cup b = I$ and $b' \cup c = I$ implies $a' \cup c = I$ for all triplets $a, b, c \in \mathcal{L}$, then the lattice $\mathcal{L}$ is Boolean.

Since transitivity of the conditional is necessary for a logic, the relation $a' \cup b = I$ cannot represent the conditional whenever the lattice is not Boolean, that is in all quantum mechanical lattices.

One might conclude from this that the failure to define a consistent conditional in "Quantum - Logic" is due to our insistence that we have only two truth values for the yes-no experiments and that a many-valued logic would present no such difficulty. We shall show next that a many-valued logic does not save the situation either.

5. The conditional in a many-valued "Quantum Logic"

The idea of using a logic with more than two truth values for quantum mechanics was first proposed by Reichenbach[4]. He postulated that elementary propositions about quantum mechanical systems should admit three truth values: True, false and undetermined.

In view of the fact, however, that the state of a system attributes to each yes-no experiment a probability function p(a) with $0 \leq p(a) \leq 1$, it seems more natural, once one has passed beyond the ordinary double-valued logic, to consider "Quantum - Logic" as an infinite-valued logic. Such a system was developed by T. Lukasiewicz[9]. In this system the truth values are represented by real numbers between 0 and 1. The value 0 represents falsehood while 1 represents the truth. Intermediate values stand for various degrees of certainty. It is possible to give a system of axioms and to prove consistency for such a logic. The conditional has the following truth values: If we denote by [P] the truth value of the proposition P, then

$$[P \rightarrow Q] = \begin{cases} 1 \text{ for } [P] \leq [Q] \\ 1 - [P] + [Q] \text{ for } [Q] < [P], \end{cases}$$

and

$$[\bar{P}] = 1 - [P].$$

We may now ask the question whether in Quantum Logic it is possible to <u>define</u> the yes-no experiment $P \rightarrow Q$ as being that measurable quantity which, for a given state, assumes the above values $[P \rightarrow Q]$.

In quantum mechanics, in Hilbert space, the yes-no experiments are represented by projection operators and if this proposed definition of $P \rightarrow Q$ is possible, then it must be possible to show the following property:

Given any two projection operators P and Q in Hilbert space H there exists a third projection R such that for any state $\psi \in H$ $(\|\psi\| = 1)$

$$(\psi, R\psi) = \begin{cases} 1 \text{ for } (\psi, P\psi) \leq (\psi, Q\psi) \\ 1 - (\psi, Q\psi) + (\psi, P\psi) \text{ for } (\psi, Q\psi) < (\psi, P\psi) \end{cases}$$

We shall now show that this impossible.

Let us assume that P and $Q \neq P$ are projections of one dimensional range and assume ψ to lie in the range of Q, so that $Q\psi = \psi$, and hence $(\psi, Q\psi) = 1$. We have then $(\psi, P\psi) < 1$ and therefore $(\psi, R\psi) = 1$. This shows that R contains Q.

Consider next a state in the range of P, so that $(\psi, P\psi) = 1$ and $(\psi, Q\psi) \leq 1$. We find then that

$$(\psi, R\psi) = (\psi, P\psi) = 1$$

and we conclude that R also contains P. Finally, we choose a state in the orthogonal complement of the plane spanned by P and Q. Thus R is the entire space. This means R is always true whatever the state. But this leads to a contradiction if $(\psi, Q\psi) < (\psi, P\psi) < 1$, because then

$$1 = (\psi, R\psi) = (\psi, P\psi) < 1.$$

Thus the definition of the conditional is not possible in a many-valued "Quantum - Logic".

We have shown this for the lattice of subspaces in a Hilbert space. However, it is also true for the lattices in general quantum mechanics although a direct proof has not yet been given. (A proof can be constructed via the representation theorem of the lattices given by Piron in his thesis[10].)

We may conclude from this that none of the obvious attempts of defining a conditional as a logical proposition can be carried through consistently in a "Quantum Logic". This state of affairs makes it very questionable whether we may properly call the lattice of general quantum mechanics a logic. If we do admit such a wide range of meaning for what constitutes a logic or a propositional calculus, there would be practically no limits for further extension of the meaning of logic. However, we could of course agree to call a system such as the lattice of quantum mechanics a logic by a kind of semantic legislation. The result would be a confusion which would make virtually impossible effective communication between thinkers of diverse tendencies. A logic should provide us with a system of general principles which allows for valid inferences in all kinds of subject matter and it should provide us with a universal language for expressing truths which we think when we abstract entirely from the special contents of our expressions. As Frege expressed it epigrammatically: Logic is not about the laws of nature but about the laws of the laws of nature[11,13].

6. The conditional in a modal logic

While it is thus seen from the preceding two sections that no consistent definition is possible of the conditional in the propositional calculus of quantum mechanics, there remains the question whether such a definition might not be possible in a logic where the conditional has an interpretation different from the Philonian conditional of primary logic. Such a theory has been developed by C. I. Lewis[17, 18]. Lewis distinguishes implication in the strict sense from material implication, the term used by Whitehead and Russell[19] for the Philonian implication. He introduces a new symbol $P \prec Q$ for this strict implication, and defines it in the terms of a new, undefined symbol $\Diamond$, designating "it is possible that". Thus $\Diamond P$ will be interpreted "it is possible that P" and correspondingly $\overline{\Diamond P}$ "it is impossible that P".

Now the statement that $P \prec Q$ means that it is impossible that P is true and Q is false, and it would be rendered in this calculus by $\overline{\Diamond(P \cap \overline{Q})}$. This corresponds formally to the Philonian implication $P \rightarrow Q$ which is equivalently expressed by $\overline{P \cap \overline{Q}}$ or $\overline{P} \cup Q$.

The calculus of modal logic is then constructed axiomatically by introducing $\Diamond$ as a new, unidentified symbol, together with the other logical constants, the rules of substitution and the rules of inference. Two rules of inference are needed in this system, viz., $\frac{P \quad P \prec Q}{Q}$ and $\frac{P \quad Q}{P \cap Q}$. The second one here is not derivable from the first one as in the ordinary propositional calculus, so it is necessary to add it as a second rule of inference.

It seems then that in this system the expression $P \prec Q$ is a proposition too, if P and Q are propositions. However, since $P \prec Q$ can be considered defined in terms of the symbol $\Diamond$, it suffices to determine the truth values of the proposition $\Diamond P$. Here the situation is as follows: If we attribute to $\Diamond P$ the truth values true or false, then the system is not more general than the ordinary system if $\Diamond P$ is considered equivalent to P and $\prec$ merely a typographical variant of $\rightarrow$. Thus a greater generality can be expected only in a many-valued logic.

In the context of the many-valued logic of Lukasiewicz, Tarski[20] has shown that the truth values of $\Diamond P$ are given by

$$[\Diamond P] = \begin{cases} 1 \text{ for } [P] \geq \frac{1}{2} \\ 2\,[P] \text{ for } [P] < \frac{1}{2}\,. \end{cases}$$

Although with this definition of $\Diamond P$ we are now able to define the truth values of the strict implication, it is not possible to utilize this system for "Quantum Logic" either. The difficulty in this case stems from the fact that in order to define the disjunction and conjunction it is necessary to have recourse to the material implication. In fact Lukasiewicz[9] gives the following definitions for their truth values:

$$[P \cup Q] = [(P \rightarrow Q) \rightarrow Q]$$

$$[P \cap Q] = [(\overline{P} \rightarrow \overline{Q}) \rightarrow \overline{Q}] = 1 - [\overline{P} \cup \overline{Q}].$$

Since we have already seen that the propositional calculus of "Quantum Logic" is inconsistent with the definition of $P \rightarrow Q$ as a proposition, the interpretation of "Quantum Logic" as a modal logic is also impossible.

7. Probability in quantum theory

The probability calculus for quantum systems is a generalization of the classical probability calculus. While the latter is derived from a measure on a Boolean algebra, the former is a similar measure but defined on a non-Boolean lattice. This has several consequences. One of the most important is a more restrictive meaning for the notion of random variable. A real valued random variable is a function $x(\Delta)$ from the Borel sets Δ of the reals to the elements of the lattice which satisfies the properties

$$x(o) = \phi,\ x(R) = I \text{ and}$$

if Δ_i are countable and disjoined then $x(\bigcup \Delta_i) = \bigcup_i x(\Delta_i)$. It follows from this definition that the range of this function is a Boolean sublattice of the entire lattice of yes-no experiments. If the lattice is realized as the projections in a Hilbert space, then this object is nothing else than a spectral measure on the real line and therefore it can be replaced, by virtue of the spectral theorem, by a self-adjoint operator. Thus the random variable as defined here is coextensive with the self-adjoint operators in Hilbert space, and it is the analogue of an observable in the lattice theoretic formulation of quantum theory.

We may formalize this situation by stating that the range H_x of the random variable x is a Boolean sublattice of the complete lattice $\mathcal{L}$ of all yes-no experiments.

In classical as well as in the quantal probability calculus, the range of a random variable is usually a subset of the set of all elements of the lattice. In quantum theory it is even required to be so since the completed set $\mathcal{L}$ of all yes-no experiments is, as we have seen, in general non-Boolean while the range of a random variable is always a Boolean subset of $\mathcal{L}$.

Thus for instance in the example of Fig. 1, the range of a random variable (that is an observable) can consist of four elements only, such as a, a', ϕ and I. But different random variables may have different ranges. Four elements b, b', ϕ and I may be the range of another random variable for the photon system.

The consequence of this new situation in the quantal probability calculus is that two random variables x and y may or may not have a joint distribution.

Failure to recognize this more general aspect of probability calculus in quantum theory has led some people to cast doubt on the consistency of the new calculus. Thus for instance in a recent publication, A. I. Fine[15] proposed that the probability calculus of quantum theory is in fact classical and that the random variables described above are what Karl Menger[16] has called "statistical random variables". Implied in this criticism is that the probability on $a \cap b$ or $a \cup b$ is usually not assigned even if it is so for the individual elements a and b. This is, however, not in agreement with the empirical content of the quantal lattice theory where the probability for $a \cap b$ can, at least in principle, be measured as we have outlined in Section 2. Thus, we disagree with the conclusion of Fine's paper.

However, a useful proposal in that paper merits acceptance, viz., that due to the more general properties of random variables in quantum theory they should be given another name, as Menger suggested in the cited paper.

Conclusion

We have shown that the attempt to interpret "Quantum - Logic", that is, the empirically-given calculus of yes-no experiments of microsystems, as a non-classical logic leads to difficulties. The central problem is the definition of the conditional as a proposition. None of the known generalizations of logic can be used for this purpose in "Quantum - Logic", and consequently no rule of inference can be formulated. It would thus be better to avoid the term logic altogether in the designation of the lattice structure of general quantum mechanics.

On the other hand, the lattice of quantum mechanics calls for a generalized probability calculus with a new meaning for random variables. This theory is as yet little developed mathematically. In particular, in view of the physical interpretation of this theory, it should be of interest to derive limit theorems for random variables as is done in classical probability theory.

This paper is contributed to this Volume in grateful recognition of Professor Wentzel's scientific influence on one of us (J. M. J.) during his formative years in Zürich, and for continued friendship during later life.

References

1) G. Birkhoff and J. von Neumann, Ann. Math. 37, 823 (1936).
2) J. M. Jauch, Foundations of Quantum Mechanics, Addison-Wesley (1968).
3) N. Bohr, Dialectica 1, 317 (1948).
4) H. Reichenbach, Philosophical Foundations of Quantum Mechanics, California Press (1941).
5) C. F. V. Weizsacker, Naturwiss., 42, 547 (1955).
6) G. Mackey, The Mathematical Foundations of Quantum Mechanics, Benjamin (1963).
7) G. Ludwig, Zs. Phys. 181, 233 (1964); Comm. Math. Phys. 4, 331 (1967).
8) Sextus Empiricus, Opera, Ed. H. Mutschmann and J. Mau, Leipzig, (1912-54).
9) T. Lukasiewicz, Aristotelic Syllogistic from the Standpoint of Modern Formal Logic, Oxford (1957).
10) C. Piron, Helvetia Physica Acta 37, 439 (1964).
11) G. Frege, Die Grundlagen der Arithmetik, Breslau (1884).
12) Karl R. Popper, Nature 219, 682 (1968).
13) W. and M. Kneale, The Development of Logic, Oxford (1966).
14) cf. e.g., S. C. Kleene, Introduction to Mathematical Logic, ch. IV, p. 82.
15) Alexander I. Fine, Philosophy of Science 35, 101 (1968).
16) Karl Menger, Proceedings of the Third Berkeley Symposium in Mathematical Statistics and Probability 2, 215 (1954).
17) C. I. Lewis, Survey of Symbolic Logic, Berkeley (1948).
18) C. I. Lewis and C. H. Langford, Symbolic Logic, New York (1932).
19) A. N. Whitehead and B. Russell, Principia Mathematica, Cambridge, 2nd ed. (1925-7).
20) A. Tarski, Logic, Semantics, Metamathematics, Oxford (1953).
21) Gy. Fahy, Ada. Sci. Math. (Szeged) 28, 269 (1967).

Received 1/24/69

The Oscillations of a Rotating Gaseous Mass in the Post-Newtonian Approximation to General Relativity

by
S. Chandrasekhar
The University of Chicago

The axisymmetric oscillations of a uniformly rotating gaseous mass are considered in the first post-Newtonian approximation to general relativity. The problem is reduced to a self-adjoint characteristic value problem in a pair of partial differential equations for the square of the oscillation frequency; and a variational expression for it is given.

1. Introduction

It is now well known that the effects of general relativity, already in the first post-Newtonian approximation, induce instabilities in gaseous masses that would be considered stable in the Newtonian theory (Chandrasekhar 1964 and 1965b; Fowler 1964). On the other hand, since rotation is known to have a stabilizing effect in the Newtonian theory (Ledoux 1945), it is clear that in an approximation in which the effects of general relativity and of rotation are *both* considered as of the first order, the two effects will be additive and rotation can suppress the relativistic instability (Fowler 1966; Durney and Roxburgh 1967). It is in fact quite simple to write down explicitly the required condition for stability when both effects are operative and are considered as of the first order (Chandrasekhar and Lebovitz* 1968, eq. [6]). However, the case, when the rotation is rapid and can no longer be considered as of the first order, is not so simple: in this case, the full post-Newtonian equations, in which framework the effects of rotation (and of internal motions quite generally) are taken into account exactly, must be used. It is the object of this paper to obtain the basic equations of this theory.

*The following misprints in this paper may be noted here. In the second line of equation (131) on page 288 a factor $(\partial U/\partial \nu)^2$ which must follow $-4V_4$ (under the integral sign and before the closing square bracket) has been omitted.

2. The equations of post-Newtonian hydrodynamics in a rotating frame of reference

The equations of hydrodynamics in the first post-Newtonian approximation can be written in the forms (Chandrasekhar 1965a, eqs. [67] and [81]; and Chandrasekhar* 1967, eqs. [7] and [8]).

$$\rho[1 + \frac{1}{c^2}(v^2+4U)]\frac{d\vec{v}}{dt} + [1 - \frac{1}{c^2}(\Pi + \frac{p}{\rho})] \text{ grad } p - \rho\,(1+\frac{2v^2}{c^2}) \text{ grad } U$$
$$- \frac{4}{c^2}(\frac{d\vec{U}}{dt} - v_\beta \text{ grad } U_\beta) + \frac{1}{c^2}\rho \text{ grad } (\frac{1}{2}\frac{\partial^2\chi}{\partial t^2} - 2\Phi) \tag{1}$$
$$+ \frac{1}{c^2}\vec{v}\,[\frac{dp}{dt} + \rho\frac{d}{dt}(\frac{1}{2}v^2 + 3U)] = 0$$

and

$$\frac{d\rho}{dt} + \rho \text{ div } \vec{v} + \frac{1}{c^2}\rho\frac{d}{dt}(\frac{1}{2}v^2 + 3U) = 0, \tag{2}$$

where χ, U, and Φ are certain potentials defined by the equations

$$\nabla^2\chi = -2U, \quad \nabla^2\vec{U} = -4\pi G\rho\vec{v}, \tag{3}$$

and

$$\nabla^2\Phi = -4\pi G\rho\phi = -4\pi G\rho(v^2 + U + \frac{1}{2}\Pi + \frac{3}{2}\frac{p}{\rho}); \tag{4}$$

also ρ denotes the density, p the pressure, $\rho\Pi$ the internal energy, $\vec{v}$ the fluid velocity in the chosen frame, and U is the Newtonian gravitational potential determined by the prevailing distribution of ρ.

For treating the equilibrium and the stability of configurations rotating uniformly with an angular velocity $\vec{\Omega}$, it is convenient to write the equations of motion (1) and (2) in a frame of reference rotating with the angular velocity $\vec{\Omega}$. We shall now consider the requisite transformations.

---

*The following misprints in this paper may be noted here. In equations (8) on page 622 and (15) on page 623 a factor ρ in front of the $1/c^2$ terms has been omitted. And a plus sign in front of the last term on the third line of equation (12) on page 623 has been omitted.

Let $\vec{x}$ and $\vec{u}$ denote the position and the velocity of a fluid element in the rotating frame. If $\vec{v}$ denotes the velocity of the same fluid element in the sta ionary frame (in which eqs. [1] and [2] are written) but resolved along the instantaneous directions of the rotating axes, then

$$\vec{v} = \vec{u} + \vec{\Omega} \times \vec{x} \tag{5}$$

and

$$\frac{d\vec{v}}{dt} = \frac{d\vec{u}}{dt} + 2\vec{\Omega} \times \vec{u} - \frac{1}{2} \operatorname{grad} [\Omega^2 (x_1^2 + x_2^2)]. \tag{6}$$

In writing equation (6), we have supposed that the x_3 -axis has been chosen along the direction of $\vec{\Omega}$.

In view of equation (5), the equation governing $\vec{U}$ becomes

$$\nabla^2 \vec{U} = -4\pi G\rho \ (\vec{u} + \vec{\Omega} \times \vec{x}), \tag{7}$$

so that we may write

$$\vec{U} = \vec{U}_* + \vec{\Omega} \times \vec{\mathfrak{D}}, \tag{8}$$

where

$$\nabla^2 \vec{U}_* = -4\pi G\rho\vec{u} \text{ and } \nabla^2 \vec{\mathfrak{D}} = -4\pi G\rho\vec{x}. \tag{9}$$

And finally the total time derivative of $\vec{U}$ which occurs in equation (1) becomes

$$\frac{d\vec{U}}{dt} = \frac{d\vec{U}_*}{dt} + \vec{\Omega} \times \frac{d\vec{\mathfrak{D}}}{dt} + \vec{\Omega} \times \vec{U}_* + \vec{\Omega} \times (\vec{\Omega} \times \vec{\mathfrak{D}}). \tag{10}$$

Now making use of the foregoing relations, we find that equation (1), after some further reductions, can be brought to the form

$$\rho\{1 + \frac{1}{c^2} [u^2 + 2 \vec{u} \cdot \vec{\Omega} \times \vec{x} + \Omega^2 (x_1^2 + x_2^2) + 4U]\} (\frac{d\vec{u}}{dt} + 2 \vec{\Omega} \times \vec{u})$$

$$- \frac{1}{2c^2} \rho (u^2 + 2 \vec{u} \cdot \vec{\Omega} \times \vec{x}) \text{ grad } [\Omega^2 (x_1^2 + x_2^2) + 4U]$$

$$- \frac{4}{c^2} \rho [\frac{d\vec{U}}{dt} + \text{grad}(\vec{\Omega} \times \vec{U} \cdot \vec{x}) + \vec{\Omega} \times \frac{d\vec{\mathcal{V}}}{dt} - u_\beta \text{grad}(\vec{U} + \vec{\Omega} \times \vec{\mathcal{V}})_\beta] \tag{11}$$

$$+ \frac{1}{c^2} (\vec{u} + \vec{\Omega} \times \vec{x}) \{\frac{dp}{dt} + 3\rho \frac{dU}{dt} + \rho\vec{u} \cdot [\frac{d\vec{u}}{dt} - \frac{1}{2} \Omega^2 \text{ grad}(x_1^2 + x_2^2)]$$

$$+ \rho \vec{\Omega} \times \vec{x} \cdot (\frac{d\vec{u}}{dt} + 2 \vec{\Omega} \times \vec{u})\}$$

$$+ \frac{1}{c^2} \rho \text{ grad } (\frac{1}{2} \frac{\partial^2 \chi}{\partial t^2} - \vec{\Omega} \times \vec{x} \cdot \text{grad } \frac{\partial \chi}{\partial t})$$

$$= - [1 - \frac{1}{c^2} (\Pi + \frac{p}{\rho})] \text{ grad } p + \rho \text{ grad } \Psi$$

where

$$\Psi = \Psi_o + \frac{1}{c^2} \Psi_1, \tag{12}$$

$$\Psi_o = U + \frac{1}{2} \Omega^2 (x_1^2 + x_2^2), \tag{13}$$

and

$$\Psi_1 = \frac{1}{4} \Omega^4 (x_1^2 + x_2^2)^2 + 2 U \Omega^2 (x_1^2 + x_2^2) + 2 \Phi$$
$$- 4\Omega^2 (x_1 \mathcal{V}_1 + x_2 \mathcal{V}_2) - \frac{1}{2} (\vec{\Omega} \times \vec{x} \cdot \text{grad})^2 \chi. \tag{14}$$

In equation (11) we have suppressed the distinguishing subscript to $\vec{U}_*$: it has to be understood that $\vec{U}$ is defined in terms of $\vec{u}$ (instead of $\vec{v}$, as formerly).

The equation of continuity (2) in the rotating frame takes the form

$$\frac{d\rho}{dt} + \rho \text{ div } \vec{u} + \frac{1}{c^2} \rho [\frac{1}{2} \frac{du^2}{dt} - \frac{1}{2} \vec{u} \cdot \text{grad } \Omega^2 (x_1^2 + x_2^2)$$
$$+ \vec{\Omega} \times \vec{x} \cdot (\frac{d\vec{u}}{dt} + 2\vec{\Omega} \times \vec{u}) + 3\frac{dU}{dt}] = 0. \tag{15}$$

3. The equations governing axisymmetric oscillations

When no internal motions are present in the rotating frame considered and the conditions are further stationary, equation (11) gives

$$\left[1 - \frac{1}{c^2}\left(\Pi + \frac{p}{\rho}\right)\right] \text{grad } p = \rho \text{ grad } \Psi . \tag{16}$$

From this equation it follows that surfaces of constant p, constant ρ, and constant Ψ all coincide. Accordingly, we may write

$$p \equiv p(\Psi) \text{ and } \rho \equiv \rho(\Psi) . \tag{17}$$

A further consequence of equation (16) is that if on the bounding surface S of the configuration, on which the pressure p vanishes, the density ρ also vanishes, then, the normal component of grad p on S vanishes. In our later considerations (see ¶ 4 below) we shall assume that these conditions prevail.

a. The equations governing small oscillations. We shall now suppose that an initial equilibrium configuration, constructed in conformity with equation (16), is slightly perturbed and that the ensuing motions are described in terms of a Lagrangian displacement of the form

$$\vec{\xi}(\vec{x})e^{\lambda t} , \tag{18}$$

where λ is a parameter whose characteristic values are to be determined. The corresponding linearized version of equation (11) which governs the perturbed motion is seen to be

$$\begin{aligned} &\rho\left\{1 + \frac{1}{c^2}\left[\Omega^2(x_1^2 + x_2^2) + 4U\right]\right\}(\lambda^2\vec{\xi} + 2\lambda\vec{\Omega}\times\vec{\xi}) \\ &- \frac{1}{c^2}\rho\lambda(\vec{\Omega}\times\vec{x}\cdot\vec{\xi}) \text{ grad } \left[\Omega^2(x_1^2 + x_2^2) + 4U\right] \\ &- \frac{4}{c^2}\rho\lambda\left[\lambda\vec{U} + \text{grad } (\vec{\Omega}\times\vec{U}\cdot\vec{x}) + \vec{\Omega}\times\Delta\vec{\mathfrak{B}} - \xi_\beta \text{ grad } (\vec{\Omega}\times\vec{\mathfrak{B}})_\beta\right] \end{aligned} \tag{19}$$

$$+ \frac{1}{c^2} \lambda\vec{\Omega} \times \vec{x} \; [\Delta p + 3\rho\Delta U - \frac{1}{2} \rho\vec{\xi} \cdot \text{grad } \Omega^2 \; (x_1^2 + x_2^2)$$

$$+ \rho\vec{\Omega} \times \vec{x} \cdot (\lambda\vec{\xi} + 2\vec{\Omega} \times \vec{\xi})]$$

$$+ \frac{1}{c^2} \rho\lambda \text{ grad } (\frac{1}{2} \lambda\delta\chi - \vec{\Omega} \times \vec{x} \cdot \text{grad } \delta\chi) \qquad (19)\ \text{(cont'd.}$$

$$= -[1 - \frac{1}{c^2} (\Pi + \frac{p}{\rho})] \text{ grad } \Delta p + \frac{1}{c^2} \Delta(\pi + \frac{p}{\rho}) \text{ grad } p$$

$$+ \Delta\rho \text{ grad } \Psi + \rho \text{ grad } \Delta\Psi,$$

where δQ and ΔQ denote, respectively, the Eulerian and the Lagrangian changes in a quantity Q resulting from the deformation caused by the displacement $\vec{\xi}$. The two changes are related by

$$\Delta Q = \delta Q + \vec{\xi} \cdot \text{grad } Q. \qquad (20)$$

Also, in equation (19), $\vec{U}$ is now defined in terms of $\vec{\xi}$, i.e.

$$\nabla^2\vec{U} = -4\pi G\rho\vec{\xi}. \qquad (21)$$

It should be noted that the first variation of equation (11) gives, in the first instance, for the right-hand side of equation (19) an expression in which the Lagrangian changes which occur are replaced by the corresponding Eulerian changes. But it can be readily verified that by virtue of the relation (20) that obtains between the two changes and of the form of equation (16) governing equilibrium, this replacement of the Eulerian by the Lagrangian changes is justified.

b. The case of axisymmetric oscillations. In this paper we shall restrict ourselves to the case when the initial equilibrium configuration is axisymmetric about the axis of rotation and further that this axisymmetry is not violated by the perturbation. In cylindrical polar coordinates, $\varpi = \sqrt{(x_1^2 + x_2^2)}$, $z = x_3$, and ϕ, this assumption of continued axisymmetry implies that the components ξ_ϖ, ξ_z, and ξ_ϕ of $\vec{\xi}$ are all independent of the azimuthal angle ϕ; and that, therefore, all the other changes, Eulerian and Lagrangian, are similarly independent of ϕ.

Since

$$\vec{\Omega} \times \vec{x} \cdot \text{grad} = \Omega \frac{\partial}{\partial \phi} \,, \tag{22}$$

it follows that on the present assumption of axisymmetry the expression (14) for Ψ_1 becomes

$$\Psi_1 = \frac{1}{4}\Omega^4\varpi^4 + 2U\Omega^2\varpi^2 + 2\Phi - 4\Omega^2 \; (x_1\mathfrak{D}_1 + x_2\mathfrak{D}_2) . \tag{23}$$

Under the assumed conditions of axisymmetry we can also write

$$\mathfrak{D}_1 = x_1\mathfrak{D}(\varpi, z) \text{ and } \mathfrak{D}_2 = x_2\mathfrak{D}(\varpi, z) , \tag{24}$$

where $\mathfrak{D}(\varpi,z)$ is governed by the equation

$$\frac{\partial^2\mathfrak{D}}{\partial\varpi^2} + \frac{3}{\varpi}\frac{\partial\mathfrak{D}}{\partial\varpi} + \frac{\partial^2\mathfrak{D}}{\partial z^2} = -4\pi G\rho . \tag{25}$$

It is convenient to write

$$\mathfrak{D}_\varpi = \varpi\mathfrak{D} = \vec{\mathfrak{D}} \cdot \vec{1}_\varpi , \tag{26}$$

so that

$$\vec{\Omega} \times \vec{\mathfrak{D}} = \Omega D_\varpi \, \vec{1}_\phi , \tag{27}$$

where $\vec{1}_\varpi$ and $\vec{1}_\phi$ are unit vectors in the principal directions of increasing ϖ and ϕ, respectively.

Since we are also assuming that the axisymmetry of the configuration is preserved during the perturbed motion, we may write, similarly,

$$\begin{gathered} \delta\mathfrak{D}_1 = x_1\delta\mathfrak{D}(\varpi, z) , \quad \delta\mathfrak{D}_2 = x_2\delta\mathfrak{D}(\varpi, z) , \\ \delta\mathfrak{D}_\varpi = \varpi\delta\mathfrak{D} , \quad \text{and} \quad \vec{\Omega} \times \delta\mathfrak{D} = \Omega\delta\mathfrak{D}_\varpi\vec{1}_\phi , \end{gathered} \tag{28}$$

where $\delta\mathfrak{D}$ is determined in terms of $\delta\rho$ by the equation

$$\left(\frac{\partial^2}{\partial\varpi^2} + \frac{3}{\varpi}\frac{\partial}{\partial\varpi} + \frac{\partial^2}{\partial z^2}\right) \delta\mathfrak{D} = -4\pi G\delta\rho . \tag{29}$$

With the aid of the foregoing relations pertaining to $\vec{\mathfrak{D}}$ and $\delta\vec{\mathfrak{D}}$, we can readily verify that

$$-\xi_\beta \text{ grad } (\vec{\Omega}\times\vec{\mathfrak{D}})_\beta = -\Omega\xi_\phi \text{ grad } \mathfrak{D}_\varpi + \frac{\Omega}{\varpi}\xi_\varpi\mathfrak{D}_\varpi \vec{1}_\phi, \tag{30}$$

and

$$\begin{aligned}\vec{\Omega} \times \Delta\vec{\mathfrak{D}} &= \Omega(\delta\mathfrak{D}_\varpi + \vec{\xi} \cdot \text{grad } \mathfrak{D}_\varpi) \vec{1}_\phi - \frac{\Omega}{\varpi}\xi_\phi\mathfrak{D}_\varpi\vec{1}_\varpi \\ &= \Omega\Delta\mathfrak{D}_\varpi \vec{1}_\phi - \frac{\Omega}{\varpi}\xi_\phi\mathfrak{D}_\varpi\vec{1}_\varpi .\end{aligned} \tag{31}$$

Finally, we may also note that by making use of equation (16) governing equilibrium (in zero order) it can be directly verified that

$$\Delta p + 3\rho\Delta U - \frac{1}{2}\rho\vec{\xi} \cdot \text{grad } \Omega^2\varpi^2 = \delta p + 3\rho\delta U + 4\rho\vec{\xi} \cdot \text{grad } U. \tag{32}$$

Now inserting the relations (30)-(32) in equation (19), we obtain

$$\begin{aligned}&\rho[1 + \frac{1}{c^2}(\Omega^2\varpi^2 + 4U)] (\lambda^2\vec{\xi} + 2\lambda\vec{\Omega} \times \vec{\xi}) + \frac{1}{2c^2}\rho\lambda^2 \text{ grad } \delta\chi \\ &+ \frac{1}{c^2}\lambda\varpi\Omega \vec{1}_\phi [\delta p + 3\rho\delta U + 4\rho\vec{\xi} \cdot \text{grad } U + \rho\tilde{\varpi}\Omega(\lambda\xi_\phi + 2\Omega\xi_\varpi)] \\ &- \frac{1}{c^2}\lambda\rho\tilde{\varpi}\Omega\xi_\phi \text{ grad } (\Omega^2\varpi^2 + 4U) \\ &- \frac{4}{c^2}\rho\lambda [\lambda\vec{U} - \Omega \text{ grad } (\varpi U_\phi) + \Omega\Delta D_\varpi \vec{1}_\phi - \frac{\Omega}{\varpi}\xi_\phi D_\varpi\vec{1}_\varpi \\ &\qquad - \Omega\xi_\phi \text{ grad } D_\varpi + \frac{\Omega}{\varpi}\xi_\varpi D_\varpi\vec{1}_\phi] = \vec{L}\vec{\xi},\end{aligned} \tag{33}$$

where

$$\begin{aligned}\vec{L}\vec{\xi} = &- [1 - \frac{1}{c^2}(\Pi + \frac{p}{\rho})] \text{ grad } \Delta p + \frac{1}{c^2}\Delta(\Pi + \frac{p}{\rho}) \text{ grad } p \\ &+ \Delta\rho \text{ grad } \Psi + \rho \text{ grad } \Delta\Psi.\end{aligned} \tag{34}$$

Returning to the equation of continuity (15), we now observe that the linearized version of this equation gives

$$\Delta\rho = -\rho \operatorname{div} \vec{\xi} - \frac{1}{c^2}\rho \left[-\frac{1}{2}\vec{\xi} \cdot \operatorname{grad} \Omega^2\varpi^2 + 3\Delta U + \varpi\Omega \left(\lambda\xi_\phi + 2\Omega\xi_\varpi\right)\right]. \tag{35}$$

It is manifest from equation (34) that by virtue of the assumed axisymmetry, the ϕ-component of $\vec{L}\vec{\xi}$ vanishes. Therefore by considering the ϕ-component of equation (33), we obtain

$$\begin{aligned} &\rho \left[1 + \frac{1}{c^2}\left(\Omega^2\varpi^2 + 4U\right)\right] \left(\lambda^2\xi_\phi + 2\lambda\Omega\xi_\varpi\right) \\ &+ \frac{1}{c^2}\lambda\varpi\Omega \left[\delta p + 3\rho\delta U + 4\rho\vec{\xi} \cdot \operatorname{grad} U + \rho\varpi\Omega \left(\lambda\xi_\phi + 2\Omega\xi_\varpi\right)\right] \\ &- \frac{4}{c^2}\lambda\rho \left(\lambda U_\phi + \Omega\Delta\mathfrak{D}_\varpi + \frac{\Omega}{\varpi}\xi_\varpi\mathfrak{D}_\varpi\right) = 0. \end{aligned} \tag{36}$$

To zero order in $1/c^2$, equation (36) gives (cf. Chandrasekhar and Lebovitz 1968, eq. [16])

$$\lambda\xi_\phi = -2\Omega\xi_\varpi + O(1/c^2). \tag{37}$$

From equation (21) we may now infer that

$$\lambda U_\phi = -2\Omega U_\varpi. \tag{38}$$

Note that since $\vec{U}$ occurs only in terms that are explicitly post-Newtonian, the inference of relation (38) from equation (37) is justified.

To order $1/c^2$, equation (36) now gives

$$\begin{aligned} \lambda\xi_\phi = &-2\Omega\xi_\varpi + \frac{4\Omega}{c^2}\left(\Delta\mathfrak{D}_\varpi + \frac{\xi_\varpi}{\varpi}\mathfrak{D}_\varpi - 2U_\varpi\right) \\ &- \frac{1}{c^2}\varpi\Omega \left(\frac{1}{\rho}\delta p + 3\delta U + 4\vec{\xi} \cdot \operatorname{grad} U\right). \end{aligned} \tag{39}$$

Making use of equation (39) to eliminate ξ_ϕ, we find that the $\tilde{\omega}$- and the z-components of equation (33) give

$$\rho\left[1 + \frac{1}{c^2}(\Omega^2\varpi^2 + 4U)\right](\lambda^2 + 4\Omega^2)\xi_{\varpi} + \frac{\lambda^2}{c^2}\rho\left(\frac{1}{2}\frac{\partial\delta\chi}{\partial\varpi} - 4U_{\varpi}\right)$$
$$+ \frac{4\Omega^4}{c^2}\rho\tilde{\omega}^2\xi_{\varpi} + \frac{2\Omega^2}{c^2}\varpi(\delta p + 3\rho\delta U)$$
$$- \frac{8\Omega^2}{c^2}\rho\left[\varpi\frac{\partial U_{\varpi}}{\partial\varpi} - U_{\varpi} + \delta\mathfrak{B}_{\varpi} + 2\xi_{\varpi}\left(\frac{\partial\mathfrak{B}_{\varpi}}{\partial\varpi} + \frac{\mathfrak{B}_{\varpi}}{\varpi} - \varpi\frac{\partial U}{\partial\varpi}\right)\right. \tag{40}$$
$$\left. + \xi_z\left(\frac{\partial\mathfrak{B}_{\varpi}}{\partial z} - \varpi\frac{\partial U}{\partial z}\right)\right] = (\vec{L}\vec{\xi})_{\varpi},$$

and

$$\rho\left[1 + \frac{1}{c^2}(\Omega^2\varpi^2 + 4U)\right]\lambda^2\xi_z + \frac{\lambda^2}{c^2}\rho\left(\frac{1}{2}\frac{\partial\delta\chi}{\partial z} - 4U_z\right)$$
$$- \frac{8\Omega^2}{c^2}\rho\left[\varpi\frac{\partial U_{\varpi}}{\partial z} + \xi_{\varpi}\left(\frac{\partial\mathfrak{B}_{\omega}}{\partial z} - \varpi\frac{\partial U}{\partial z}\right)\right] = (\vec{L}\vec{\xi})_z. \tag{41}$$

Returning to equation (35), we may, in view of the zero-order relation (37), now write

$$\Delta\rho = -\rho\left(\text{div}\,\vec{\xi} + \frac{1}{c^2}\Delta Q\right) \tag{42}$$

where

$$\Delta Q = 3\,\Delta U - \frac{1}{2}\vec{\xi}\cdot\text{grad}\,\Omega^2\varpi^2 = \Delta\left(3U - \frac{1}{2}\Omega^2\varpi^2\right). \tag{43}$$

(Note that the Eulerian change in $\Omega^2\varpi^2$ is zero.) Since

$$\text{dw}\vec{\xi} = \frac{\partial\xi_{\tilde{\omega}}}{\partial\varpi} + \frac{\partial\xi_z}{\partial z}, \tag{44}$$

equation (42) expresses the Lagrangian change in ρ in terms only of ξ_{ϖ} and ξ_z. The corresponding changes in p and $(\Pi + p/\rho)$ follow from the isentropic conditions

$$\frac{\Delta p}{p} = \gamma\frac{\Delta\rho}{\rho} \quad \text{and} \quad \Delta\left(\Pi + \frac{p}{\rho}\right) = \frac{\Delta p}{\rho}, \tag{45}$$

where γ denotes the appropriate adiabatic exponent. Thus,

$$\Delta p = -\gamma p \left(\operatorname{div} \vec{\xi} + \frac{1}{c^2} \Delta Q\right)$$

and (46)

$$\Delta\left(\Pi + \frac{p}{\rho}\right) = -\frac{\gamma p}{\rho} \left(\operatorname{div} \vec{\xi} + \frac{1}{c^2} \Delta Q\right).$$

Again from equations (5), (6), and (37) we deduce

$$\Delta v^2 = -\vec{\xi} \cdot \operatorname{grad} \Omega^2 \varpi^2 = -\Delta(\Omega^2 \varpi^2). \tag{47}$$

From equations (46) and (47) it now follows that (cf. eq. [4])

$$\begin{aligned} \Delta\phi &= \Delta v^2 + \Delta U + \frac{1}{2} \Delta(\Pi + p/\rho) + \Delta(p/\rho) \\ &= \Delta(U - \Omega^2\varpi^2) - \frac{1}{2} \frac{p}{\rho} (3\gamma - 2) \operatorname{div} \vec{\xi} + O(1/c^2). \end{aligned} \tag{48}$$

The corresponding Eulerian change in Φ is given by (cf. Chandrasekhar 1965b, eq. [28])

$$\begin{aligned} \delta\Phi = G &\int_V \rho(\vec{x}')\phi(\vec{x}')\xi_\alpha(\vec{x}') \frac{\partial}{\partial x'_\alpha} \frac{1}{|\vec{x}-\vec{x}'|} d\vec{x}' \\ &+ G \int_V \frac{\rho(\vec{x}')\Delta\phi(\vec{x}')}{|\vec{x}-\vec{x}'|} d\vec{x}'. \end{aligned} \tag{49}$$

The Eulerian change in U to $O(1/c^2)$ is given by

$$\delta U = G \int_V \rho(\vec{x}')\xi_\alpha(\vec{x}') \frac{\partial}{\partial x'_\alpha} \frac{1}{|\vec{x} - \vec{x}'|} d\vec{x}' - \frac{1}{c^2} G \int_V \frac{\rho(\vec{x}')\Delta Q(\vec{x}')}{|\vec{x}-\vec{x}'|} d\vec{x}'; \tag{50}$$

and the required Eulerian change in χ is given by

$$\delta\chi = -G \int_V \rho(\vec{x}')\xi_\alpha(\vec{x}') \frac{\partial}{\partial x'_\alpha} |\vec{x}-\vec{x}'| \, d\vec{x}'. \tag{51}$$

And finally, from equation (23) we deduce

$$\delta\Psi = \delta U + \frac{1}{c^2} (2\Omega^2\varpi^2\delta U + 2\delta\Phi - 4\Omega^2\varpi\delta\mathfrak{D}_\varpi), \tag{52}$$

where $\delta\mathcal{D}_{\varpi}$ is defined in equations (28) and (29). (It is sufficient to know $\delta\chi$ and $\delta\Psi$ to zero order in $1/c^2$.)

We have now expressed the changes in the various quantities which occur in equations (34), (40), and (41) in terms of ξ_{ϖ} and ξ_z. Therefore, this pair of equations together with the appropriate boundary conditions on ξ_{ϖ} and ξ_z define the required characteristic value problem for λ^2.

4. The variational expression for λ^2

It can be verified that the characteristic value problem for λ^2 specified by equations (34), (40), and (41) is a self-adjoint one and that the result of multiplying equations (40) and (41) by ξ_{ϖ} and ξ_z, respectively, summing, and integrating over the volume V occupied by the fluid is to obtain an expression for λ^2 which provides a variational base for determining it. After a lengthy reduction the expression for λ^2 one obtains is

$$
\begin{aligned}
\lambda^2 \Bigg\{ & \int_V \rho\left[1 + \frac{1}{c^2}\left(\Omega^2\varpi^2 + 6U + \Pi + \frac{p}{\rho}\right)\right] (\xi_{\varpi}^2 + \xi_z^2)\, d\vec{x} \\
& - \frac{1}{2c^2}\int_V \delta\chi \,\mathrm{div}\,(\rho\vec{\xi})\, d\vec{x} - \frac{4}{c^2}\int_V \rho\, U_z \xi_z\, d\vec{x} \\
& - \frac{1}{\pi G c^2}\int_V \left[\frac{U_{\varpi}^2}{\varpi^2} + (\mathrm{grad}\, U_{\varpi})^2\right] d\vec{x} \Bigg\} \qquad (53) \\
= - & \int_V \left(1 + \frac{2}{c^2}U\right)\gamma p\, (\mathrm{div}\vec{\xi})^2\, d\vec{x} \\
& - 2\int_V \rho\left\{ 1 + \frac{1}{c^2}\left[2U + \Pi + (\gamma+1)\frac{p}{\rho}\right] \right\} (\vec{\xi}\cdot \mathrm{grad}\,\Psi)\Big\}\,\mathrm{div}\,\vec{\xi}\, d\vec{x} \\
& - \int_V \left\{ \frac{d\rho}{d\Psi} + \frac{1}{c^2}\left[\left(2U + \Pi + \frac{p}{\rho}\right)\frac{d\rho}{d\Psi_0} + \rho\,\frac{d(\Pi + p/\rho)}{d\Psi_0} + \rho\right] (\vec{\xi}\cdot\mathrm{grad}\Psi)^2 d\vec{x} \right. \\
& + G\int_V\int_V \frac{\mathrm{div}[\rho(\vec{x})\vec{\xi}(\vec{x})]\;\mathrm{div}\,[\rho(\vec{x}')\vec{\xi}(\vec{x}')]}{|\vec{x}-\vec{x}'|}\, d\vec{x}d\vec{x}' \\
& + \frac{2}{c^2}\int_V \rho\left(2U + \Pi + \frac{p}{\rho}\right)\vec{\xi}\cdot \mathrm{grad}\,\delta U\, d\vec{x} + \frac{4}{c^2}\int_V \delta p\, \vec{\xi}\cdot \mathrm{grad}\, U\, d\vec{x} \\
& - \frac{1}{c^2}\int_V \rho(\delta U)^2 d\vec{x} - \frac{6}{c^2}\int_V \delta U(\gamma p\,\mathrm{div}\,\vec{\xi} + \vec{\xi}\cdot \mathrm{grad}\, p)\, d\vec{x}
\end{aligned}
$$

(Equation [53] continues on following page)

(Equation [53] continued from preceding page)

$$- 4\Omega^2 \int_V \rho[1 + \frac{1}{c^2}(2\Omega^2\varpi^2 + 6U + \Pi + \frac{p}{\rho})]\, \xi_\varpi^2\, d\vec{x}$$

$$+ \frac{4\Omega^2}{c^2}\int_V \rho\vec{\xi} \cdot \mathrm{grad}\,(\varpi^2\delta U)\, d\vec{x} + \frac{4\Omega^2}{c^2}\int_V \varpi\delta\mathfrak{B}_\varpi\, \mathrm{div}\,(\rho\vec{\xi})\, d\vec{x}$$

$$- \frac{4\Omega^2}{c^2}\int_V (\delta p + 3\rho\delta U)\, \varpi\xi_\varpi\, d\vec{x} - \frac{4\Omega^2}{\pi G c^2}\int_V [\frac{U_\omega^2}{\varpi^2} + (\mathrm{grad}\, U_\varpi)^2]\, d\vec{x}$$

$$+ \frac{16\Omega^2}{c^2}\int_V \rho(\frac{\partial D_\varpi}{\partial\varpi} + \frac{\mathfrak{B}_\varpi}{\varpi} - \tilde{\varpi}\frac{\partial U}{\partial\varpi})\, \xi_\varpi^2\, d\vec{x}$$

$$+ \frac{16\Omega^2}{c^2}\int_V \rho(\frac{\partial\mathfrak{B}_\varpi}{\partial z} - \varpi\frac{\partial U}{\partial z})\, \xi_\varpi\xi_z\, d\vec{x}$$

$$+ \frac{8\Omega^2}{c^2}\int_V \left\{ \xi_\varpi[\frac{\partial}{\partial\varpi}(\varpi U_\varpi) + \delta\mathfrak{B}_\varpi] + \xi_z\frac{\partial}{\partial z}(\varpi U_\varpi)\right\} d\vec{x}.$$

5. Concluding remarks

For configurations with even moderate central condensations, the surfaces of constant Ψ can be well approximated by the same surfaces appropriate for the Roche model (cf. Chandrasekhar 1967*b*). On such an approximation, the variational expression (53) can be used to obtain estimates for λ^2 in the manner described by Chandrasekhar and Lebovitz (1968, ¶ VIII) in the Newtonian context. We postpone such applications of the present theory to a future occasion.

The research reported in this paper has in part been supported by the Office of Naval Research under Contract Nonr-2121(24) with the University of Chicago.

References

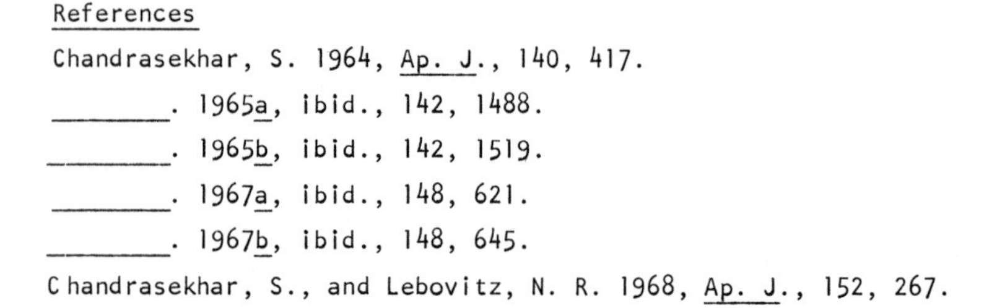

Chandrasekhar, S. 1964, *Ap. J.*, 140, 417.
________. 1965*a*, ibid., 142, 1488.
________. 1965*b*, ibid., 142, 1519.
________. 1967*a*, ibid., 148, 621.
________. 1967*b*, ibid., 148, 645.
Chandrasekhar, S., and Lebovitz, N. R. 1968, *Ap. J.*, 152, 267.

References (continued)

Durney, B. R., and Roxburgh, I. W. 1967, *Proc. Roy. Soc. London, A*, 296, 189.

Fowler, W. 1964, *Rev. Mod. Phys.*, 36, 545.

________. 1966, *Ap. J.*, 144, 180.

Ledoux, P. 1945, *Ap. J.*, 102, 143.

Received 3/19/69.

General Methods for Integrating the Relativistic Equation of Spin Motion, and Their Application to Some Cases of Experimental Interest*

by

M. Fierz

Eidg. Technische Hochschule, Zurich

and

V. L. Telegdi

The University of Chicago

1. Introduction

Various methods are at present under study to measure the "anomaly"

$$a = (g/2 - 1)$$

of the muon magnetic moment. All these methods rest on the gradual transformation (from longitudinal to transverse) which the polarization of the muon undergoes as the latter moves on a periodic (or quasi-periodic) orbit in a magnetic field B in the laboratory. These orbits are obtained by integrating the Lorentz equation†

$$\dot{\underline{u}} = \underline{\underline{F}} \cdot \underline{u} \qquad (\text{i.e. } \dot{\vec{u}} = \vec{u} \times \vec{B}) \tag{1}$$

for a specific $\vec{B}$. The techniques for integrating (1) analytically or numerically constitute a highly developed discipline, and we may assume in what follows that $\underline{u}(\tau)$ can be obtained for any arrangement of experimental interest.

---

*This paper is an extended version of work done by the authors while at CERN (1960) and circulated as SC Internal Report 60-9.

†See the Appendix for all questions of notation.

To obtain the spin motion, one must solve

$$\dot{\underline{s}} = (1 + a)\ \underline{\underline{F}} \cdot \underline{s} + a(\underline{s} \cdot \underline{\underline{F}} \cdot \underline{u})\ \underline{u} \tag{2}$$

with $\underline{u}(\tau)$ and $\underline{\underline{F}}(\tau)$ given. Let us recall that in Eq. (2) $s = (s^\circ, \vec{s})$ is a unit (space-like) 4-vector of polarization so defined that it satisfies

$$\underline{s} \cdot \underline{u} = 0. \tag{3}$$

This equation simply means that in any instantaneous rest-frame (R) of the particle, where $\underline{u} = (1,0)$, $\underline{s}$ reduces to the ordinary polarization (or spin, or magnetic moment) vector $\vec{s}$:

$$\underline{s}(R) = (0, \vec{s}). \tag{3'}$$

In thinking about relativistic spin problems, it is always helpful to consider the spin motion in a succession of such frames (R), because it is only then that one may interpret $\vec{s}$ in an intuitive way. A possible way of solving such problems is to compute from the laboratory fields $\underline{\underline{F}}(L)$ the magnetic field $\vec{B}(R)$ by Lorentz transformation. This way, however, has some pitfalls due to the peculiarities of relativistic kinematics (Thomas precession, etc.). These peculiarities are altogether absorbed into the formalism if one uses only covariant concepts throughout, and for this reason the direct solution of Eq. (2) appears generally preferable in all but the simplest cases. In deriving such solutions, one is in practice not so much interested in computing the <u>absolute</u> motion of $\underline{s}$ (with respect to some frame fixed in the laboratory) as in the relative motion of $\underline{s}$ with respect to $\underline{u}$ and $\dot{\underline{u}}$ (and other kinematic variables). One thus will seek to solve for the components of $\underline{s}$ in some co-moving coordinate frame adapted to the problem at hand.

In what follows, we shall discuss some general methods of solving Eq. (2) with the help of such frames, and apply them to derive the spin motion in some of the proposed experimental arrangements. It will turn out that all the useful frames are automatically rest-frames R (although with particular orientation of their space axes!), so that the final results are easily amenable to intuitive interpretation.

2. Michel's tetrad[1]

One notices that for a = 0, Eq. (2) reduces to

$$\dot{\underline{s}} = \underline{\underline{F}} \cdot \underline{s} \tag{4}$$

i.e. for a "normal" particle (having g = 2) $\underline{s}$ follows the same equation as $\underline{u}$. Let us eliminate the leading term $\underline{\underline{F}} \cdot \underline{s}$ in Eq. (2) by choosing a frame in which $\underline{s}$ would stay fixed if one had a = 0. Clearly such a frame is provided by a tetrad of unit 4-vectors $e^{(i)}$ (i = 0...4) which all move in accordance with Eq. (2), i.e. for which

$$\dot{e}^{(i)} = \underline{\underline{F}} \cdot \underline{e}^{(i)}. \tag{5}$$

The components $\xi^{(i)}$ of $\underline{s}$ in such a basis are defined by

$$\underline{s} \equiv \xi^{(i)}\, \underline{e}^{(i)}. \tag{6}$$

We choose $\underline{e}^{(o)} = \underline{u}$, satisfying Eq. (5) automatically and implying $\xi^o = 0$. The remaining space-like components $\{\xi^{(\alpha)}\}$ (α = 1,2,3) form a vector $\vec{\xi}$, which obeys

$$\dot{\vec{\xi}} = 0 \tag{4'}$$

for a = 0. To obtain $\dot{\vec{\xi}}$ for the general case, we substitute Eq. (6) into Eq. (2):

$$\dot{\xi}^{(i)}\underline{e}^{(i)} + \xi^{(i)}\underline{\dot{e}}^{(i)} = \xi^{(j)}(1 + a)\ \underline{\underline{F}} \cdot \underline{e}^{(j)} + a(s \cdot F \cdot u)\ \underline{e}^{o}$$

$$\dot{\xi}^{(i)}\underline{e}^{(i)} = a\xi^{(j)}\underline{\underline{F}}\ \underline{e}^{(j)} + a(s \cdot F \cdot u)\ e^{o} \tag{2'}$$

and multiplying with $\underline{e}^{(\alpha)}$ $(\alpha = 1,2,3)$:

$$\dot{\xi}^{(\alpha)} = -a\left(\underline{e}^{(\alpha)} \cdot \underline{\underline{F}} \cdot \underline{e}^{(\beta)}\right)\ \xi^{(\beta)} \tag{2''}$$

(put $j = \beta$, because $\xi^{o} = 0$). It is convenient to introduce the notation

$$\underline{e}^{(\alpha)} \cdot \underline{\underline{F}} \cdot \underline{e}^{(\beta)} = \tilde{F}^{\alpha\beta} = -\tilde{B}^{\gamma}. \tag{7}$$

The $\tilde{F}^{\alpha\beta}$ represent the components of the magnetic field $\tilde{B}$ seen in the tetrad; because the latter is a succession of R-frames, the $\tilde{E}$ do not contribute to the precession. With this notation Eq. (2'') becomes

$$\dot{\vec{\xi}} = -a\vec{\xi} \times \tilde{B} = \dot{\vec{\Omega}} \times \vec{\xi} \tag{8}$$

where $\vec{\Omega} = a\tilde{B}$ represents the rotation of $\vec{\xi}$ with respect to the tetrad. In terms of the laboratory variables $t, \vec{v}, \vec{B}$ and $\vec{E}$ one writes

$$d\vec{\xi}/dt = \gamma^{-1}\ \vec{\Omega} \times \vec{\xi}$$

$$\gamma^{-1}\ \vec{\Omega} = a\left(\vec{B} = \frac{\gamma}{\gamma+1}\ (\vec{B}\cdot\vec{v})\vec{v} - \vec{v} \times \vec{E}\right). \tag{9}$$

In the <u>absence</u> of electric fields (in the laboratory) one may always put

$$\underline{e}^{(o)} = \underline{u}, \text{ and } \underline{e}^{(1)} = \gamma(v,\hat{v}) \equiv (u,\gamma\hat{u})$$

because this choice automatically satisfies the requirements of

Eq. (5). This is, in general, not so for the choice[2]

$$e^{(2)} = (0,\hat{n}) \equiv (0,\dot{\hat{u}}) \equiv (0,\dot{\vec{u}}/\dot{\vec{u}}),$$

$$e^{(3)} = (0,\hat{b}) \equiv (0,\hat{u} \times \hat{n}) = (0,\hat{u} \times \dot{\hat{u}}). \tag{10}$$

For instance, in the presence of an external longitudinal field $\vec{B}_{11} = \hat{u}(\vec{B} \cdot \hat{u})$ and for a flat orbit perpendicular to a field $\vec{B}_\perp$, $\hat{n}$ and $\hat{t}$ do not rotate about $\vec{B}_{11}$ as $\vec{e}^{(2)}$ and $\vec{e}^{(3)}$ should to satisfy Eq. (5). We make the unique ansatz

$$\begin{aligned} e^{(2)} &= (0,\ \hat{n}\cos\phi - \hat{b}\sin\phi), \\ e^{(3)} &= (0,\ \hat{n}\sin\phi + \hat{b}\cos\phi), \\ \phi &= \phi(\tau), \end{aligned} \tag{11}$$

and determine $\phi(\tau)$ from Eq. (5). This time-dependence of ϕ will supply part of the motion of $\vec{e}^{(2)}$ and $\vec{e}^{(3)}$ which is not automatically provided by $\hat{n}(\tau)$ and $\hat{b}(\tau)$ through the equation of motion of $\hat{u}$, e.g. the action of a $\vec{B}_{11}$. From Eq. (2), with $i = 2$, we get in the case that $\vec{E} = 0$ in the laboratory

$$\dot{\phi} = \hat{n} \cdot (\hat{b} \times \vec{B} - \dot{\hat{b}}). \tag{12}$$

One has

$$\hat{n} \cdot \hat{b} \times \vec{B} = B_{11}$$

$$\dot{\hat{n}} = \ddot{\vec{u}}/\dot{u} - \hat{n}(\ddot{\vec{u}} \cdot \hat{n})/\dot{u}$$

$$\dot{\hat{b}} = \hat{u} \times \dot{\hat{n}}$$

$$\begin{aligned} -\hat{n} \cdot \dot{\hat{b}} &= -\hat{n} \cdot (\hat{u} \times \ddot{\vec{u}})/\dot{u} = -\frac{\dot{\vec{u}} \cdot (\vec{u} \times \ddot{\vec{u}})}{u\,\dot{u}\,\dot{u}} \equiv S \\ &= -\hat{n} \cdot \hat{u} \times [\dot{\vec{u}} \times \vec{B} + \vec{u} \times \dot{\vec{B}}]/\dot{u} \\ &= -B_{11} + (\hat{n} \cdot \dot{\vec{B}})/B_\perp \quad (B_\perp = \dot{u}/u) \end{aligned} \tag{12a}$$

$$\dot{\phi} = B_{11} + S, \text{ or} \tag{12b}$$

$$\dot{\phi} = \hat{n} \cdot \dot{\vec{B}}/B_{\perp}. \tag{13}$$

We see from Eqs. (12a) and (12b) that $\dot{\phi} = 0$ only if the orbit has no helicity ($S = 0$) and if $B_{11} = 0$. Note that $\hat{n} \cdot \dot{\vec{B}}$ is the component of $d\vec{B}/d\tau = (\vec{u} \cdot \vec{\nabla})\vec{B}$ along $\hat{n}$, and arises when $\vec{B}$ is not time-dependent in the laboratory, but changes in absolute orientation along the orbit.

We now use Eq. (7) to calculate the components of the magnetic field B acting in the R-frame defined by the tetrad:

$$\begin{aligned} B_1 &= -\tilde{F}^{23} = -\underline{e}^{(2)} \cdot \underline{\underline{F}} \cdot \underline{e}^{(3)} = \hat{n} \cdot \hat{b} \times \vec{B} = B_{11} \\ B_2 &= -\tilde{F}^{31} = -\underline{e}^{(3)} \cdot \underline{\underline{F}} \cdot \underline{\dot{e}}^{(1)} = \gamma\, B_{\perp} \sin\phi \\ B_3 &= +\tilde{F}^{21} = \underline{e}^{(2)} \cdot F \cdot \underline{\dot{e}}^{(1)} = -\gamma\, B_{\perp} \cos\phi. \end{aligned} \tag{14}$$

As is obvious from Eq. (5), we can represent the components of $\tilde{B}$ by two fixed vectors: $\tilde{B}_{11} = B_{11}$, along $\hat{u}$; and $\tilde{B}_{\perp} = \gamma\, B_{\perp}$, along $-\hat{b}$. The accompanying figure illustrates this convenient representation.

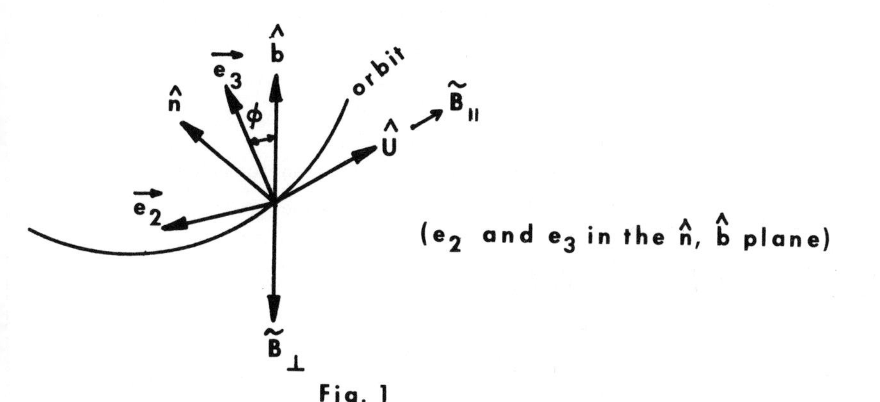

Fig. 1

To express the polarization precession with respect to the physically convenient frame $\hat{u}$, $\hat{n}$, $\hat{b}$, we make the following remark: $\vec{e}_2$ and $\vec{e}_3$ rotate rigidly about $\hat{u}$ at an angular velocity $\omega_{\hat{u}} = \dot{\phi}$, while $\vec{\xi}$ rotates with respect to $\vec{e}_1$ and $\vec{e}_2$ with an angular velocity $\Omega_1 = a\tilde{B}_{11}$. $\vec{\xi}$ rotates about $\hat{b}$ (in the $\hat{u}$, $\hat{n}$ plane) with a velocity $\Omega_3 = \Omega_{\hat{b}} = a\tilde{B}_\perp$. The precession vector $\vec{\Omega}$ is hence completely specified by

$$\boxed{\begin{aligned} \Omega_{\hat{u}} &= \dot{\phi} + a\tilde{B}_{11} = \hat{n} \cdot \dot{\vec{B}}/B_- + aB_{11} \\ \Omega_{\hat{b}} &= a\tilde{B}_\perp = aB_\perp\gamma, \ \dot{\phi} = S + B_{11} \end{aligned}} \qquad (15)$$

in terms of the laboratory fields (but in units of proper time). Clearly $\Omega_{\hat{u}}$ is responsible for any erection of the polarization out of the $\hat{u},\hat{n}$ plane.

$B_\perp$, B_{11} and S in Eq. (15) are in general functions of the proper time τ, i.e. they vary along the orbit. The latter will be in most cases either periodic or quasiperiodic. The relevant frequency is here always given by $B_\perp$; the spin components with respect to the frame $(\hat{u},\hat{n},\hat{b})$ will change slowly with τ, as a is a small quantity (for muons and electrons), when both S and B_{11} are small compared to $B_\perp$. In that case it will be legitimate to replace $\Omega_{\hat{u}}$ and $\Omega_{\hat{b}}$ by their time averages $\langle\Omega_{\hat{u}}\rangle$, $\langle\Omega_{\hat{b}}\rangle$ over an orbit period, and to consider these averages as slowly varying.

Examples ($\vec{E} = 0$ always)

A. Circular orbit in plane ⊥ to $\vec{B}$ const.

This represents, of course (neglecting focusing fields or gradients) the idea behind the basic (g-2) experiments.

$$\begin{aligned} &B_{11} = 0; \ \dot{\vec{B}} = 0; \ \dot{\phi} = 0 \rightarrow \Omega_{\hat{u}} = 0, \ \xi_3 = \text{const.} \\ &\Omega_{\hat{b}} = aB\gamma \end{aligned} \qquad \text{(Ex. 1}$$

Note that the particle frequency $\omega_p = B$ in our units, so that

$$\Omega_{\hat{b}} = a\gamma\omega_p,$$

i.e. the Mendowitz-Case[3] result.

B. Same as A, but with a current-carrying wire at orbit center

The idea here (due to M. Schwartz and J. Steinberger) is to trap initially longitudinally polarized muons in orbits around a wire. The (g-2) precession produces a transverse polarization, and the tangential magnetic field induces an erection of the polarization out of the orbit plane. An experiment of this kind was performed at CERN.[4]

$$B_{11} = b = \frac{1}{2\pi\rho} = \text{wire field; } S = 0;\ \dot{\phi} = h \text{ [see Eq. 12b]}$$

$$B_{\perp} = B$$

$$\Omega_{\hat{u}} = (1 = a)b = (1 + a)\left(\frac{h}{B}\right)\omega_p \qquad \text{(Ex. 2)}$$

$$\Omega_{\hat{b}} = aB\gamma = \gamma a\ \omega_p$$

C. Same as B, Ex. 2, but Orbit not concentric with wire

$\vec{B} = \vec{B}_1 + \vec{h};\ B_1 = (0,0,B_1);\ \vec{b} = \vec{J} \times \vec{r}/(2\pi\rho^2);\ \vec{J} = (0,0,J)$

$\vec{r}$ = radius vector, ρ = distance in (x,y) plane from the wire. According to the discussion in the last paragraph preceding the examples, we have to determine the $\langle\Omega_i\rangle$.

It follows from the equation of motion that

$$u_z - A_z = u_z + (J/2\pi)\ln(\rho/\rho_o) = u_o \qquad \text{(Ex. 3)}$$

is a constant. We shall now assume that $(u/u_o) \ll 1$ so that we may neglect terms quadratic on this quantity. In addition, the constant of integration ρ_o is so defined that

$$\langle \ln(\rho/\rho_o)\rangle = 0. \qquad \text{(Ex. 4)}$$

In taking the means, we may use the orbits for J = 0. If we insert into $S = (\vec{u}\ \dot{\vec{u}}\ \ddot{\vec{u}})/u\ \dot{u}^2$

$$u_x = -u_{\perp} \sin \omega t,\ u_y = u_{\perp} \cos \omega t,\ u_z = u_o - (J/2\pi)/\ln(\rho/\rho_o)$$

we find that in taking the mean the terms linear in J vanish because of Eq. (Ex. 4) and of the periodicity of the orbit.

The mean value of B_{11} is compensated by <S> [compare Eq. (ex. 3)], so that

$$\langle\dot{\phi}\rangle = \langle b_{11}\rangle = (1/T)\int_0^T d\tau\ (\vec{h}\cdot\vec{u})/u = \left(\frac{1}{2\pi R}\right)\int_0^{2\pi R}(\vec{h}\cdot\vec{d\ell}). \qquad \text{(Ex. 5)}$$

Here T = period of the orbit, R its radius. One sees from Eq. (Ex. 5) immediately that

$$\langle\dot{\phi}\rangle = J/2\pi R \equiv h. \qquad \text{(Ex. 6)}$$

Thus to this approximation the effect of the wire field is the same as in Example B.

D. Helical motion in a uniform field $\vec{B}$

Pitch angle $\beta = \sin^{-1}(u_z/u)$; $B_{11} = B\sin\beta$, $B_\perp = B\cos\beta$

$v_\perp = v\cos\beta$

$\dot{\vec{B}} = 0$, $\dot{\phi} = 0$ (here $S = -B_{11}$);

$$\left.\begin{aligned}\Omega_{\hat{u}} &= a\sin\beta\cdot B\\ \Omega_{\hat{b}} &= a\,\gamma\cos\beta\cdot B\end{aligned}\right\}\ |\Omega| = aB\gamma(v_\perp) = a\gamma(v_\perp)\omega_p \qquad \text{(Ex. 7}$$

Notice that in this case there does develop a component of polarization along b, and that in Eq. (Ex. 7) the argument of is $v_\perp$ not v as in Example A.[5] Both of these effects have to be taken into account in any proposed experiment.

E. Motion in the Hine-Lederman spiral

As is well known particle beams can be trapped in a toroidal region, like in storage rings. One can show that the trapping property is preserved if one "wraps the ring around a cylinder", i.e. into a cylindrical helix. The particles then "walk up a staircase". Initially longitudinally polarized muons suffer a polarization change which depends on the anomaly a.

$$\begin{aligned}
&\text{Pitch angle } \beta = \sin^{-1}(u_z/u); \ \omega_p = B/\cos\beta \\
&B = B_\perp, \text{ but } S \neq 0; \ \hat{n} \cdot \vec{\dot{B}}/B_\perp = B_\phi \ \omega_p/B = B \operatorname{tg}\beta \\
&\Omega_{\hat{u}} = B \operatorname{tg}\beta \\
&\Omega_{\hat{b}} = a\, B\, \gamma .
\end{aligned} \tag{Ex. 8}$$

The component $\Omega_{\hat{u}}$, not dependent on the anomaly, is objectionable. It can be eliminated by providing a $B_{11} = -S = -B \operatorname{tg}\beta$, e.g. by mounting a solenoidal winding all along the helical cavity of the screw magnet.

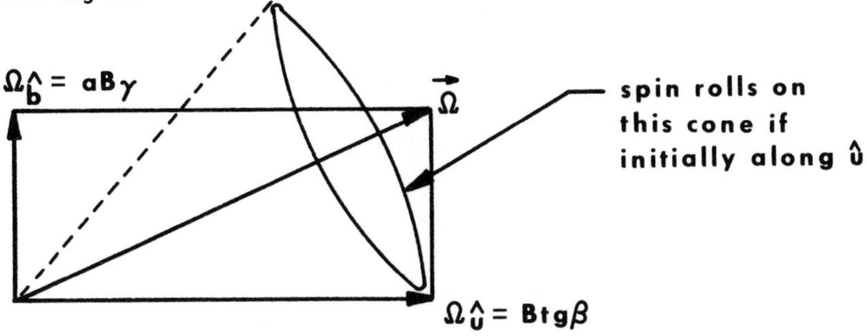

Fig. 2

3. Direct solution in terms of kinematic variables

Though the introduction of a basis $\{\underline{e}^{(i)}\}$, where each vector moves according to Eq. (5), is very elegant, it is simpler and more straightforward to use a basis $\{\underline{\varepsilon}^{(i)}\}$ which is linked to the kinematics of the orbit.

We put

$$\begin{aligned}
\underline{\varepsilon}^{(1)} &= (u, \gamma\, \hat{u}) \\
\underline{\varepsilon}^{(2)} &= (0,\ \hat{n}) \\
\underline{\varepsilon}^{(3)} &= (0,\ \hat{t})
\end{aligned} \tag{16}$$

that is, we start with the basis of Eq. (10).

Now writing

$$\underline{s} = \eta^{(i)} \ \varepsilon^{(i)} \tag{17}$$

we insert this into the equation of motion (4) and find

$$\begin{aligned} \dot{\eta}_1 &= -a\gamma\ B_\perp\ \eta_2 \\ \dot{\eta}_2 &= +a\gamma\ B_\perp\ \eta_1 + \{(1 + a)B_{11} + S\}\ \eta_3 \\ \dot{\eta}_3 &= \qquad\qquad - \{(1 + a)B_{11} + S\}\ \eta_2 \end{aligned} \tag{18}$$

where S is the helicity of the orbit:

$$S = \frac{\vec{u} \cdot (\dot{\vec{u}} \times \ddot{\vec{u}})}{u \quad \dot{u}^2} \equiv (\hat{u}\ \hat{n}\ \dot{\hat{n}}). \tag{19}$$

The effect due to $B_{11} + S$ is just transformed away by using the rotating system [Eq. (11)], because

$$\dot{\phi} = B_{11} + S$$

[compare Eq. (12)].

The geometrical meaning of the helicity S can be seen in the following way: the hodograph of our particle is a curve on the sphere $|\vec{v}|$ = constant; this sphere we may use as unit-sphere, i.e. we use $\hat{u}$ instead of $\vec{v}$.

At each point of the hodograph we may draw on the sphere a circle osculating the hodograph, that is a circle that is tangent to it and has the same curvature:

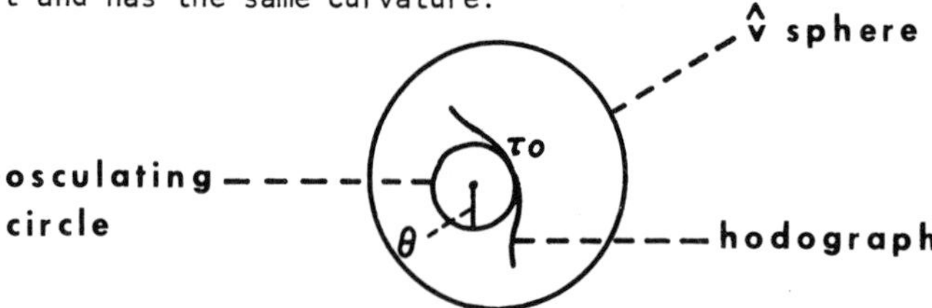

The angular aperture of the circle we call θ.

If the hodograph were always to coincide with the circle, and the velocity-point were to move around it with constant angular velocity ω, then the orbit would be a screw with a pitch angle $\beta = 90° - \theta$. For such a helical orbit

$$\hat{u} \times \hat{n}$$

is a unit vector, tangent to the sphere and normal to the circle.

$\dot{\hat{n}}$ is a vector normal to the circle and lying in its plane. Its absolute value is ω. From this follows immediately

$$(\dot{\hat{n}}\ \hat{v}\ \hat{n}) = -\omega \cos\theta = -\omega \sin\beta \tag{20}$$

(the minus sign comes because $\hat{u}$, $\hat{n}$, $\dot{\hat{n}}$ form a left-system).

In τ_o the hodograph and the circle coincide. This means: to each point of the orbit belongs an osculating screw, whose helicity [Eq. (19)] gives $S(\tau)$.

Appendix on notation

1. Vector and Tensor Algebra

We denote by $\vec{v}$ a 3-vector, of components v^α ($\alpha = 1,2,3$); $v = v$;

by $\underline{v}$ a 4-vector, of components v^i ($i = 0,1,2,3$); $v = (v^o, \vec{v})$;

by $\underline{\underline{T}}$ a second rank skew tensor, of components T^{ik} ($i,k = 0,1,2,3$); $\underline{\underline{T}} = (T^{o\alpha}, T^{\alpha\beta})$; e.g. the e.m. field tensor $\underline{\underline{F}} = -(\vec{E},\vec{B})$.

We adopt a metric of signature (+ - - -). The dot ($\cdot$) between quantities represents the contraction over neighboring indices with this metric: $\underline{\underline{T}} \cdot \underline{v} = \Sigma_k T^{ik} v^k = T^{io} v^o - \Sigma_\alpha T^{i\alpha} v^\alpha = \underline{w} = \{w^i\}$. The symbol ^ indicates a 3-dimensional unit vector: $\hat{v} = \vec{v}/v$.

2. Symbols

The ordinary velocity is indicated as $\vec{v}$, the 4-velocity as $\underline{u}$. One has

$$\underline{u} = (u^o, \vec{u})$$
$$= \gamma(1,\vec{v}), \quad \gamma = (1 = v^2)^{-1/2}, \ c = 1.$$

Clearly, $\hat{u} \equiv \hat{v}$. $\underline{u} \cdot \underline{u} = +1$.
The dot above a symbol represents the derivative with respect to the proper time $\tau = t/\gamma$, e.g.

$$\dot{\underline{u}} = \frac{d}{d\tau}\underline{u} = \gamma \frac{d}{dt}\underline{u}.$$

All frequencies in this note are computed in units of (inverse) proper time. This is particularly convenient because no conversion to real time is needed so long as only comparison of frequencies are made.

$\parallel$ means parallel, and $\perp$ perpendicular, always with respect to the velocity $\vec{v}$ in the laboratory frame. Thus $\vec{B}_{11} = (\vec{B} \cdot \hat{v})\hat{v}$, $\vec{B}_{\perp} = \vec{B} - \vec{B}_{11}$.

($\dot{}$) above a symbol means that the quantity refers to the instantaneous rest frame R[$\underline{u}$(R) = (1,0)] of a particle of 4-velocity $\underline{u}$.

3. Units

We put (e/m) = 1 and c = 1 throughout the text. With this notation, the Larmor frequency of a particle in a magnetic field B is $\omega_L = B/\gamma$. Using the proper time scale (see above) this frequency, now called particle frequency, is $\omega_p = B$.

References

1. Bargmann, Michel and Telegdi, Phys. Rev. Letters 2, 435 (1959).
2. The reader will recognize $(\hat{u},\hat{n},\hat{b})$ as the Frenet triad; for a discussion of this, see e.g. H.S.M. Coxeter, Introduction to Geometry, p. 322, J. Wiley, New York (1961).
3. H. Mendowitz and K. M. Case, Phys. Rev. 97, 33 (1955).
4. G. Charpak, F.J.M. Farley, R.L. Garwin, T. Muller, J.D. Sens, V.L. Telegdi, C.M. York and A. Zichichi, Proc. of the 1960 Rochester Conference, p. 776.
5. This fact, well known to physicists at CERN since 1960, has recently been rediscovered by G.R. Henry and J.E. Silver. It provides a correction of 7 ppm to the anomaly of the electron as measured at Michigan. (A. Rich, private communication).

Received 5/29/69

Theory of the Eigenvalues of the S-Matrix*

by

M. L. Goldberger

and

C. E. Jones†

Palmer Physical Laboratory, Princeton University

The general analytic properties of the eigenvalues of the S-matrix describing n coupled two-body channels is discussed. The important role of singularities of the eigenvalues resulting from the coincidence of eigenvalues for complex energies is treated in detail. It is shown that considerable insight into the underlying dynamics, in particular, the importance of high threshold channels on lower ones can be obtained from an eigenphase analysis of data. A kind of Levinson theorem is derived.

1. Introduction

One of the reasons for the complexity of the physics of strongly interacting systems is that the coupling between many channels must be considered in any realistic discussion. Although it is sometimes possible to ignore in some sense channels other than the one of interest, it is important to understand the possible effects on the problem of other channels. The most economical description of a system of coupled channels is in terms of the eigenvalues of the S-matrix. Unitarity is most simply expressed in terms of such eigenvalues; resonance poles whose characteristics frequently dominate the behavior of scattering amplitudes are believed to be generally associated with a definite eigenamplitude. The latter circumstance leads to the important property of factorization of residues in a many-channel Breit-Wigner resonance formula.

---

*Work supported by the U.S. Air Force Office of Scientific Research, under contract Number AF 49(638)-1545.

†Present address Department of Physics, Mass. Inst. of Technology.

In addition to their utility in abstract theoretical discussions, eigenamplitudes may play an important role in quantitative dynamical calculations. For example in connection with the concept of compositeness of all strongly interacting particles, use is frequently made of some form of Levinson's theorem which relates phase shifts at threshold and at infinite energy. This theorem can be stated in simple form for eigenphase shifts and a study of the properties of such quantities may give insight into the idea of compositeness.

There has been very little discussion in the literature of S-matrix eigenamplitudes.[1,2] It is the purpose of this paper to elucidate some of their properties and to show how their behavior may lead to a qualitative understanding of the underlying dynamics.

We treat here the problem of many coupled two-body channels with non-degenerate thresholds. The assumption of non-degenerate thresholds is a non-trivial complication. Presumably, eigenamplitudes will play an important role in any treatment of three or more body channels. Fortunately, nature has been kind in providing us with many situations where a quasi-two body description is not unreasonable.

Our starting point is the S-matrix for a given angular momentum with the elements labeled by the various channels. We assume that the undiagonalized amplitudes have simple threshold and interaction cuts on the real axis (although the location of the latter will play little role in the subsequent discussion), as well as bound state and resonance poles. It will be useful to consider analytic variations in the strength of the interaction and we shall therefore assume that the amplitudes are analytic in the "coupling" or the angular momentum (in some region) as needed. Our discussion then--assuming only analytic properties--includes potential theory and is presumably also broad enough to include interesting relativistic problems.

In Section 2 we examine the analytic structure of the eigenamplitudes. In addition to the branch points present in the undiagonalized amplitudes, one also finds singularities which we call crossing points that occur when two or more eigenamplitudes coincide. The crossing-point singularities are simple in nature and eigenamplitudes interchange their roles when analytically continued around these points. Certain general characteristics of crossing points are discussed, including their location.

Section 3 contains a proof of the important statement that no crossing-point singularities can be present on the real energy axis except possibly at thresholds. Crossing-point singularities at thresholds correspond to the coincidence of only two eigenamplitudes.

The general structure of the eigenphases is described in Section 4. This is most easily discussed in terms of a gradual variation of the interaction strength from zero. The manner in which the crossing-point singularities move around the thresholds under this coupling variation is crucial in determining a kind of Levinson's theorem for the eigenphase shifts. The failure, in general, of two eigenphase shifts to coincide at a given (real) energy is due to a repulsion phenomenon similar to the splitting of degenerate energy levels in molecular physics; this repulsion gives a virtually unique general structure to the eigenphase shifts in the presence of a number of bound states and resonances. The configuration of the eigenphase shifts is among other things an indicator of the role of higher threshold channels in producing a low-lying resonance.

A summary of results and a discussion of possible applications is given in Section 5.

2. Analytic structure of eigenamplitudes

We discuss the problem of scattering in a state of given angular momentum, ℓ, with n spinless two-body channels. Physical scattering processes which take place in the energy interval between the $j^{\underline{th}}$ and $(j + 1)^{\underline{st}}$ thresholds $(j \leq n)$ is described by a j x j S-matrix, denoted by S. Unless explicitly stated otherwise, all subsequent matrices are j x j corresponding to the j channels open at the energy under consideration. The j x j scattering amplitude matrix T is defined in terms of S by

$$S = I + 2i\, \rho^{1/2}\, T\, \rho^{1/2} \tag{2.1}$$

where I is a j x j unit matrix and ρ is a diagonal j x j phase space matrix with elements ρ_{km} given by

$$\rho_{km} = \nu_k^{\ell+1/2}\delta_{km}. \qquad (2.2$$

In Eq. (2.2), ν_k is the square of the center-of-mass momentum in the kth channel and δ is the ordinary Kronecker delta. Physical unitarity at energies in the jth interval leads to

$$\frac{T-T^{\dagger}}{2i} = T\rho T^{\dagger}. \qquad (2.3$$

This relation may also be written as

$$(T^{\dagger})^{-1} - (T)^{-1} = 2i\rho. \qquad (2.4$$

We recall that with appropriate choice of phases S may be assumed symmetric and therefore T is also.

We assume the usual analytic structure for T including a right-hand cut starting at the lowest threshold and a left-hand cut. Each element of T when analytically continued below the lowest threshold $\nu_1 = 0$ is real and we can write T^{-1} in the following form:

$$T^{-1} = Y - i\rho, \qquad (2.5$$

where each element of Y is real below the $(j+1)^{st}$ threshold down to the start of the left-hand cut. The elements of Y also have right hand branch points at the thresholds ν_k, $k > j$.

We proceed to examine the analytic properties of the eigenvalues of S. It is somewhat more convenient to work with the eigenvalues of the matrix $\rho^{-1/2}\,T^{-1}\,\rho^{-1/2} + iI$ denoted by t, which from Eq. (2.5) is given by

$$t = \rho^{-1/2}\,Y\,\rho^{-1/2}. \qquad (2.6$$

The eigenvalues λ of this matrix are related to the eigenvalues Λ of S by

$$\Lambda = 1 + 2i\ (\lambda - i)^{-1}. \tag{2.7}$$

The eigenvalues λ are determined by the characteristic equation

$$F(\lambda,\nu) = \det\ (t - \lambda I) = 0, \tag{2.8}$$

where we imagine that the angular momentum and coupling strength are held fixed. The variable ν stands for any of the ν_k, only one of which, by energy conservation, may be considered independent. The function $F(\lambda,\nu)$ is a polynomial in λ of the $j^{\underline{th}}$ degree whose coefficients have, in general, the same analyticity as the elements of the T-matrix namely the usual right and left-hand cuts. The implicit function theorem then states the λ's as functions of energy have the same analytic structure as the elements of T but may have, in addition, branch points occurring where two or more eigenvalues coincide. At these new singularities, which we call crossing points, the derivative $d\lambda(\nu)/d\nu$ fails to exist. The derivative is given by

$$\frac{d\lambda(\nu)}{d\nu} = - \left.\frac{\partial F(\lambda,\nu)}{\partial\nu}\right|_{\lambda=\lambda(\nu)} \left[\frac{\partial F(\lambda,\nu)}{\partial\lambda} \right]^{-1}_{\lambda=\lambda(\nu)} \tag{2.9}$$

where $\lambda(\nu)$ is determined by the equation $F(\lambda(\nu),\nu) = 0$. The coincidence of two or more roots means a multiple root of the characteristic equation, a zero of $\partial F/\partial\lambda|\lambda=\lambda(\nu)$ and thus the derivative of λ may fail to exist. We emphasize that a coincidence of eigenvalues is a necessary but not a sufficient condition for a singularity. In fact in the sections that follow we shall find that there are regions where there are no crossing-point singularities even though eigenvalues do coincide.

We turn now to a consideration of the nature of the crossing-point singularities and their location in the complex energy plane. Fortunately, since $F(\lambda,\nu)$ is a polynomial in λ whose coefficients are analytic functions of ν, we can invoke the theory of algebraic functions[3] in order to describe the behavior of the eigenvalues $\lambda(\nu)$ near a crossing point singularity. Suppose that q eigenvalues coincide at the crossing point $\nu = \nu_c$. In the neighborhood of ν_c we may write

$$\lambda(\nu) = \lambda_c + \sum_{k=0}^{\infty} \alpha_k (\nu - \nu_c)^{\frac{k+p}{q}}, \tag{2.10}$$

where λ_c is the q-fold degenerate eigenvalue, the α_k are constants and p is a positive integer ≥ 1. Evidently the function $\lambda(\nu)$ has a q-sheeted Riemann surface with respect to the crossing-point singularity at $\nu = \nu_c$. The q degenerate eigenvalues at ν_c are specified away from $\nu = \nu_c$ by the q determinations of

$$(\nu - \nu_c)^{\frac{k+p}{q}}$$

in Eq. (2.10). Thus as the point $\nu = \nu_c$ is circled, the q eigenvalues undergo a cyclic permutation.

Before going further, we illustrate the general discussion by consideration of a two-channel problem. The characteristic equation $F(\lambda,\nu) = 0$ is quadratic in this case and the eigenvalues λ are given by

$$\lambda = \frac{\frac{Y_{11}}{\rho_1} + \frac{Y_{22}}{\rho_2} \pm \sqrt{\Delta}}{2}$$

$$\Delta = \left(\frac{Y_{11}}{\rho_1} - \frac{Y_{22}}{\rho_2}\right)^2 + \frac{4Y_{12}^2}{\rho_1\rho_2}, \tag{2.11}$$

where we have used the fact that Y is a symmetric matrix. Crossing-point singularities can occur when the discriminant $\Delta = 0$. Clearly the singularities are of square root type and the eigenvalues interchange as we continue around the point where $\Delta = 0$.

Still within the framework of the two-channel case, let us see what can be said about the location of the crossing points. We note that Δ is a real analytic function with a left-hand cut and a finite right-hand cut along the real axis from $\nu_1 = 0$ to $\nu_2 = 0$. From this real analyticity it follows that the zeroes of Δ occur symmetrically with respect to the real axis. Because of the absence in Δ of the second threshold cut the number of crossing points in the eigenamplitudes is twice the number of zeroes of Δ which occur on the first and second sheets with respect to the second threshold cut. (See Fig. 1.)

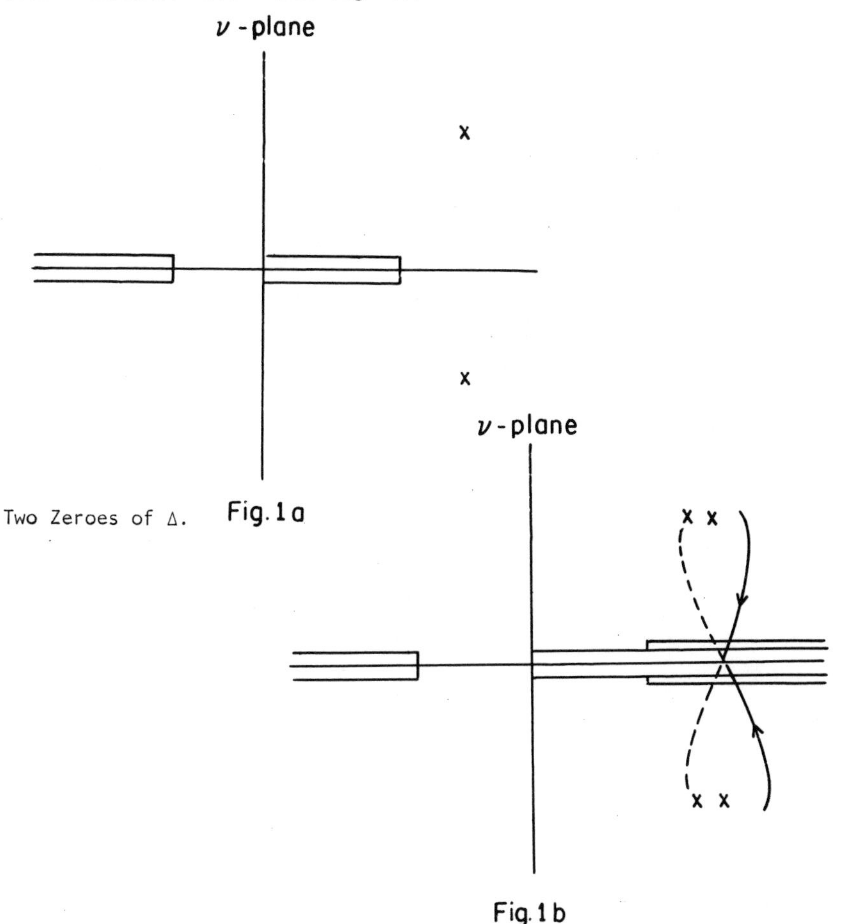

Two Zeroes of Δ. Fig. 1a

Fig. 1b

Four Crossing Points in Eigenamplitudes.

The possible presence of crossing points on the real axis is treated in Section 3.

To proceed with the discussion of the j channel case we use the generalized discriminant Δ_j defined in terms of the j eigenvalues λ_j by

$$\begin{aligned}\Delta_j = (\lambda_1 - \lambda_2)^2 (\lambda_1 - \lambda_3)^2 (\lambda_1 - \lambda_4)^2 \ldots (\lambda_1 - \lambda_j)^2 \\ \times (\lambda_2 - \lambda_3)^2 (\lambda_2 - \lambda_4)^2 \ldots (\lambda_2 - \lambda_j)^2 \\ \times (\lambda_{j-1} - \lambda_j)^2.\end{aligned} \tag{2.12}$$

The discriminant may also be expressed in terms of the coefficients of the characteristic equation as follows:[4]

$$\Delta_j = \frac{(-1)^{j(j-1)/2}}{a_0} D(F,F'), \tag{2.13}$$

where D(F,F') is called Sylvester's determinant and is defined by

$$D(F,F') = \begin{array}{l} \left.\begin{array}{l} \\ \\ \\ (j-1) \text{ rows} \\ \\ \\ \end{array}\right\{ \\ \\ \left.\begin{array}{l} \\ \\ j \text{ rows} \\ \\ \\ \end{array}\right\{ \end{array} \left| \begin{array}{llllll} a_0 & a_1 & \ldots & a_{j-2} \; a_{j-1} \; a_j & |\leftarrow (j-2) \text{ zeroes} \longrightarrow & \\ & a_0 & \ldots & & a_j & \\ & & a_0 & \ldots & & a_j \\ ja_0 & (j-1)a_1 & \ldots & a_{j-1} & |\leftarrow (j-1) \text{ zeroes} \longrightarrow & \\ & ja_0 & & & a_{j-1} & \\ & & & ja_0 & & a_{j-1} \end{array} \right| \tag{2.14}$$

and

$$F(\lambda,\nu) = a_0\lambda^j + a_1\lambda^{j-1} + \ldots + a_j. \tag{2.15}$$

It is well-known from the theory of equations that the coefficients a_k in the characteristic equation, Eq. (2.15) can be expressed as sums of the principal minors of the original matrix $t = \rho^{-1/2} \gamma \rho^{-1/2}$. For example, a_k is made up of sums of products of k elements of t (see Eq. (2.17)). Evidently then Δ_j is real between the $j^{\underline{th}}$ and $(j + 1)^{\underline{st}}$ thresholds where each element of D(F,F') is real. Below the lowest threshold, the elements of t are all pure imaginary (because $\rho \to i|\rho|$). This means that the elements of Sylvester's determinant are alternately purely real or purely imaginary. We write D schematically below the lowest threshold as

$$D(F,F') = \begin{vmatrix} 1 & i & 1 & i & 1 & \cdots & \\ & 1 & i & 1 & i & 1 & \cdots \\ & & & \cdots & & & \\ 1 & i & 1 & i & & \cdots & \\ & 1 & i & 1 & i & & \cdots \end{vmatrix} \quad \begin{matrix} \left.\vphantom{\begin{matrix}1\\1\end{matrix}}\right\} (j-1) \text{ rows} \\ \\ \left.\vphantom{\begin{matrix}1\\1\end{matrix}}\right\} j \text{ rows} \end{matrix} \qquad (2.16a$$

To see that D is real, multiply the first, third, etc., columns by i and divide by the corresponding factors outside to obtain

$$D(F,F') = \frac{1}{i^{(j-1)}} \begin{vmatrix} i & i & i & i & \cdots \\ & 1 \; -1 & 1 \; -1 & \cdots \\ & & \cdots & & \\ i & i & i & i & \cdots \\ & 1 \; -1 & 1 \; -1 & \cdots \end{vmatrix} \qquad (2.16b$$

which is clearly real.

Thus Δ_j has the singularity structure shown in Fig. 2 and again as in the two-channel problem, the zeroes of Δ_j are located symmetrically with respect to the real energy axis. Furthermore there are twice as many crossing-point singularities in the eigen-amplitudes because the latter contain the branch point $\nu_j = 0$ absent in Δ_j.

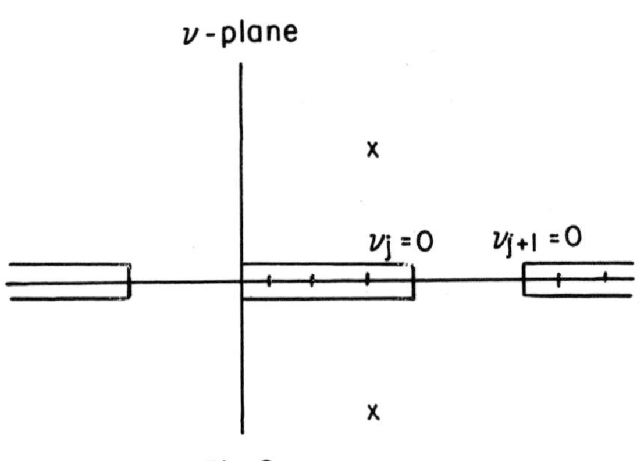

Fig. 2a

Cut Structure of Δ_j with Two Symmetric Zeroes Indicated.

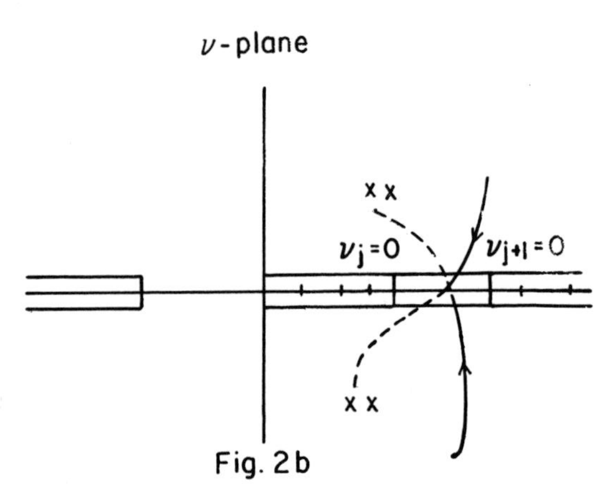

Fig. 2b

Four Crossing-Point Singularities in Eigenamplitudes.

In the singularity structure of the eigenamplitudes shown in Fig. 2b we have used the fact that the eigenvalues λ are all pure imaginary below the lowest threshold down to the start of the left-hand cut. This follows from the fact that the matrix t is (-i) times a real symmetric matrix (which thus has real eigenvalues) in this region.

We turn now to a discussion of the threshold properties of the eigenvalues λ_k. We shall show that if there are no crossing-point singularities at $\nu_j = 0$, the λ_k are analytic functions of $(\nu_j)^{1/2}$ in this neighborhood, with the exception of one eigenvalue which is proportional to ρ_j^{-1}. This latter eigenvalue we refer to as the threshold eigenvalue λ_1 since the corresponding characteristic value Λ_1 of the S-matrix is unity at $\nu_j = 0$ and the eigenphase is a multiple of π.

To study the eigenvalues in the neighborhood of $\nu_j = 0$, we study in detail the characteristic equation, $F(\lambda,\nu) = 0$ in this region. We have

$$F(\lambda,\nu) = |t-\lambda I| = (-1)^j[\lambda^j - \lambda^{j-1} \sum (PM)_1 + \lambda^{j-2} \sum (PM)_2 + \ldots + (-1)^j (PM)_j]. \tag{2.17}$$

The coefficients $\sum (PM)_k$ denotes the sum of all determinants formed from square arrays within the matrix t of order k whose principal diagonals lie along the principle diagonal of t. These determinants are called principal minors (PM) of t. In terms of the matrix t, Eq. (2.6) we have

$$\sum (PM)_k = \sum_{\substack{i<\ell<m<\ldots<r \\ k \text{ indices}}} \begin{vmatrix} \dfrac{Y_{ii}}{\rho_i} & \dfrac{Y_{i\ell}}{(\rho_i\rho_\ell)^{1/2}} & \cdots & \dfrac{Y_{ir}}{(\rho_i\rho_r)^{1/2}} \\ \dfrac{Y_{i\ell}}{(\rho_i\rho_\ell)^{1/2}} & \dfrac{Y_{\ell\ell}}{\rho_\ell} & & \\ \dfrac{Y_{ir}}{(\rho_i\rho_r)^{1/2}} & & \cdots & \dfrac{Y_{rr}}{\rho_r} \end{vmatrix} \tag{2.18}$$

In order to isolate the threshold eigenvalue, which we expect to be proportional to ρ_j^{-1}, we multiply the characteristic equation, (2.17), by $(\rho_j)^j$ and define $\bar{\lambda} = \rho_j \lambda$. The equation for $\bar{\lambda}$ is then

$$(\rho_j)^j \, |t - \lambda I| = (-1)^j [\bar{\lambda}^j - b_1 \bar{\lambda}^{j-1} + b_2 \rho_j \bar{\lambda}^{j-2} + \ldots + (-1)^j b_j \rho_j^{\,j-1}] = 0 \qquad (2.19)$$

where

$$b_k = \rho_j \sum (PM)_k . \qquad (2.20)$$

If $\rho_j = 0$, we find

$$(-1)^j [\bar{\lambda}^j - b_i(0)\, \bar{\lambda}^{j-1}] = 0 \qquad (2.21)$$

with $b_1(0) = Y_{jj}(0)$. Thus if $Y_{jj}(0) \neq 0$, there are j-1 zero roots and one non-zero root, namely $\bar{\lambda}_1 = Y_{jj}(0)$ corresponding to the threshold eigenvalue λ_1 proportional to ρ_j^{-1} as $\rho_j \to 0$.

It is easy to find an exact power series for $\bar{\lambda}_1$ in the neighborhood of the point $\nu_j = 0$ or $\rho_j = 0$. We need simply rewrite the characteristic equation, Eq. (2.19), as

$$\bar{\lambda} = b_1 - b_2 \left(\frac{\rho_j}{\bar{\lambda}}\right) + b_3 \left(\frac{\rho_j}{\bar{\lambda}}\right)^2 + \ldots + (-1)^j \, b_j \left(\frac{\rho_j}{\bar{\lambda}}\right)^{j-1} \qquad (2.22)$$

and observe from the definition of the b_k, Eq. (2.18), that they have the form

$$b_k = b_k(0) + c_k \rho_j + d_k \nu_j + \ldots \qquad (2.23)$$

in the neighborhood of $\nu_j = 0$. We may then solve Eq. (2.22) by iteration, setting first $\nu_j = 0$ to obtain the leading term $\bar{\lambda}_1 = b_1(0)$. In this fashion we generate a power series in the variable $\nu_j^{1/2}$.

We turn now to a consideration of the other eigenvalues, the remaining j-1 solutions of the characteristic equation, (2.19). Since all of the coefficients in Eq. (2.19) are analytic functions of $\nu_j^{1/2}$, with our assumption of no crossing-point singularities at $\nu_j = 0$

it follows from the implicit function theorem that all solutions to Eq. (2.19) must be analytic functions of $(\nu_j)^{1/2}$. On the other hand, if we multiply the characteristic equation $F(\lambda,\nu) = 0$, Eq. (2.17) by ρ_j and let $\rho_j \to 0$, we get an equation for the (j-1) non-infinite λ eigenvalues at $\nu = 0$.

$$(-1)^j[-\lambda(0)^{j-1}\, b_1(0) + \lambda(0)^{j-2}\, b_2(0) + \ldots + (-1)^j b_j(0)] = 0. \tag{2.24}$$

Now the (j-1) $\lambda(0)$'s given by Eq. (2.24) are generally finite and from our previous argument the corresponding $\bar{\lambda}$'s (recall $\bar{\lambda} = \rho_j\lambda$) must be analytic in $(\nu_j)^{1/2}$, consequently we have

$$\lambda_k = \nu_j{}^{\ell} \sum_{p=1}^{\infty} \beta_{pk}(\nu_j)^{\frac{p}{2}},\ k \neq 1, \tag{2.25}$$

which expension is valid in a neighborhood of the point $\nu_j = 0$. We can equally well write

$$\lambda_k = \sum_{p=1}^{\infty} \beta_{pk}(\nu_j)^{\frac{p-1}{2}},\ k \neq 1$$

It is a remarkable fact that although the individual matrix elements of S generally do possess singularities as a function of $(\nu_j)^{1/2}$ (through factors of $\rho_j^{1/2} \sim [\nu_j^{1/2}]^{1/2}$), the eigenvalues λ_k generally do not (aside from the pole in λ_1). The important question of what happens where crossing points occur at threshold is studied in the next Section.

We bring to a close this discussion of the singularity structure of the eigenvalues of S with a few words about how the cuts may be drawn connecting the crossing point singularities. In Fig. 2c we have redrawn Fig. 2b, designating by 0 the crossing point singularity reached by going onto the second sheet (with respect to the branch point at $\nu_j = 0$) from below the cut and by $\otimes$ that reached by passage from above. The two crossing points on the physical sheet are designated by x. The points x, 0 and x, $\otimes$ shown separately are of course actually at the same points (numerically) but on different sheets. The two singularities on the physical sheet may be joined by the cut C_1 or alternatively by the cuts C_2 and C_2' which extend from the physical to the unphysical sheet and the choices may be dictated by convenience. In fact that the cuts may be drawn as indicated follows from the reality properties of the λ's. Using the reality properties one may verify that in circling either set of the cuts drawn in Fig. 3 one arrives back at the same value.

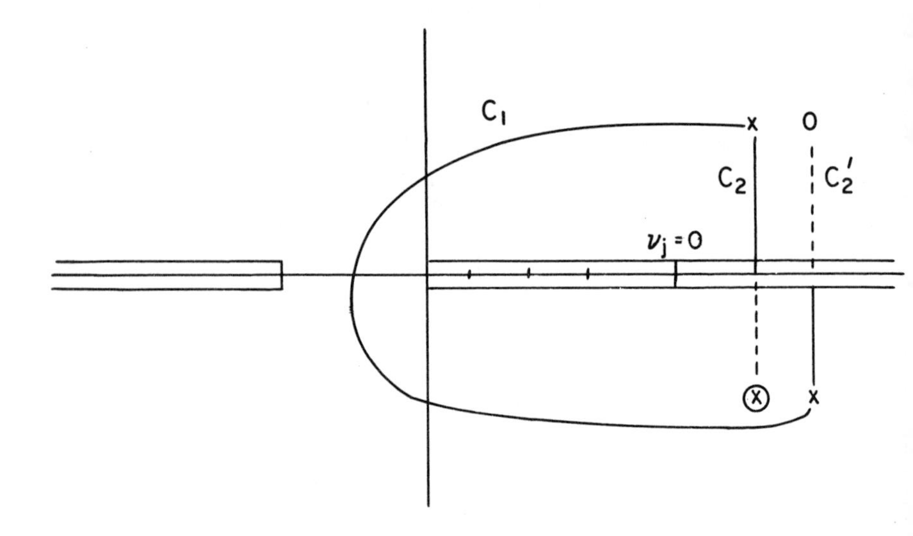

Fig. 2c

Alternate Ways of Connecting Crossing-Point Singularities.

3. Absence of crossing-point singularities on the real axis except at thresholds

If two eigenvalues of the matrix t coincide at a non-threshold value of the energy on the real axis, we have an important illustration of the fact that coincident of eigenvalues is a necessary but not sufficient condition for a crossing point singularity. We shall, in fact, prove that on the real axis, with the exception of thresholds, no crossing-point singularities can occur even if two (or more) eigenvalues coincide.

We recall that t is a j x j matrix with energy dependent elements and for real physical values of the energy between the $j^{\underline{th}}$ and $(j+1)^{\underline{st}}$ thresholds, t is a real symmetric matrix with all eigenvalues real. Suppose that two or more eigenvalues are equal at the point $\nu_j = \nu_c$, located in the energy region just described. We now ask whether such a point can be a singularity of the coincident eigenvalues $\lambda_k(\nu_j)$. If such a crossing point singularity exists, we can draw a cut from $\nu_j = \nu_c$ along the real axis in such a direction to make the λ_k real analytic functions. The discontinuity across this presumed cut is given in the usual way by

$$\lim_{\varepsilon \to 0} [\lambda_k(\nu_j + i\varepsilon) - \lambda_k(\nu_j - i\varepsilon)] = 2i \ \text{Im} \ [\lambda_k(\nu_j + i\varepsilon)]. \qquad (3.1)$$

But since the eigenvalues are necessarily real on the real energy axis, $\text{Im}\lambda_k = 0$, and thus there is no branch point at $\nu_j = \nu_c$.[5] This is not to say that eigenvalues cannot accidentally coincide on the real axis, but when they do, no singularity results. We'll return to this point later in connection with phase shift repulsion.

The question of coincident eigenvalues at $\nu_j = 0$ requires a special discussion. Recall the form of the matrix t:

$$t = \begin{pmatrix} \frac{Y_{11}}{\rho_1} & \frac{Y_{12}}{\sqrt{\rho_1\rho_2}} & \cdots & \frac{Y_{ij}}{\sqrt{\rho_1\rho_j}} \\ \frac{Y_{12}}{\sqrt{\rho_1\rho_2}} & \frac{Y_{22}}{\rho_2} & \cdots & \\ \vdots & & & \\ \frac{Y_{ij}}{\sqrt{\rho_1\rho_j}} & & \cdots & \frac{Y_{jj}}{\rho_j} \end{pmatrix} \qquad (3.2)$$

Clearly the argument given in the preceding paragraph will not work because t is not real symmetric below the point $\nu_j = 0$ due to presence of factors $(\rho_j)^{1/2}$. These threshold singularities must be carefully disentangled from true crossing-point singularities that may occur at threshold.

It is evident from the discussion in Section 2 that in the absence of crossing-point singularities, the $\overline{\lambda}_k (= \rho_j \lambda_k)$ are analytic functions of $(\nu_j)^{1/2}$ and for $k \neq 1$, so are the λ_k. By working in the complex $\xi = (\nu_j)^{1/2}$ plane we may effectively remove the threshold singularities and can study the location of the crossing-point singularities in the ξ - plane, in particular near $\xi = 0$.

First, we show that crossing-point singularities at $\xi = 0$, if any, must correspond to the coincidence of eigenvalues in pairs. It is convenient to use the eigenvalues $\overline{\lambda}$. The matrix whose eigenvalues are $\overline{\lambda}$ is $\rho_j t$:

$$\rho_j t = \begin{pmatrix} Y_{11}\dfrac{\rho_j}{\rho_1} & \dfrac{Y_{12}}{\sqrt{\rho_1 \rho_2}}\,\rho_j & \cdots & Y_{1j}\sqrt{\dfrac{\rho_j}{\rho_1}} \\ \dfrac{Y_{12}}{\sqrt{\rho_1 \rho_2}}\,\rho_j & Y_{22}\dfrac{\rho_j}{\rho_2} & \cdots & \\ \vdots & & & \\ Y_{1j}\sqrt{\dfrac{\rho_j}{\rho_1}} & \cdots & & Y_{jj} \end{pmatrix} . \qquad (3$$

The matrix $\rho_j t$ is obviously real and symmetric as a function $\xi^{1/2}$ near $\xi = 0$. Then employing the same argument used in terms of the variable ν_j for real energies away from threshold, we can now conclude that the eigenvalues $\overline{\lambda}$ must be analytic functions of $\xi^{1/2}$ near $\xi = 0$. Algebraic function theory used in Section 2 thus enables us to conclude that any crossing-point singularities at $\xi = 0$ involve an interchange of two eigenvalues.

We have shown that crossing-point singularities may exist at thresholds and if they do occur, they must have a certain form. We now study an extremely important case in which a crossing-point singularity definitely does exist in the threshold eigenvalue λ_1. An understanding of this situation is important for the discussion of the eigenphases given in the next section.

A singularity does appear in λ_1 when $Y_{jj} = 0$ at $\nu_j = 0$. We note that it follows from Eq. (2.24), since $b_1(0) = Y_{jj}(0)$, that there are in this case generally only (j-2) finite roots λ and thus two others must be infinite. The previously given expansion for $\overline{\lambda}$, in the neighborhood of $\nu_j = 0$, Eq. (2.22), now diverges; we start again to find an exact expansion of the two eigenvalues which become infinite at $\nu_j = 0$ when $Y_{jj} = 0$ at this point.

We begin by writing the characteristic equation for the $\overline{\lambda}$'s, Eq. (2.19), in the form

$$(\overline{\lambda}-\overline{\lambda}_1)(\overline{\lambda}-\overline{\lambda}_2) = \rho_j[b_3(\frac{\rho_j}{\overline{\lambda}}) - b_4(\frac{\rho_j}{\overline{\lambda}})^2 + \ldots - (-1)^j (\frac{\rho_j}{\overline{\lambda}})^{j-2}], \qquad (3.4)$$

where

$$(\overline{\lambda}-\overline{\lambda}_1)(\overline{\lambda}-\overline{\lambda}_2) = \overline{\lambda}^2 - b_1\overline{\lambda} + b_2\rho_j,$$

so that

$$\overline{\lambda}_{1,2} = \frac{b_1 \pm (b_1^2 - 4b_2\rho_j)^{1/2}}{2}. \qquad (3.5)$$

The coefficients b_1 and b_2 near $\nu_j = 0$ have the form (See Eq. (2.23))

$$\begin{aligned} b_1 &= c_1\rho_j + d_1\nu_j + \ldots \\ b_2 &= b_2(0) + c_2\rho_j + d_2\nu_j + \ldots \end{aligned} \qquad (3.6)$$

since $b_1(0) = Y_{jj}(0) = 0$ by hypothesis. Note also in this case that

$$b_2(0) = -\sum_{i<j} Y_{ij}{}^2/\rho_i,$$

a negative definite quantity unless all the Y_{ij}'s are zero at $\nu_j = 0$.

We shall use Eq. (3.4) to construct an exact expansion of the two λ-roots which become infinite at $\nu_j = 0$. The zero$\underline{th}$ order approximation for the corresponding $\bar{\lambda}$'s follows from Eq. (3.5) which yields near $\nu_j = 0$,

$$\bar{\lambda}_{1,2} \sim \text{const. } \nu_j^{1/2(\ell+1/2)}, \quad \ell \leq 1 \tag{3.7a}$$

$$\bar{\lambda}_{1,2} \sim \text{const. } \nu_j \quad , \quad \ell \geq 2. \tag{3.7b}$$

It is easy to see that an iteration scheme based on these starting values substituted into the right-hand side of Eq. (3.4) converges. There is a marked difference, however, between the cases $\ell \leq 1$ and $\ell \geq 2$. In the former case, the expansion involves powers of $\xi^{1/2}$ and hence there are crossing-point singularities at $\xi = 0$. In the $\ell \geq 2$ case, however, only integral powers of ξ occur in the expansion, hence there are no crossing-point singularities right at $\xi = 0$.

There are, of course, generally crossing-point singularities in the latter situation and they may be quite near the point $\xi = 0$ or $\nu_j = 0$ even though the singularity disappears when $Y_{jj} = 0$ at $\nu_j = 0$. We illustrate this point in the two channel problem with with $\ell = 2$. For two channels the $\bar{\lambda}_{1,2}$ given by Eq. (3.5) are exact. We now assume that $b_1(0)$ is small but non-zero. The crossing points are found by setting the discriminant equal to zero; we keep only the first few powers of ν_j.

$$b_1^2 - 4b_2\rho_j = b_1^2(0) + 2d_1 b_1(0)\nu_j + d_1^2\nu_j^2 + \nu_j^{5/2}\,[2c_1 b_1(0) - 4b_2(0)] = 0. \tag{3.8}$$

To lowest order in $b_1(0)$ the location of the crossing-point singularities is

$$\nu_j = -\frac{b_1(0)}{d_1} \pm i\,\frac{[2c_1 b_1(0) - 4b_2(0)]^{1/2}}{d_1}\left(-\frac{b_1(0)}{d_1}\right)^{5/2} \tag{3.9}$$

We take $d_1 < 0$ and $b_1 > 0$ (a normal resonance situation in the one-channel problem). As noted earlier, $b_2(0) < 0$ if $b_1(0) = 0$ and we expect $b_2(0)$ to be negative if $b_1(0)$ is sufficiently small, which we assume. Thus the two crossing-point singularities given by (3.9) are symmetrically placed with respect to the real axis, in accord with our general theory, and are near $\nu_j = 0$ if $b_1(0)$ is small.

4. Properties of eigenphase shifts and "Levinson's Theorem"

We have completed our study of the general analytic structure of the eigenvalues of the S-matrix and the properties of crossing-point singularities. We are now in position to examine the physically interesting subject of the properties of the eigenphase shifts.

A general outline for this section is as follows: (A) We first derive a relation between the products of eigenvalues of the S-matrix in two adjacent threshold intervals. This establishes a basis for the transfer of resonance poles from one threshold region to another as the effective coupling is changed. (B) A general repulsion property for the eigenphase shifts is deduced which shows that eigenvalues will not generally coincide on the real energy axis. (C) It is shown that the sum of two eigenphase shifts that have a crossing-point singularity at threshold is free of the singularity. (D) The essentially unique structure of the eigenphases which results as a consequence of (B) and (C) is discussed along with a kind of Levinson's theorem. (E) It is shown how the gradual change of eigenphases can be understood from the standpoint of analytic continuation.

A. In the energy interval between the $j^{\underline{th}}$ and $(j + 1)^{\underline{st}}$ thresholds we have to do with a $j \times j$ S-matrix denoted by $S^{(j)}$. The matrix obtained by striking out the $j^{\underline{th}}$ row and the $j^{\underline{th}}$ column is called $S^{(j-1)}$, when analytically continued below the threshold $\nu_j = 0$ onto the interval between $\nu_{j-1} = 0$ and $\nu_j = 0$ is the physical S-matrix corresponding to j-1 open channels. We seek the relationship between the eigenvalues $\tilde{\Lambda}_j$ of $S^{(j-1)}$. We recall that $\det S^{(j)} = \Lambda_1\Lambda_2\ldots\Lambda_j$ and that $\det S^{(j-1)} = \tilde{\Lambda}_1\tilde{\Lambda}_2\ldots\tilde{\Lambda}_{j-1}$. To relate these determinants we remark that the jj element of $[S^{(j)}]^{-1}$ is given by

$$[S^{(j)}]^{-1}_{jj} = \frac{\det S^{(j-1)}}{\det S^{(j)}} \,, \qquad (4.1)$$

from which we obtain

$$[S^{(j)}]^{-1}_{jj} \, \Lambda_1 \Lambda_2 \ldots \Lambda_j = \tilde{\Lambda}_1 \tilde{\Lambda}_2 \ldots \tilde{\Lambda}_{j-1}. \qquad (4.2)$$

It is also useful to write the analog of Eq. (4.2) in terms of the eigenvalues of our t-matrix. We obtain

$$(\lambda_1 - i)(\lambda_2 - i)\ldots(\lambda_j - 1) = (\frac{Y_{jj}}{\rho_j} - i)(\tilde{\lambda}_1 - i)(\tilde{\lambda}_2 - i)\ldots(\tilde{\lambda}_{j-1} - i) \qquad (4.3)$$

where the $\tilde{\lambda}$ are the eigenvalues of the t-matrix associated with $S^{(j-1)}$. A resonance pole of the S-matrix is to be associated with a simple zero of one of the factors $\lambda_k - i$; such a zero generally leads to a simple pole in each element of $S^{(j)}$. We see from Eq. (4.3) that corresponding to the vanishing of $\lambda_k - i$ one of the factors $\tilde{\lambda}_\ell - i$ must be zero unless $Y_{jj}/\rho_j - i = 0$ at the point in question. Thus, barring the latter continuency, resonance poles can be naturally passed down from one of the j λ eigenvalues region to one of the (j-1) $\tilde{\lambda}$ eigenvalues as the coupling is increased.[6] The vanishing of $Y_{jj}/\rho_j - i$ can be ruled out as unlikely since it corresponds in a sense to a pole state of the $j^{\underline{th}}$ channel alone.

It is perhaps worth noting at this point, and we return to it later, that a resonance pole in one eigenvalue also appears in any other eigenvalue with which it coincides at a crossing-point singularity. On any given Riemann sheet with respect to the singularity, the pole occurs in only one eigenvalue; there are no higher order poles in the S-matrix resulting from the crossing points.

B. In section 3, we concluded that no crossing-point singularities can occur on the real energy axis for physical values of the energy. Coincidence of two or more eigenvalues for such energies does not lead to a singularity. This result implies a splitting of coincident eigenvalues when the coupling strength or angular momentum is varied which is reminiscent of the repulsion of degenerate eigenvalues in molecular systems.

To see this, we recall from section 2 that the discriminant Δ_j is a real analytic function, the vanishing of which corresponds to a coincidence of two or more eigenvalues λ_k. Consider the coincidence of two eigenvalues for some real value of $\nu = \nu_0$; since this cannot correspond to a singularity in the eigenvalues, we must have $\Delta_j \sim (\nu-\nu_0)^2$ in the neighborhood of ν_0. If the effective coupling strength is varied, this double zero of Δ_j will generally split into two simple zeroes which move off the real axis to complex conjugate points since Δ_j is real analytic. Thus we expect $\Delta_j = f^2(\nu)(\nu-\nu_c)(\nu-\nu_c^*)$ where $f(\nu)$ is regular and non-zero in a domain including a portion of the real axis and the nearby points ν_c and ν_c^*. We assume also that there are no other crossing-point singularities in this domain. We can then write for the two eigenvalues which coincide at ν_c and ν_c^*

$$\begin{aligned} \lambda_i &= f(\nu)[(\nu-\nu_c)(\nu-\nu_c^*)]^{1/2} + g(\nu) \\ \lambda_k &= -f(\nu)[(\nu-\nu_c)(\nu-\nu_c^*)]^{1/2} + g(\nu) \end{aligned} \tag{4.4}$$

where $g(\nu)$ is regular in the domain of interest. This follows because the trace of t (the sum of all eigenvalues) is clearly analytic and from our assumption that the sum of the remaining j-2 eigenvalues are analytic and hence $\lambda_1 + \lambda_2$ is analytic. It is clear that if ν_c is near the real axis the repulsion effect is present. This is illustrated in Fig. 3.

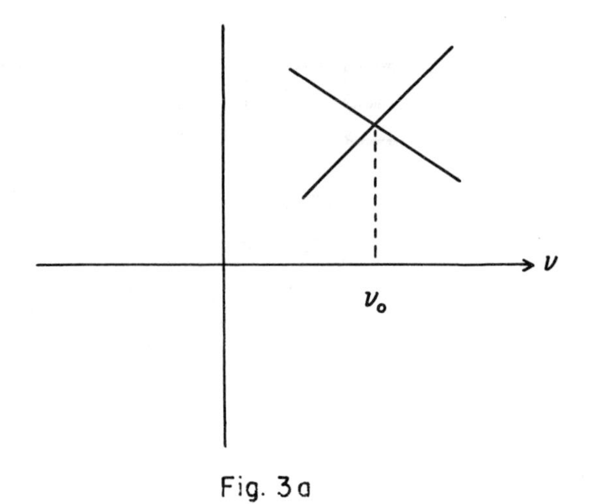

Fig. 3a

Coincident Eigenvalues on Real Axis

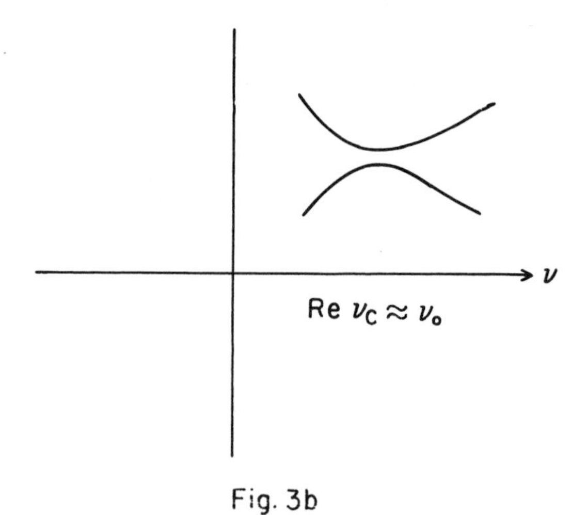

Fig. 3b

Repulsion of Eigenvalues

C. We now demonstrate that the sum of the j eigenphase shifts $\delta_0, \delta_1, \ldots \delta_j$ in the interval between the $j^{\underline{th}}$ and $(j+1)^{\underline{st}}$ thresholds has simple properties. Since the S-matrix eigenvalue $\Lambda_k = \exp(2i\delta_k)$, we deduce from Eq. (2.7) that

$$\sum_{k=1}^{j} \delta_k = \frac{i}{2} \ln \frac{(\lambda_1 - i)(\lambda_2 - i)\ldots(\lambda_j - i)}{(\lambda_1 + i)(\lambda_2 + i)\ldots(\lambda_j + i)} \,. \tag{4.5}$$

We recall that the vanishing of one of the $\lambda_k - i$ corresponds to a pole of the S-matrix. As long as we are in the $j^{\underline{th}}$ threshold region we do not expect any poles of the S-matrix to approach the real axis from an unphysical sheet since if there are lower threshold open channels any such potentially discrete state will decay, i.e. it will be a resonance. This means that there cannot be any logarithmic singularities corresponding to resonance poles reaching the real energy axis. Although crossing point singularities may exist near the real axis above threshold for individual λ_k's, such singularities are absent in the argument of the logarithm in Eq. (4.5). This follows from the fact that the numerator and denominator in (4.5) may be expressed in terms of the characteristic equation $F(\lambda,\nu)$ as $F(\pm i,\nu)$; the coefficients of λ in $F(\lambda,\nu)$ have no singularities except at threshold. Even though λ_j becomes infinite at threshold this presents no difficulty. We need only multiply numerator and denominator by $[\rho_j]^j$ which effectively replaces the λ's by $\overline{\lambda}$'s, all of which are finite at $\nu_j = 0$. We note also that because of its relation to the characteristic equation, the argument of the logarithm in (4.5) is an analytic function of $\xi = (\nu_j)^{1/2}$ so that crossing points at threshold also do not appear in (4.5).

We see, then, that although individual δ_k's may vary dramatically due to repulsion phenomena resulting from nearby crossing-point singularities, the sum $\sum_{k=1}^{j} \delta_k$ is a particularly smooth function of both energy and coupling strength, a fact which we now exploit.

D. Although the physical S-matrix changes its dimension as each new channel opens up, it is still a continuous function of the energy because of threshold factors. This in turn means that the eigenphase shifts can be chosen as continuous functions as has been recently discussed by D. Gross.[7] This is illustrated in Fig. 4.

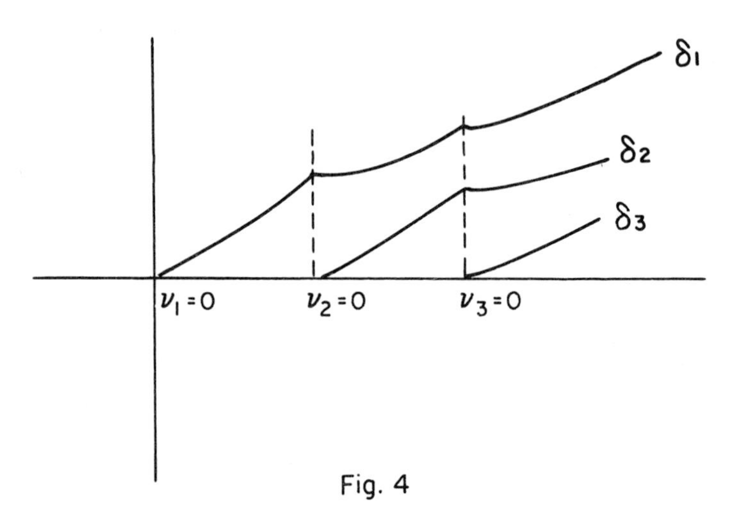

Fig. 4

Continuous Eigenphase Shifts

At each threshold one eigenphase is zero (or a multiple of π) corresponding to the one infinite λ eigenvalue discussed in section 2, Eq. (2.22). We note that the eigenphases are not analytic at the various thresholds and one cannot continue them from one threshold region to another.

Using the properties deduced so far we trace the general energy dependence of the eigenphase shifts as we gradually increase the effective coupling strength from zero up to where resonances and ultimately bound states appear. We do this in detail first for two channels and then show how the techniques are generalizable to an arbitrary number of channels.

For weak coupling or angular momentum $\ell \to \infty$ we assume that all eigenphase shifts are near zero (See Fig. 5a). The sequence of figures, 5a - 5e, show the analytic continuation of the phase shifts as a resonance above $\nu_2=0$ comes down through the second threshold and finally out through the lowest threshold to become a bound state. We describe each step in some detail:

Fig. 5a, weak coupling limit. This is our starting configuration and serves to define our eigenphase shifts with no freedom to make additions of multiples of π since everything will be determined by analytic continuation from here on.

Fig. 5b. A resonance occurs above $\nu_2 = 0$. Note that the large variation <u>must</u> occur in the upper phase shift because of the repulsion effect.

Fig. 5c and 5d. We consider a case in which the phase shift shift δ_1 is now above π at $\nu_2 = 0$ and the fact that it has risen through $\pi/2$ below $\nu_2 = 0$ is evidence for the resonance having moved to the lower region. The configuration in Fig. 5c shows the situation but is not acceptable. The dotted line represents $\delta_2 + \pi$ which we see implies a coincidence of eigenvalues. Repulsion theory leads to Fig. 5d. Although the individual δ_1 and δ_2 are dramatically different in Figs. 5c and 5d, we emphasize that their sum is, in fact, the same in both figures. The fact that $\delta_1 + \delta_2$ in Fig. 5d undergoes a drop by π from $\nu_2 = 0$ to $\nu_2 = \infty$ is essentially an expression of the fact that a "bound" state (i.e. a resonance which would have been a bound state in the absence of coupling to channel 1) has passed below $\nu_2 = 0$.

Fig. 5e. A true bound state emerges and $\delta_1(\nu_2=0) = \pi$. The results in the familiar way from a logarithmic singularity of δ_1 which comes out of the lowest threshold.

One can also imagine a bound state configuration shown in Fig. 6. In this case the resonance pole was presumably associated more closely with the first threshold channel.

One can imagine a succession of bound states emerging as a result of resonances being passed down from higher channels.

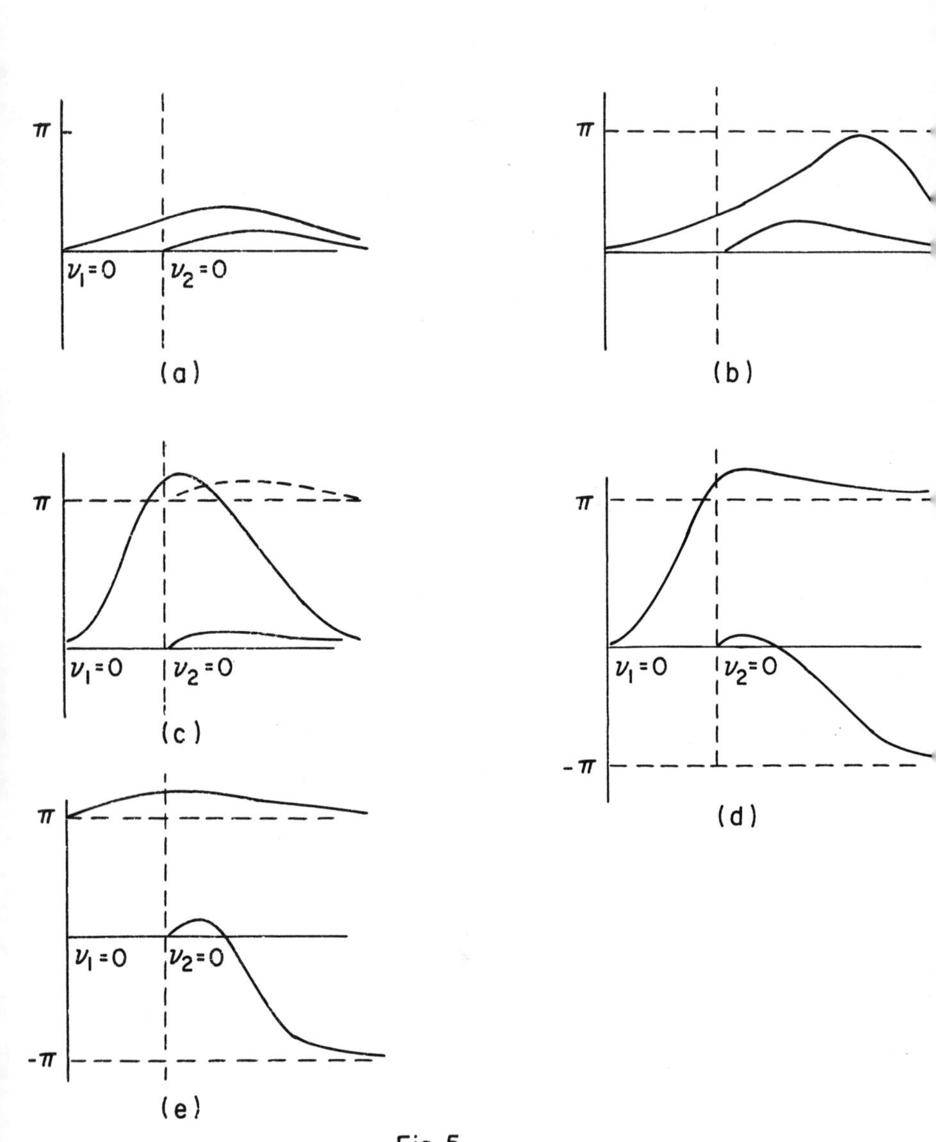

Fig. 5

Eigenphase Shifts for Two Channels

See Text for Explanation

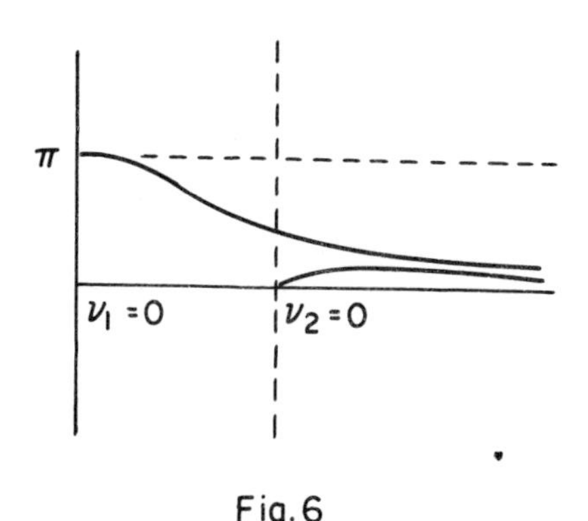

Fig. 6
Bound State Associated Primarily with Channel One

In Fig. 7a we show schematically the situation where two bound states have emerged and in Figs. 5b three, in Fig. 7c, n (even) and in Fig. 7d, n (odd).

This procedure can be readily generalized to many channels and to situations where the bound states have or have not actually emerged and to cases where there are resonances which arise from both channels 1 and 2, say, in the two channel problem. In every case the configuration is uniquely determined by: (1) Continuity of the phase shifts; (2) $\sum_k \delta_k = 0$ at infinite energy; (3) a knowledge of whether resonance poles were passed down from a higher channel or not. The phase shifts we obtain are those which result from analytic continuation in the effective coupling strength starting with the weak coupling configuration shown in Fig. 5a. In the latter connection we note that the arrangement of phase shifts obtained by subtracting π from δ_1 and adding π to δ_2 in Fig. 5d, shown in Fig. 8, while not, of course, experimentally distinguishable from 5d, would not come about naturally by analytic continuation. There cannot in general be any logarithmic singularities emerging at $\nu_2 = 0$ to force δ_2 up to π at $\nu_2 = 0$.

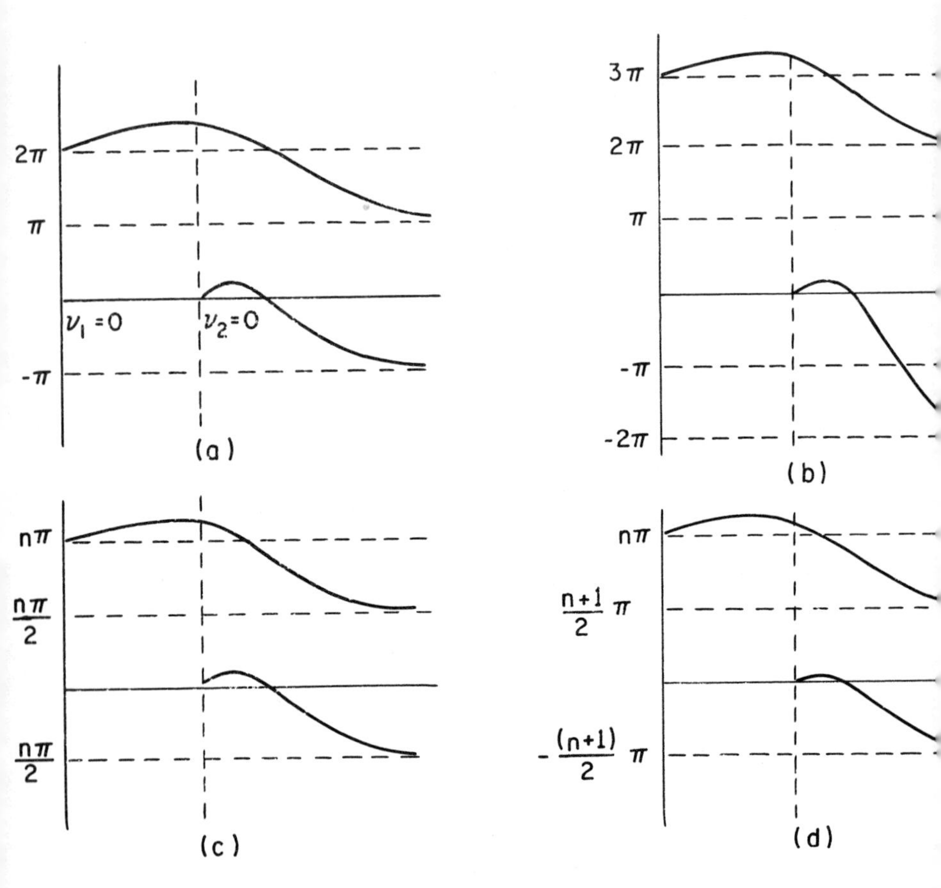

Fig. 7

Two (a), Three (b), n(even) (c), n(odd) (d) Bound States from Channel 2

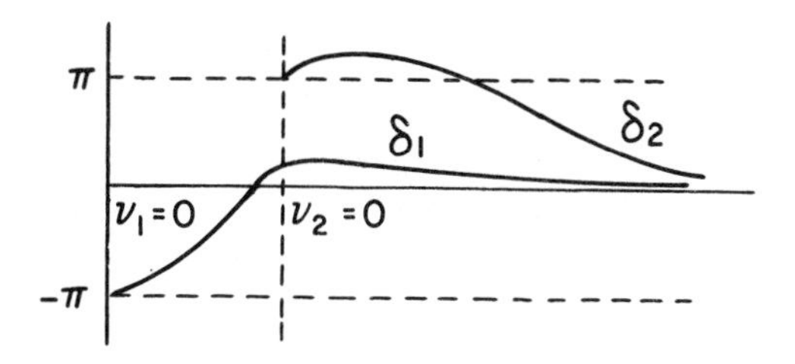

Fig. 8

Phase Shift Configuration Not Obtainable by Analytic Continuation from Weak Coupling

We note that by virtue of our insistence upon analytic continuation in effective coupling strength we are led to the following extremely simple version of Levinson's theorem:

$$\delta_1(\nu_1=0) = N_B\pi, \tag{4.6}$$

where N_B is the number of bound states. This result is true independent of the number of channels. Our version of this important theorem is consistent with the familiar form involving the sum of the eigenphases:

$$\sum_{k=1}^{n} [\delta_1(\nu_k = 0) - \delta_k(\infty)] = N_B\pi, \tag{4.7}$$

with $\delta_k(\nu_k = 0) = m\pi$, where $\delta_k(\nu=0)$ is the $k^{\underline{th}}$ threshold phase shift. For us,

$$\sum_{k=1}^{n} \delta_k(\infty) = 0 \text{ and } \delta_k(\nu_k = 0) = 0 \text{ for } k \neq 1.$$

In concluding this sub-section we show some typical phase shift configurations for a three-channel problem. Starting again from the weak coupling situation we imagine that a resonance develops above the highest threshold, see Fig. 9a. This resonance can again appear only in the highest phase shift. As the coupling is increased so that the resonance moves into the second threshold region, we assume the third channel threshold phase drops to $-\pi$ whereas δ_1 levels off at $+\pi$, just as in the two-channel case, and is shown in Fig. 9b. As the resonance moves down into the first energy interval, no significant change can take place because of repulsion theory, Fig. 9c. If it should happen that the resonance originated in some sense in the second energy interval and then moved down into the first, we find that the second threshold phase shift undergoes the rapid change in the second interval, as shown in Fig. 9d. An intrinsically first-channel resonance is shown in Fig. 9e. If the coupling were further increased in this case so that a bound state emerged, the phase shift δ_1 would drop from π to zero, whereas δ_2 and δ_3 would go from zero to zero. In Fig. 9f we show a situation in which two resonances originating in the third channel have moved down, one into the first interval and one into the second. If a third resonance moves down below the third threshold we obtain either the configuration shown in Fig. 9g if there is one resonance in each of the lower intervals or that of Fig. 9h if there are two resonances in the first interval. The rapid variation of δ_2 in above $\nu_3 = 0$ reflects the normal drop of a resonance phase shift which because of the repulsion effect must be reflected in a lower phase shift.

E. We now show in detail how the remarkable change in phase shifts described by Fig. 5d can take place in an analytic fashion as a resonance passes down from one threshold region to another. For simplicity we consider a two channel s-wave problem although the results are quite general. From Eq. (2.11) we have

$$\lambda_\pm = \frac{\rho_2 \dfrac{Y_{11}}{\rho_1} + Y_{22} \pm \left[\left(\rho_2 \dfrac{Y_{11}}{\rho_1} - Y_{22}\right)^2 + 4\, \dfrac{Y_{12}^2}{\rho_1}\, \rho_2\right]^{1/2}}{2\rho_2} . \qquad (4.8)$$

The dramatic behavior of the two eigenphase shifts comes about by a

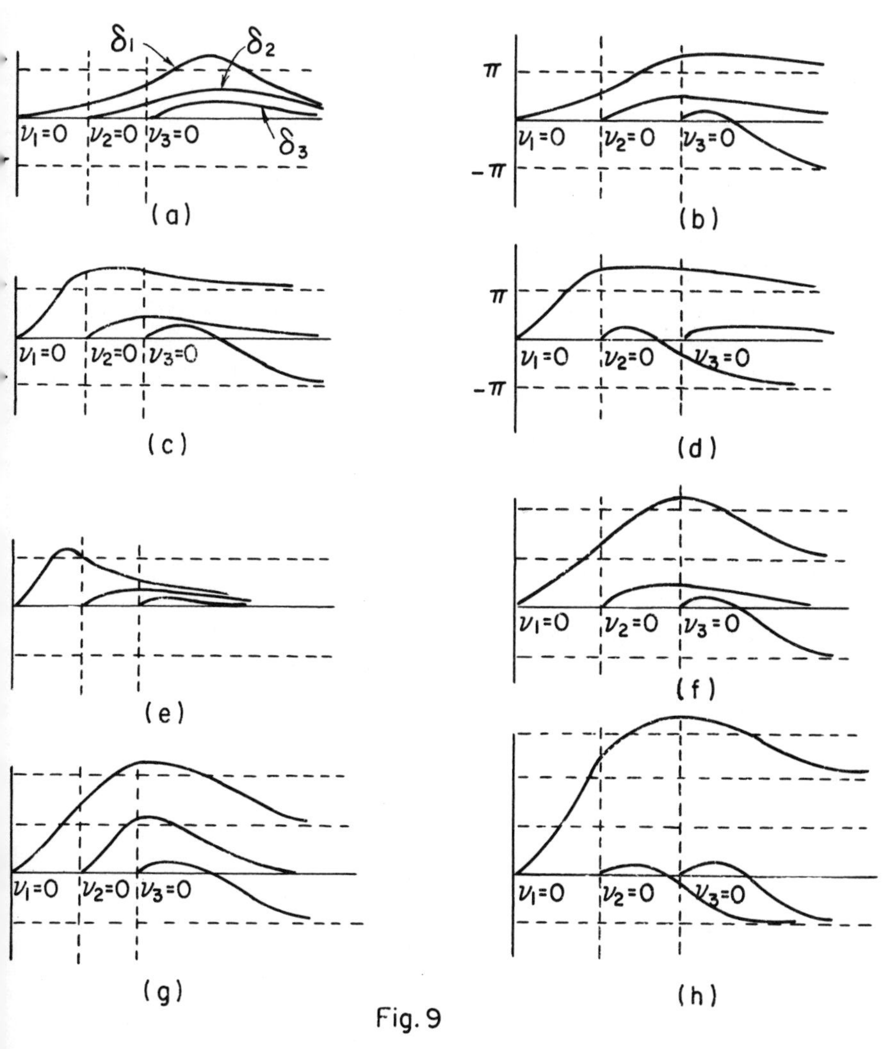

Fig. 9

Eigenphase Shifts for Three Channels

See Text for Explanation

crossing-point singularity moving up to the threshold $\nu_2 = 0$ which in fact happens when $Y_{22}(\nu_2=0) = 0$. When $Y_{22}(\nu_2=0) = 0$, both λ's become infinite (see section 3). We expand all of the quantities in (4.9) about $\nu_2 = 0$ and keep only lowest order terms; we imagine Y_1, ρ_1, Y_{12} and Y_{22} are constants. Thus we write

$$\lambda_{\pm} = \frac{c\xi + a \pm [(c\xi - a)^2 + \eta\xi]^{1/2}}{2\xi} \tag{4.9}$$

where $Y_{22} = a$, $Y_1/\rho_1 = c$, $4Y_{12}^2/\rho_1 = \eta$ and $\rho_2 = \nu_2^{1/2} = \xi$. Originally we imagine a to be small and positive and then we analytically continue to a negative. For a one channel problem this is characteristic of a bound state emerging from threshold.

The crossing-point singularities located at ξ_c arise from the vanishing of the discriminant in Eq. (4.9);

$$\xi_c = \frac{-(\eta-2ac) \pm [(\eta-2ac)^2 - 4a^2c^2]^{1/2}}{2c} . \tag{4.10}$$

For a small and positive, the two crossing points lie on the negative real ξ axis. One of them is roughly given by $\xi_c \approx -a^2/\eta$ while the other is $\xi_c \approx -2\eta/c$. Clearly, if a is continued to negative values by means of a small counter-clockwise semi-circular loop about the origin, the crossing point near the origin in the ξ-plane will encircle the origin counter-clockwise and return to a point on the negative real axis.

For $a > 0$, the threshold eigenphase shift, δ_2, is given in terms of λ_+ by

$$\delta_2 = \frac{i}{2} \ln \frac{\lambda_+ - i}{\lambda_+ + i} . \tag{4.11}$$

Since $\lambda_+ \to +\infty$ as $\xi \to 0$ we may define the phase of the argument of the logarithm in (4.11) so that $\delta_2 = 0$ in this limit. We show in Fig. 10 an Argand diagram with the complex numbers $\lambda_+ \pm i$ for $a > 0$ and $\xi > 0$.

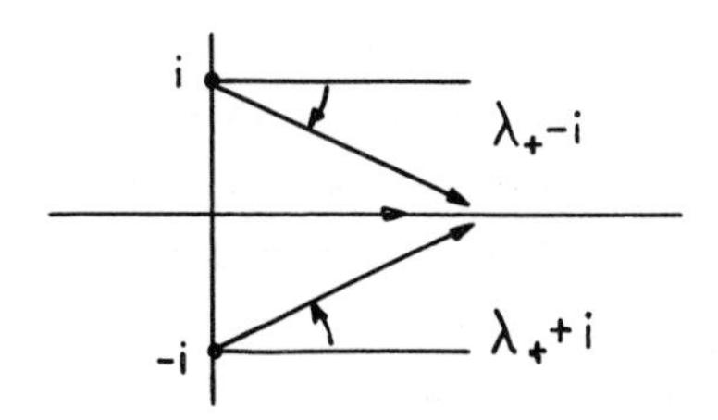

Fig. 10

Argand Diagram for Eq. (4.11)

The angle for the vector $\lambda_+ - i$ is measured downwards (i.e. a negative angle) from the horizontal and that for $\lambda_+ + i$ is measured upwards (i.e. a positive angle) as shown. The magnitude of both angles is simply δ_2 which clearly goes to zero as $\lambda_+ \to +\infty$. The other phase shift, δ_1, is given by

$$\delta_1 = \frac{i}{2} \ln \frac{\lambda_- - i}{\lambda_- + i} , \tag{4.12}$$

and its Argand diagram is similar to that shown in Fig. 10. In this case, however, as $\xi \to 0$, λ_- is finite and $\delta_1(\nu_2=0)$ is some finite angle, not in general a multiple of π. As a approaches zero, $\delta_1(\nu_2=0)$ generally increases and finally becomes π when a = 0 and the crossing point is at $\xi_c = 0$.

Let us now follow this process, assuming that the initial configuration of the phase shifts is as shown in Fig. 11, for a > 0 but small.

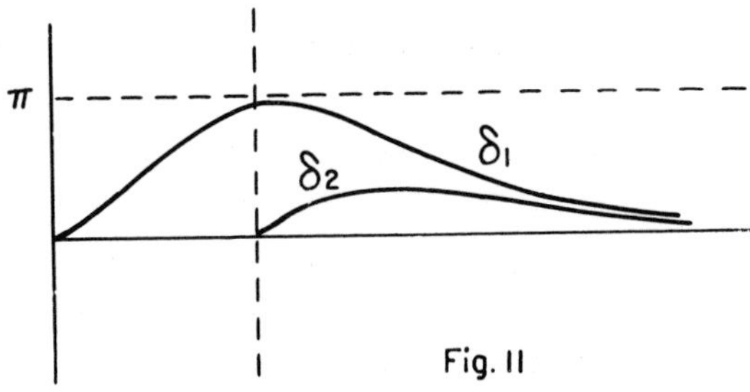

Fig. 11

Initial Phase Shift Configuration before Analytic Continuation in "a"

It is a matter of indifference what the value of $\delta_1(\nu_2=0)$ is so long as it is less than π. Now as we noted above, if a goes from positive to negative via a counter-clockwise loop, ξ_c will encircle the origin in the ξ-plane in a counter-clockwise loop. Then the path in the -plane for a > 0 which deforms into the correct path for determining δ_1 and δ_2 after the continuation to a < 0 has been effected is shown in Fig. 12. This, of course, because the crossing-point singularity ξ_c about which we have looped "undoes the loop" as the continuation to a < 0 is carried out.

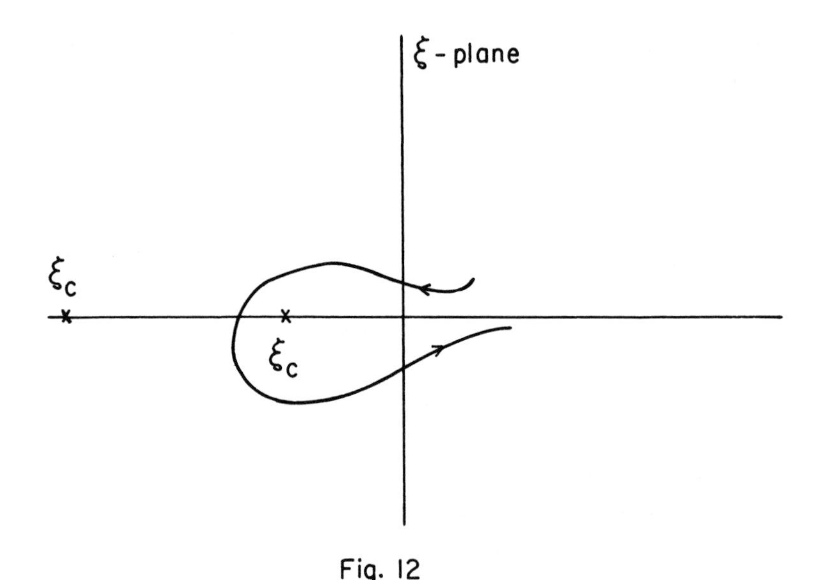

Fig. 12

Path in ξ-plane for Study of Crossing-Point Effects

Let us now trace the behavior of δ_2 on the path shown in Fig. 12. It is clear from Fig. 10 that the small loop about the origin causes the vector $\lambda_+ + i$ to execute nearby a complete circle clockwise about the point $-i$. This moves us onto a new sheet of the logarithm and $(-\pi)$ must be added to any phase shifts computed from Eq. (4.11) in a straightforward manner. Looping the crossing point has the effect of changing λ_+ to λ_- and hence along the real axis we obtain

$$\tilde{\delta}_2 = \delta_1 - \pi \qquad (4.13)$$

where the tilde indicates δ_2 continued along the path in Fig. 12; it is the "new" δ_2 appropriate when $a < 0$. Since δ_1 goes to zero at infinity, $\tilde{\delta}_2$ goes from zero to $-\pi$ along the loop path.

In the case of δ_1, Eq. (4.12), the first loop about the origin gives no big change because λ_- is finite at $\xi = 0$. As the crossing point singularity is circled, λ_- goes into λ_+; then as one moves under the point $\xi = 0$, the vector $\lambda_+ - i$ does a clockwise loop about the point i. This adds $(+\pi)$ to the phase shift and we move onto a new branch of the logarithm. We have then

$$\tilde{\delta}_1 = \delta_2 + \pi. \qquad (4.14)$$

Since $\delta_2 \to 0$ at infinite energy, we see that on the path shown in Fig. 12, $\tilde{\delta}_1$ goes from around π to (precisely) π. The Argand diagram showing the initial and final configurations for δ_1 and δ_2 are shown in Fig. 13.

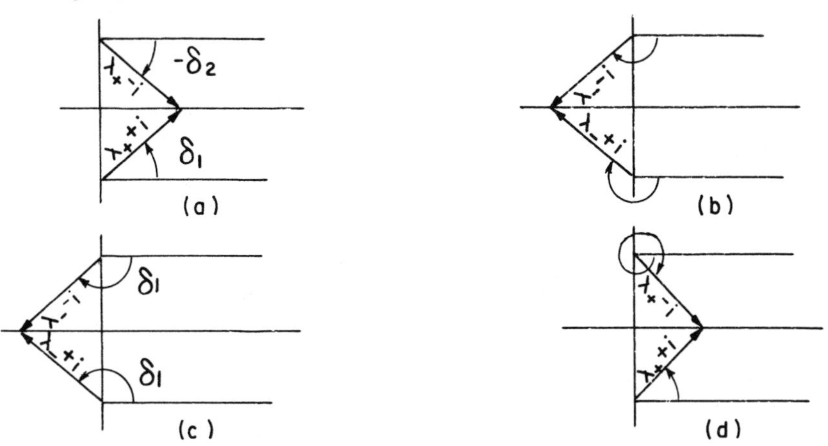

Fig. 13

Argand Diagrams Showing Initial (a) and Final (b) Configuration for δ_2 Taken Around Path of Fig. 12. The Arrows in (b) show the sense in which Angles are to be measured. (c) and (d) show the same things for δ_1.

We see, therefore, that by analytic continuation in the effective coupling which causes $a = Y_{22}$ to change sign near $\xi = 0$ the phase shift configuration shown in Fig. 11 goes over into Fig. 14 (which is the same as 5d).

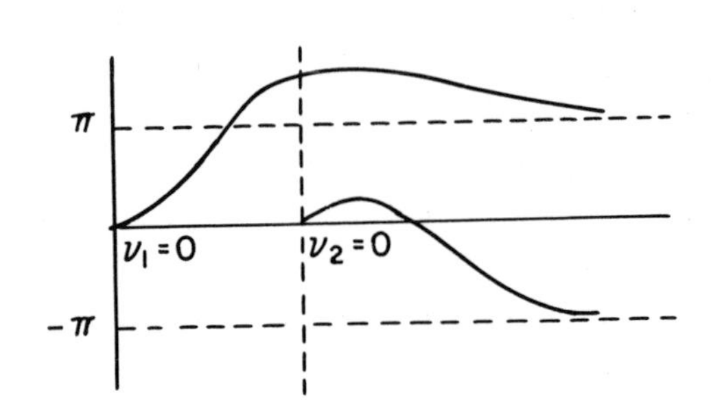

Fig. 14

Phase Shift Configuration Corresponding to a Resonance Passing Below ν_2=0

We note from Eq. (4.13) and (4.14) that

$$\tilde{\delta}_1 + \tilde{\delta}_2 = \delta_1 + \delta_2$$

so that the sum remains essentially unchanged as we vary the parameters over a small region.

5. Summary and conclusions

We have studied the general structure of eigenphase shifts for a system of n two-body channels with non-degenerate thresholds. It is clear that an analysis of experimental data in terms of these phase shifts can give valuable information about the underlying dynamics in that their behavior is an accurate reflection of the importance of higher channels. Particularly striking is the fact that a bound state situation which is customarily thought of as leading to a phase shift falling from π at the lowest threshold may actually lead to a phase shift in the lowest energy interval which rises from π (or what is experimentally indistinguishable, rising from zero); the drop by π required by Levinson's theorem is reflected in the behavior of an eigenphase starting at a higher threshold. Such a behavior could, for example, reconcile the suggestion of Chew[8] in connection with the I = 0, J = 0, $\pi - \pi$ phase shift being associated with a bound state and the experimental indication about this phase shift increasing below the ρ threshold.

We have not explored the possibility of using eigenphase shifts in dynamical calculations. These appear superficially complicated in that the branch points associated with crossing point singularities add to the already complex singularity structure of amplitudes. On the other hand, the discontinuities across cuts joining the crossing points are expressible in terms of the eigenphases and in practice the additional complexity may not be too severe. Realistic calculations involving more than two-body channels may very well be more readily considered in the eigenphase framework since the theoretical concept of these quantities is well defined.

Acknowledgements

It is a pleasure to thank Dr. M. Froissart and Professor E. P. Wigner for several helpful discussions.

References

1. See, however, R. C. Hwa and D. Feldman, Annals of Physics 21, 21, 453 (1963); also
2. D. Atkinson, K. Dietz, and D. Morgan, Annals of Physics 37, 77 (1966).
3. See, for example, W. F. Osgood, Functions of a Complex Variable (G. E. Stechert, New York, 1938) Chapter VII, P. 176.
4. J. V. Uspensky, Theory of Equations (McGraw Hill, New York, 1948).
5. We are indebted to Dr. M. Froissart and Dr. F. J. Dyson who independently produced this argument in support of our conjecture which was based on the explicitly soluble two channel problem.

References (continued)

6. The precise manner in which the resonance moves to a lower threshold region involves the well-known shadow pole phenomenon as discussed, for example, by R. J. Eden and J. R. Taylor, Phys. Rev. 133, B1575 (1964). For our purposes we need not go into this matter.
7. D. J. Gross, University of California preprint, unpublished.
8. G. F. Chew, Phys. Rev. Letters 16, 60 (1966).

Received 2/14/69

Current Propagator and Hadron Production
in Electron-Positron Collisions

by

J. J. Sakurai
The University of Chicago

1. Introduction

In the past eight years the hypothesis of vector meson dominance has been successfully applied to various electromagnetic processes involving hadrons.[1] As is well known, the basic ingredients of vector meson dominance is conveniently summarized by postulating the "current-field identity"

$$j_\mu^{(em)} = \lambda_\rho \rho_\mu^\circ + \lambda_\omega \omega_\mu + \lambda_\phi \phi_\mu \qquad (1$$

where $j_\mu^{(em)}$ is the hadronic part of the electromagnetic current density, and ρ_μ°, ω_μ and ϕ_μ are the renormalized ρ°, ω and ϕ fields.[2] It is then natural to discuss the propagator of the current $j_\mu^{(em)}$ in much the same way as field theoritsts in the "old days" discussed the propagators of various meson fields. This paper is devoted to studying the properties of the propagator of the current $j_\mu^{(em)}$, hereafter referred to simply as the "current propagator."

We could, of course, discuss the current propagator without reference to the vector mesons or the vector meson fields. One might wonder why a Källén-type representation for the current propagator was not studied extensively fifteen years ago. There are essentially three reasons for this. First, it is only in recent years that currents (rather than Heisenberg fields) came to play dominant roles in elementary particle physics. We may mention in this connection that some of the recent developments in current algebra--viz. the spectral-function (Weinberg) sum rules[3] and hard-pion techniques[4]--directly rely on the current propagator and its SU(2), SU(3) and SU(2) ⊗ SU(2) generalizations. Second, and more important, in old-fashioned theories where the current is visualized as being made up of bilinear densities of "fundamental fields," the spectral representation for the current propagator is expected to be highly divergent. In contrast, the expected behavior of the current propagator is much more convergent in theories that take seriously the current-field identity (1). Third, the weight function that appears in the spectral representation of the current propagator is directly related to the cross section for hadron production in electron-positron collisions, which, thanks to electron-positron colliding beam facilities, can now be measured experimentally. The existing electron-positron storage rings at Novosibirsk and Orsay have already been used to study hadron production in the ρ, ω and ϕ regions; the colliding beam facilities now being constructed at Frascati and Cambridge (Mass.) will hopefully extend the measurement of the hadronic cross section up to center-of-mass energies of 4 - 6 GeV.

In this paper we show how the properties of the current propagator can be explored in detail by performing electron-positron colliding beam experiments. In particular we demonstrate that precise measurements of the hadronic cross section in electron-positron collisions may answer the following intriguing questions:

(1) Does the current propagator fall off with q^2 in a manner understandable from simple-minded vector meson dominance?

(2) How large is the hadronic contribution to (g - 2) for the muon?

(3) Is the hadronic contribution to charge renormalization finite?

(4) Is the Schwinger term finite?

(5) Is the "bare mass" of the ρ meson finite?

Even though most of the materials of this paper are neither original nor new, it is hoped that the physical emphasis found here is somewhat novel and refreshing.

2. Spectral representation

The hadronic part of the electromagnetic current density is given in the SU(3) notation by

$$j_\mu^{(em)} = j_\mu^3 + \frac{1}{\sqrt{3}} j_\mu^8. \tag{2}$$

Let us first focus our attention on the Fourier transform of the vacuum expectation value of the time-ordered product

$$\Delta_{\mu\nu}^{\alpha\beta}(q) = i\int d^4x \; e^{-iq\cdot x} <0|\; T(j_\mu^\alpha(x) j_\nu^\beta(0))\,|0> \tag{3}$$

It is tempting to identify this expression with the covariant propagator for the current density j_μ^α when β is set equal to α. However, using the commonly accepted commutation relations[5]

$$\begin{aligned} [j_o^\alpha(x),\; j_o^\beta(x')]_{x_o=x_o'} &= if_{\alpha\beta\gamma} j_o^\gamma(x)\,\delta^{(3)}(\vec{x}-\vec{x}'), \\ [j_o^\alpha(x),\; j_k^\beta(x')]_{x_o=x_o'} &= if_{\alpha\beta\gamma} j_k^\gamma(x)\,\delta^{(3)}(\vec{x}-\vec{x}') - ic_{\alpha\beta}\partial_k\delta^{(3)}(\vec{x}-\vec{x}') \end{aligned} \tag{4}$$

(with $j_o^\alpha = -ij_4^\alpha$), it is readily seen that $\Delta_{\mu\nu}$ defined by (3) is not covariant. Contracting (3) with q_μ, we obtain

$$\begin{aligned} q_\mu \Delta_{\mu\nu}^{\alpha\beta}(q) &= \int d^4x\, e^{-iq\cdot x} <0|\,[j_o^\alpha(x),\; j_\nu^\beta(0)]_{x_o=0}\,|0>\; \delta(x_o) \\ &= \begin{cases} 0 \text{ for } \nu=4 \\ <0|c_{\alpha\beta}|0> q_k \text{ for } \nu=k. \end{cases} \end{aligned} \tag{5}$$

With $<0|c_{\alpha\beta}|0> \neq 0$ the right-hand side of (5) is not covariant, hence $\Delta_{\mu\nu}^{\alpha\beta}$ cannot be a covariant tensor of rank two. Note that the trouble arises because of the famous Schwinger term[6] in the expression for the time-space commutator. Since a general argument can be given for the nonvanishing of a diagonal ($\alpha=\beta$) Schwinger term ($<0|c_{\alpha\alpha}|0> \neq 0$), it is inconvenient to deal with the expression (3). If we, however, work with a new propagator defined by

$$\Delta_{\mu\nu}^{\alpha\beta}(q) = i\int d^4x\, e^{-iq\cdot x} <0|T(j_\mu^\alpha(x)\; j_\nu^\beta(0))\,|0> + <0|c_{\alpha\beta}|0>\; \delta_{\mu4}\delta_{\nu4}, \tag{6}$$

then covariance is restored. In the following we mean by $\Delta_{\mu\nu}^{\alpha\beta}$ the covariant expression (6) rather than the noncovariant expression (3).

Setting aside for a moment the question of convergence, we write down a spectral representation for the diagonal (α=β) propagator $\Delta_{\mu\nu}^{(\alpha\alpha)}(q)$

$$\Delta_{\mu\nu}^{\alpha\alpha}(q) = \int \frac{\rho^{\alpha\alpha}(m^2)}{q^2+m^2-i\varepsilon} \left(\delta_{\mu\nu} + \frac{q_\mu q_\nu}{m^2}\right) dm^2, \quad (\alpha \text{ not summed}) \tag{7}$$

where we have assumed that the vector current is conserved, as is the case for α=1, 2, 3 and 8. If we are talking about the full (i.e. isoscalar and isovector) electromagnetic current, we may work with

$$\Delta_{\mu\nu}^{(em)} = \int \frac{\rho^{(em)}(m^2)}{q^2+m^2-i\varepsilon} \left(\delta_{\mu\nu} + \frac{q_\mu q_\nu}{m^2}\right) dm^2 \tag{8}$$

where

$$\rho^{(em)}(m^2) \equiv \rho^{33}(m^2) + \frac{1}{3}\rho^{88}(m^2). \tag{9}$$

The weight function $\rho^{\alpha\alpha}(m^2)$ can be written explicitly in terms of a hadronic matrix element $<A|j_\mu^\alpha(0)|0>$ as follows:

$$\left(\delta_{\mu\nu} - \frac{q_\mu q_\nu}{q^2}\right)\rho^{\alpha\alpha}(-q^2) = (2\pi)^3 \sum_A \delta^{(4)}(q-p_A)<0|j_\mu^\alpha(0)|A><A|j_\nu^\alpha(0)|0>.$$

Clearly, if a convergent spectral representation of the form (6) is to be written down, it is essential that the weight function satisfies the property

$$\int \frac{\rho^{\alpha\alpha}(m^2)}{m^2} dm^2 < \infty.$$

Furthermore, if we require that the current propagator fall off with q^2 like a free field propagator [in the sense that the $\delta_{\mu\nu}$ part of $\Delta_{\mu\nu}^{\alpha\alpha}$ goes as (finite constant)/q^2], we obtain an even stronger result

$$\int \rho^{\alpha\alpha}(m^2)\, dm^2 < \infty .$$

As will be discussed in later sections, the convergence conditions (11) and (12) are completely nontrivial. In old-fashioned theories in which the high-energy behavior of the quantum electrodynamics of hadrons is assumed to resemble the high-energy behavior of "pure" quantum electrodynamics (i.e. the quantum electrodynamics of leptons and photons), the integrals (11) and (12) are, respectively, linearly and quadratically divergent with m^2 (or, equivalently, quadratically and quartically divergent with the center-of-mass energy of the hadronic system).

3. Colliding-beam cross section

The matrix element $<A|j_\mu^\alpha|0>$ with α set equal to 3 or 8 is the amplitude for an isovector or isoscalar time-like photon to turn into a hadronic state A. It is therefore directly measurable in a colliding-beam reaction[7]

$$e^+ + e^- \to A. \qquad (1\text{:}$$

Likewise, the matrix elements $<A|j_\mu^\alpha|0>$ with α=1, 2, 4, 5 and their axial-vector analogs are in principle measurable in electron-neutrino collisions. Returning now to the spectral function, it can be derived from (10) that the spectral function ρ^{33} and ρ^{88} are directly expressible in terms of the total production cross sections for $e^+ + e^- \to T = 1$ and $T = 0$ hadronic systems:[8]

$$\sigma^{(T=1)}(s) = \frac{16\pi^3\alpha^2}{s^2}\ \rho^{33}(m^2)\big|_{m^2=s},$$
$$\sigma^{(T=0)}(s) = (16/3)\ \frac{\pi^3\alpha^2}{s^2}\ \rho^{88}(m^2)\big|_{m^2=s}\ . \qquad (14)$$

These cross sections have been studied at the Novosibirsk and Orsay storage rings up to center-of-mass energies of about 1030 MeV. The isovector channel is dominated by the ρ peak, and the pion-pion cross section is very well represented by the ρ dominance expression with finite width corrections[9]

$$\sigma(e^+e^- \to \pi^+\pi^-) = \frac{\pi\alpha^2}{3}\,\frac{(s-4m_\pi^2)^{3/2}}{s^{5/2}}\;\frac{m_\rho^4[1+0.48(\Gamma_\rho/m_\rho)]^2}{(m_\rho^2-s)^2+[m_\rho\Gamma_\rho(k/k_\rho)^3(m_\rho/\sqrt{s})]^2},$$

$$k = \frac{1}{2}(s-4m_\pi^2)^{1/2}, \quad k_\rho = \frac{1}{2}(m_\rho^2-4m_\pi^2)^{1/2}$$

with $\Gamma_\rho \approx 110$ MeV, $m_\rho \approx 770$ MeV.[10] The isoscalar channel is dominated by ω and ϕ, and within 30% the spectral-function (Weinberg[3]) sum rule [which can be derived from the commutation relations (4) provided that the Schwinger term is a c-number]

$$\int \frac{\rho^{33}}{m^2}\,dm^2 = \int \frac{\rho^{88}}{m^2}\,dm^2$$ (

or, equivalently,

$$\frac{1}{3}\int s\sigma^{(T=1)}(s)\,ds = \int s\sigma^{(T=0)}(s)\,ds,$$ (

is satisfied by considering the ρ, ω and ϕ contributions only.[10, 11]

What lie beyond ρ, ω and ϕ are anybody's guess at this writing. Some Regge-pole models based on an infinite sequence of subsidiary or daughter trajectories predict an infinite number of vector mesons, but those models do not make predictions on the coupling strengths of the "higher" vector mesons to the photon. Also important is the question of possible continuum contributions (e.g. nonresonant πA_1 states) to the total hadronic cross section.

For the purpose of the present paper we are mainly concerned with the asymptotic behavior of the colliding-beam cross section. For example, the requirements (11) and (12) mentioned earlier can be re-written in terms of the hadron production cross section in electron-positron collisions as follows

$$\int \frac{\rho^{(em)}}{m^2}\,dm^2 < \infty \;\Rightarrow\; s^2\sigma(e^+e^- \to \text{hadrons}) \to 0$$

$$\int \rho^{(em)}\,dm^2 < \infty \;\Rightarrow\; s^3\sigma(e^+e^- \to \text{hadrons}) \to 0$$ (1

We thus see that the finiteness of $\int [\rho^{(em)}/m^2]dm$ and $\int \rho^{(em)}dm^2$ imposes very stringent conditions on the asymptotic behavior of the colliding-beam cross section.[12]

4. Muon magnetic moment

The hadronic correction to the anomalous magnetic moment, (g-2), of the muon arises from the Feynman diagram shown in Fig. 1.

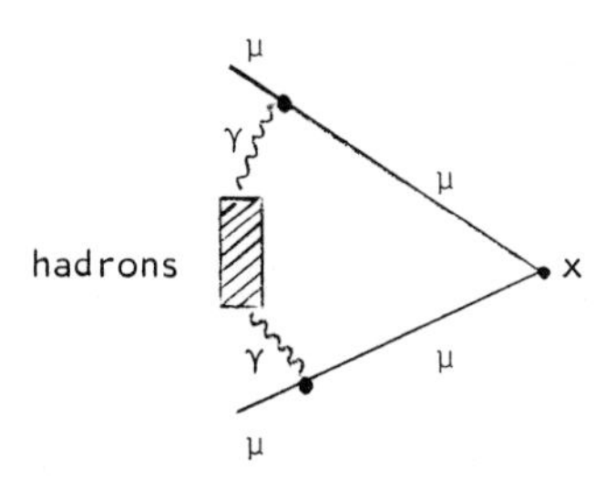

Fig. 1 The hadronic correction to the muon magnetic moment

It is straightforward to show that the contribution of this diagram can be expressed in terms of the spectral function $\rho^{(em)}$ as follows:[13]

$$\frac{1}{2}(g-2)_{had.} = 4\alpha^2 \int \frac{\rho^{(em)}(m^2)G(m^2)}{m^4}\, dm^2 ,$$
$$G(m^2) = \int_0^1 dx \frac{x^2(1-x)}{x^2 + (m^2/m_\mu^2)(1-x)} \tag{19}$$

With $m^2 \gg m_\mu^2$, $G(m^2)$ goes as $1/m^2$, hence the finiteness of $\frac{1}{2}(g-2)_{had.}$ depends on

$$\int \frac{\rho^{(em)}(m^2)}{m^6} < \infty . \tag{20}$$

This is a very weak requirement. Even if the colliding-beam cross section behaves asymptotically as in point-particle quantum electrodynamics, the convergence of the integral (20) is guaranteed. This, of course, is related to the fact that the vacuum polarization contribution to g-2 in the pure quantum electrodynamics of leptons and photons is finite.

It is important to notice that the expression (19) heavily weights relatively low mass states. If we just keep the ρ, ω and ϕ

contributions, we obtain, from the experimentally measured cross sections,[14]

$$\frac{1}{2}\,(g-2)_{\mathrm{had.}} \approx (5 - 7) \times 10^{-8} \qquad (\text{'}$$

which is below the error of the most recent measurement of the muon magnetic moment.

5. Hadronic corrections to charge renormalization

In the pure quantum electrodynamics of leptons and photons the change in the electric charge of a charged particle brought about by the vacuum polarization due to lepton pairs is logarithmically divergent. It is therefore of interest to ask whether this is still the case for the hadronic part of vacuum polarization.

Suppose the photon propagator in the absence of hadrons is given by $\delta_{\mu\sigma}/q^2$. We write the modified propagator in the presence of the electromagnetic couplings of hadrons (to order e^2) as

$$\frac{\delta_{\mu\sigma}}{q^2} + \frac{e^2}{q^2}\,\delta_{\mu\nu}\Pi_{\nu\lambda}(q)\,\delta_{\lambda\sigma}\,\frac{1}{q^2}\;. \qquad (2$$

The polarization tensor $\Pi_{\nu\lambda}$ must have the same absorptive part as the current propagator; in addition it must satisfy the gauge condition

$$q_\nu \Pi_{\nu\lambda} = 0 \qquad (2$$

which guarantees the vanishing rest mass of the photon. Thus

$$\Pi_{\nu\lambda} = \int \frac{\rho^{(em)}(m^2)}{q^2+m^2-i\varepsilon}\,(\delta_{\nu\lambda} + \frac{q_\nu q_\lambda}{m^2})\,dm^2 - \delta_{\nu\lambda}\int \frac{\rho^{(em)}(m^2)}{m^2}\,dm^2\;. \qquad (2$$

For small values of q the photon propagator is seen to be modified as follows

$$\frac{\delta_{\mu\sigma}}{q^2} \rightarrow [\,1 - e^2 \int \frac{\rho^{(em)}(m^2)}{m^4}\,dm^2\,]\,\frac{\delta_{\mu\sigma}}{q^2}\;. \qquad (2$$

As usual, the change in the coefficient of the photon propagator can be absorbed in the definition of the renormalized charge. So the fractional change in the electric charge due to the hadronic contribution can be expressed in terms of the directly observable quantity $\rho^{(em)}(m^2)$:

$$\frac{\delta e^2}{e^2} = -\int \frac{\rho^{(em)}(m^2)}{m^4} dm^2 \tag{26}$$

This integral is even more convergent than $\int [\rho^{(em)}(m^2)/m^2]dm^2$ whose finiteness is guaranteed by the existence of the spectral representation (8). Thus, in theories that take seriously the current-field identity, there is no divergence in the hadronic part of charge renormalization, a point first emphasized by Kroll, Lee and Zumino.[2]

Suppose we assume that the integral (26) can be saturated by ρ, ω and ϕ. We then obtain

$$\frac{\delta e^2}{e^2} = -0.4\%, \tag{27}$$

a fantastically small fraction. This is an order of magnitude smaller than the "rigorous" upper limit obtained by Kroll, Lee and Zumino,[2] $|\delta e^2/e^2| < 3\%$, under the assumption that the spectral integral is dominated at the two-pion and three-pion thresholds.

6. The magnitude of the Schwinger term

The coefficient $c_{\alpha\alpha}$ (or the vacuum expectation value of $c_{\alpha\alpha}$ if $c_{\alpha\alpha}$ is a q-number) that appears in the basic time-space commutation relation (4) can be related to the spectral function as follows:

$$<0|c_{\alpha\alpha}|0> = \int \frac{\rho^{\alpha\alpha}(m^2)}{m^2} dm^2. \tag{28}$$

A "modern" way to prove this "old" result[6] exploits the powerful theorem of Bjorken[8]

$$\begin{aligned}\lim_{q_0 \to \infty,\ \vec{q}\ \text{fixed}} \int e^{-iq\cdot(x-x')} <\beta|T(A(x)B(x')|\alpha> d^4x \\ = \frac{i}{q_0}\int e^{-i\vec{q}\cdot(\vec{x}-\vec{x}')} <\beta|[A(x),B(x')]|\alpha>_{x_0=x_0'} d^3x + \mathcal{O}\left(\frac{1}{q_0^2}\right).\end{aligned} \tag{29}$$

Applying this to $\Delta_{4k}^{\alpha\alpha}(q)$, we obtain

$$\lim_{q_o \to \infty,\ \vec{q}\ \text{fixed}} q_o q_k \int \frac{\rho^{\alpha\alpha}(m^2)}{m^2(q^2+m^2-i\varepsilon)} dm^2$$

$$= \frac{i}{q_o} \int e^{-i\vec{q}\cdot(\vec{x}-\vec{x}')} <0|[j_o^\alpha(x), j_k^\alpha(x')]|0>_{x_o=x'_o} d^3x, \quad (3$$

which immediately leads to the sum rule (28) when (4) is used.

It is worth while to recall that the coefficient of the Schwinger term in the quark density algebra based on

$$j_\mu^\alpha(x) = \lim_{\varepsilon\to 0} i\bar{q}(x-\tfrac{\varepsilon}{2})\gamma_\mu \frac{\lambda_\alpha}{2} q(x+\tfrac{\varepsilon}{2}) \quad (3$$

diverges as[6] $1/\varepsilon^2$. In terms of m^2 this means that $\int [\rho^{\alpha\alpha}(m^2)/m^2]dm^2$ is linearly divergent with m^2, hence[8, 15]

$$\sigma(e^+e^- \to \text{hadrons}) \sim \frac{1}{s}\ . \quad (3$$

This is perhaps not surprising since perturbative quantum electrodynamics gives a cross section for $e^+ + e^- \to \mu^+ + \mu^-$ that varies as $1/s$. In other words, in the quark-density algebra the hadronic cross section in electron-positron collisions is "point-like" in the sense that the total cross section is comparable to muon pair cross section at high energies.

Another popular current algebra is the gauge-field algebra (the Lee-Weinberg-Zumino algebra[16]) based on the hypothesis that the ρ meson field appearing on the right-hand side of the current-field identity actually satisfies the canonical commutation relations of the type discussed in Wentzel's classical treatise.[17] In this algebra the coefficient of the Schwinger term is a finite c-number

$$c_{33} = m_\rho^2/f_\rho^2 \quad (3$$

where f_ρ is the (universal) ρ meson coupling constant at zero momentum transfer. In such a theory it is perhaps not too ridiculous to saturate $\int [\rho^{\alpha\alpha}(m^2)/m^2]dm^2$ by ρ, ω and ϕ, as customarily done in practical applications of the spectral-function sum rules.[3] As for the magnitude of the Schwinger term we obtain from the Orsay cross section

$$c_{33} \approx 0.024\ \text{GeV}^2 \tag{34}$$

by integrating over the ρ meson peak. As mentioned earlier, the hadronic cross section in the asymptotic region is predicted to go to zero faster than[12] $1/s^2$, and we would not expect too many hadronic events beyond ρ, ω and ϕ.

7. Possible free-field behavior for the current

In this section we speculate on the possibility that the current behaves like a free vector-meson field at infinity, or, more specifically,

$$\Delta_{\mu\nu}^{(em)}(q) \xrightarrow{q^2\to\infty} (\delta_{\mu\nu}/q^2)\int \rho^{(em)}(m^2)dm^2 + \frac{q_\mu q_\nu}{q^2}\int \frac{\rho^{(em)}(m^2)}{m^2}dm^2. \tag{35}$$

Since $\int \rho^{(em)}dm^2$ must be convergent, we have

$$\lim_{s\to\infty} s^3\sigma(e^+e^- \to \text{hadrons}) = 0. \tag{36}$$

This highly convergent behavior is expected if it is possible to approximate the spectral function by a finite sum of δ function terms (e.g. at ρ, ω and ϕ).

The ratio $\int \rho^{\alpha\alpha}(m^2)dm^2 / \int [\rho^{\alpha\alpha}(m^2)/m^2]dm^2$ has the dimension of squared mass, and it is of some interest to examine its physical meaning. To this end let us, for a moment, talk about the propagator of the vector meson field rather than that of the vector current.

In 1961 K. Johnson[18] showed that the canonical commutation relations for a neutral vector meson field coupled to a conserved current imply the spectral-function sum rules[19]

$$\begin{aligned} &\int \sigma(m^2)dm^2 = 1, \\ &\int \frac{\sigma(m^2)}{m^2}dm^2 = \frac{1}{m_o^2} \end{aligned} \tag{37}$$

for the propagator of the vector field where m_o is the bare mass of the vector meson. If analogous formulas could be derived for the (non-Abelian) ρ meson field, we would immediately write down, by virtue of the proportionality relation between $\sigma(m^2)$ and $\rho(m^2)$,

$$\frac{\int \rho^{\alpha\alpha}(m^2)dm^2}{\int [\rho^{\alpha\alpha}(m^2)/m^2]dm^2} = (m_\alpha^{(0)})^2 \qquad (3$$

where $m_\alpha^{(0)}$ is the bare mass of the neutral vector meson associated with the current j_μ^α. Unfortunately, for the physically interesting case of massive Yang-Mills field the proof of (37) (as given by Lee Weinberg and Zumino[16]) hinges on the question of whether we can throw away the vacuum expectation value of a certain singular term in $[(\partial_o\rho_k^\alpha - \partial_k\rho_o^\alpha), \rho_\ell^\beta]$ (viz., the vacuum expectation value of $\rho_k^\alpha(0)\rho_\ell^\beta(0)$). In any case it is amusing that we can talk about measuring the "bare" ρ meson mass if future experiments indicate that the total hadronic cross section in electron-positron collisions goes to zero faster than $1/s^3$.

References

1) For a review see a talk presented at the Summer Institute for Theoretical Physics, University of Colorado, by J. J. Sakurai (to be published in Lectures in Theoretical Physics, Vol. XI, Gordon and Breach).

2) To the author's best knowledge the current-field identity for the ρ meson field was first considered by R. Schroer, R. Haag and K. Nishijima (unpublished). They pointed out to the author in the spring of 1961 that a relation of the type $j_\mu^\alpha \propto \rho_\mu^\alpha$ conveniently summarizes the essential features of ρ meson universality proposed in J. J. Sakurai, [Annals of Phys. 11, 1 (1960)]. At about the same time M. Gell-Mann and F. Zachariasen [Phys. Rev. 124, 953 (1961)] wrote down a matrix element relation fully equivalent to the current-field identity. More recently, this operator identity was extensively investigated by N. M. Kroll, T. D. Lee and B. Zumino [Phys. Rev. 157, 1376 (1967)].

3) S. Weinberg, Phys. Rev. Letters 18, 507 (1967). For applications to the electromagnetic current see T. Das, V. S. Mathur and S. Okubo, Phys. Rev. Letters 18, 761 (1967); R. J. Oakes and J. J. Sakurai, Phys. Rev. Letters 19, 1266 (1967).

4) H. J. Schnitzer and S. Weinberg, Phys. Rev. 164, 1828 (1967).

5) M. Gell-Mann, Phys. Rev. 125, 1067 (1962); Physics 1, 63 (1964).

6) J. Schwinger, Phys. Rev. Letters 3, 296 (1959); T. Goto and T. Imamura, Progr. Theoret. Phys. 14, 396 (1955).

7) Throughout the paper we assume that perturbation theory is valid and consider $e^+ + e^- \to A$ only to lowest order, i.e., to order α in the amplitude, to order α^2 in the colliding beam cross section.

8) J. D. Bjorken, Phys. Rev. 148, 1467 (1966).

9) W. R. Frazer and J. Fulco, Phys. Rev. Letters 2, 365 (1959); G. J. Gounaris and J. J. Sakurai, Phys. Rev. Letters 21, 244 (1968).

10) J. E. Augustin *et. al.*, Phys. Letters 28 B, 503, 508, 513, 517 (1969).

11) This can be taken as evidence for the absence of a unitary singlet component in the electromagnetic current. See, in particular, V. S. Mathur and S. Okubo (to be published).

12) J. Dooher, Phys. Rev. Letters 19, 600 (1967).

13) C. Bouchiat and L. Michel, J. Phys. Radium 22, 121 (1961); L. Durand III, Phys. Rev. 128, 441 (1962), and 129, 2835 (1963)(E).

14) G. J. Gounaris (to be published); M. Gourdin (to be published).

15) V. N. Gribov, B. L. Joffe and I. Ya. Pomeranchuk, Phys. Letters 24 B, 554 (1967).

16) T. D. Lee, S. Weinberg and B. Zumino, Phys. Rev. Letters 18, 1029 (1967).

17) G. Wentzel, *Quantum Theory of Fields* (translated by J. M. Jauch), Interscience Publishers (1949).

18) K. Johnson, Nuclear Phys. 25, 435 (1961).

19) See also G. Källén, Helv. Phys. Acta 25, 417 (1952); H. Lehmann, Nuovo Cimento 11, 342 (1954).

Received 3/15/69

A Relativistic Complex Pole Model with Indefinite Metric*

by

T. D. Lee†

C.E.R.N., Geneva, Switzerland

A relativistic complex pole model with an indefinite metric is described. The model enables one to study in detail scattering processes involving multiple exchanges of these negative metric quanta, corresponding to Feynman diagrams with one or more closed loops. The usual Feynman integration paths have to be modified, and the general rule for such modifications is given. The implication for causality of these multiple exchange processes is discussed.

1. Introduction

In a recent paper,[1] it was shown that the S-matrix of a theory with an indefinite metric can be unitary, provided there is no stable negative metric state in the theory. This suggests that, perhaps, one could regard the usual regularized propagators for the photon γ and for the intermediate boson $W^{\pm}$ as possible physical realities, and thereby resolve the presently existing serious divergence difficulties in both electromagnetic and weak interactions.[2] Explicit calculations have been made[3] for all $O(\alpha)$ radiative corrections and some of the higher order corrections in such a modified form of quantum electrodynamics. To the order verified, the S-matrix is, indeed, found to be unitary, Lorentz-invariant and divergence-free.[4] A general characteristic of such a theory is the breakdown of the usual analyticity requirement.

---

†On leave from Columbia University.

*This research was supported in part by the U. S. Atomic Energy Commission and the National Science Foundation.

The analytic continuation of any scattering amplitude may now have rather unusual complex poles and complex branch points on the physical sheet. For the realistic case of electromagnetic and weak interactions, while such changes in analyticity have been studied for lower order processes, a systematic analysis of all higher order modifications remains a difficult task, this difficulty is further complicated by the new feature that the usual Feynman integration paths have to be detoured.[1,3] In this paper, we shall discuss a simple relativistic model. By studying this mathematical model, one may then gain more insight into some of the technical complexities of high order processes in the realistic cases.

In order to have some perspective, let us briefly review the situation in quantum electrodynamics. In such a theory with indefinite metric, the electromagnetic interaction is given by

$$e\, j_\mu (A_\mu + i\, B_\mu) \tag{1.1}$$

where j_μ denotes the usual electromagnetic current, A_μ the usual electromagnetic field and B_μ a new negative-metric vector field of mass m. The free propagator of the combined field $(A_\mu + i\, B_\mu)$ is

$$\frac{1}{q^2} - \frac{1}{q^2 + m^2} = \frac{m^2}{q^2(q^2 + m^2)} \ . \tag{1.2}$$

Through the interaction (1.1), the (renormalized) propagator of $(A_\mu + i\, B_\mu)$ becomes[3]

$$\frac{M^2(0)}{q^2[q^2 + M^2(q^2)]} \tag{1.3}$$

where $[q^2 + M^2(q^2)]^{-1}$ has a cut along the real axis and two complex poles at

$$-q^2 = m^2 \pm i\gamma m \tag{1.4}$$

on the physical sheet. These complex poles are due simply to the fact that, for free fields, the negative-metric state at $q^2 = -m^2$ is degenerate with the appropriate continuum states, such as

$$e^+e^-,\ \mu^+\mu^-,\ \pi^+\pi^-,\ 2e^+2e^-,\ \text{etc.} \tag{1.5}$$

all of which are of positive metric. This degeneracy is removed by the anti-Hermitian interaction

$$i\, e\, j_\mu\, B_\mu \tag{1.6}$$

in (1.1); such an interaction necessarily moves these degenerate states from $-q^2 = m^2$ to $-q^2 = m^2 \pm i\gamma m$ in the complex plane, and, therefore, makes it possible to have a unitary S-matrix.

The presence of these complex poles is the general property of a theory with indefinite metric but unitary S-matrix. It implies that the S-matrix must be <u>different</u> from the usual $\lim_{t\to\infty} U(t,-t)$ in the interaction representation, since the former is well-defined by using the exact outgoing and incoming eigenstates of the total Hamiltonian, but the latter does not exist, being badly divergent (i.e., exponentially) in the limit. The usual derivation of Feynman rules, therefore, fails.

However, it has been shown in Ref. 1 that there exists an identity between the S-matrix and the finite sector of $U(t,-t)$, called $U^{reg}(t,-t)$, as $t \to \infty$. Consequently, correct results of the S-matrix can nevertheless be derived by using the usual Feynman rules, provided certain contours of integrations are modified. One of the purposes of this paper is to supply some further examples in which the correctness of the modified Feynman paths can be checked in detail against calculations based on the eigenvector solution of the total Hamiltonian.

As already discussed in Ref. 1, the presence of such complex poles has an important effect on the motion of wave packets. For example, we may consider the colliding beam reaction

$$e^+e^- \to \Gamma^0 \to \mu^+\mu^- \tag{1.7}$$

at the resonant energy, where Γ^o denotes the complex pole at

$$-q^2 = M^2 = m^2 + i\gamma m.$$

There should appear an apparent "precursor". Its amplitude can be calculated by studying the outgoing wave packet χ^{out} for $t < r$, where r denotes the relative distance between μ^+ and μ^- in the center of mass system. For clarity, we shall concentrate only on the radial dependence of the wave packet. Its angular dependence is simply proportional to $1 + \cos\theta$ or $1 - \cos\theta$, depending on the helicity states of the leptons; these angular factors will be omitted in the following expressions. For a normalized incident e^+e^- wave packet χ^{in} given by

$$|\chi^{in}|^2 \sim \frac{(\delta p)}{r^2} e^{-\delta p |r+t|}, \tag{1.8}$$

the outgoing $\mu^+\mu^-$ wave packet χ^{out} is found to have an estimated intensity

$$|\chi^{out}|^2 \sim \frac{(\alpha\gamma)^2}{(\delta p) r^2} e^{-\gamma(r-t)} \tag{1.9}$$

for $r - t >> (\delta p)^{-1}$, where α is the fine structure constant and (δp) is the momentum width of the incident e^+e^- wave packet, assuming that (δp) is much larger than γ. Since $t = 0$ is chosen to be the time of collision according to a classical picture, (1.9) shows clearly the appearance of $\mu^+\mu^-$ before $t = 0$.

While the effect of a simple complex pole (or a simple complex branch point) on the physical sheet is to have a precursor which decreases exponentially with increasing r, in general one may find precursors whose intensity decreases with r according to a power law. Such possibilities will be demonstrated in section 7 by explicit examples. From these examples, one expects to have such a precursor in, e.g.,

$$e^+e^- \to 2\Gamma^o \to \mu^+\mu^- \tag{1.10}$$

at the threshold energy $-q^2 = (M + M^*)^2 \cong 4m^2$. In this case, instead of (1.9), one may expect, as an order of magnitude estimation,

$$| \chi^{out} |^2 \sim \alpha^4 \frac{1}{(\delta p)\ m\ r^2} \left(\frac{1}{r-t}\right)^3 \tag{1.11}$$

for

$$(r-t) >> (\delta p)^{-1}. \tag{1.12}$$

The mathematical reason for this form of $| \chi^{out} |^2$ stems from the unusual property that across the point $-q^2 = (M + M^*)^2$ the scattering amplitude has a non-analytic jump due to the switching from one sheet of a multi-valued function to another sheet of the same function. The details of such a switching will be discussed in section 5.

It is of interest to inquire whether such a precursor can be experimentally detected at present. This will be discussed in section 7. As we shall see, because in present day high energy physics experiments our time resolutions are only $\sim 10^{-10}$ sec., a direct test of such interesting possibilities unfortunately seems to lie quite outside immediate experimental feasibility.

2. The model[5]

In the model, there are three neutral boson fields $A(x)$, $B(x)$ and $C(x)$. As we shall see, $A(x)$ acts the same role as the photon field $A_\mu(x)$ in electrodynamics, $B(x)$ is like the negative-metric field $B_\mu(x)$ in (1.1) and $C(x)$ is introduced to simulate the continuum states (1.5). For simplicity, all three fields are assumed to be of zero spin. The Lagrangian density of the A, B, C system is given by

$$\mathcal{L}_{ABC} = \mathcal{L}_0 + \mathcal{L}_1 \tag{2.1}$$

where

$$\mathcal{L}_0 = -\frac{1}{2}\left(\frac{\partial A}{\partial x_\mu}\right)^2 - \frac{1}{2}\left(\frac{\partial B}{\partial x_\mu}\right)^2 - \frac{1}{2}\left(\frac{\partial C}{\partial x_\mu}\right)^2 - \frac{1}{2}A^2\delta^2 - \frac{1}{2}(B^2 + C^2)\, m^2, \quad (2.2)$$

$$\mathcal{L}_1 = -i\,\gamma\, m\, B\, C, \quad (2.3)$$

δ, m, γ are all real parameters, $x_\mu = (\underset{\sim}{r}, it)$ and the repeated subscript μ is summed over from 1 to 4. [Other interactions will be included later. See (2.18) for the total Lagrangian $\mathcal{L}$.]

The fields A, B and C satisfy, respectively,

$$\eta^{-1} A^\dagger \eta = A, \qquad \eta^{-1} B^\dagger \eta = -B$$

and (2.4)

$$\eta^{-1} C^\dagger \eta = C$$

where † denotes Hermitian conjugation and η is the metric; therefore, the Lagrangian $\mathcal{L}_{ABC}$ satisfies the pseudo-Hermitian condition

$$\eta^{-1} \mathcal{L}^\dagger_{ABC}\, \eta = \mathcal{L}_{ABC}. \quad (2.5)$$

A particular representation of η is

$$\eta_o = (-1)^{N_B} \quad (2.6)$$

where N_B denotes the total occupation number operators of B-quanta. In this representation $\eta = \eta_o$, one has

$$A = A^\dagger,\ B = B^\dagger,\ C = C^\dagger,$$

$$\mathcal{L}_0 = \mathcal{L}_0^\dagger \text{ but } \mathcal{L}_1 = -\mathcal{L}_1^\dagger. \quad (2.7)$$

In the following, unless specified, the metric η is not restricted to any specific representation; instead, we assume η to be of the general form[6]

$$\eta = T^{\dagger} \eta_o T \tag{2.8}$$

where T can be any non-singular matrix.

From the Lagrangian, one finds that the conjugate momenta of A, B and C are, respectively,

$$P_A = \frac{\partial A}{\partial t}, \qquad P_B = \frac{\partial B}{\partial t}$$

$$\text{and} \qquad P_C = \frac{\partial C}{\partial t}. \tag{2.9}$$

Similarly to (2.4), these conjugate momenta satisfy

$$\eta^{-1} P_A^{\dagger} \eta = P_A, \qquad \eta^{-1} P_B^{\dagger} \eta = -P_B$$

$$\text{and} \qquad \eta^{-1} P_C^{\dagger} \eta = P_C. \tag{2.10}$$

By following the usual canonical procedures, one finds that the Hamiltonian for the A, B, C system is given by

$$H_{ABC} = \frac{1}{2} [P_A^2 + (\nabla A)^2 + A^2\delta^2 + P_B^2 + (\nabla B)^2 + B^2m^2 + P_C^2 + (\nabla C)^2 + C^2m^2]$$

$$+ i \gamma m BC. \tag{2.11}$$

The quantization is determined by

$$[P_A(\underset{\sim}{r},t), A(\underset{\sim}{r}',t)] = [P_B(\underset{\sim}{r},t), B(\underset{\sim}{r}',t)] = [P_C(\underset{\sim}{r},t), C(\underset{\sim}{r}',t)]$$

$$= -i\delta^3(\underset{\sim}{r}-\underset{\sim}{r}'), \tag{2.12}$$

while all other equal-time commutators are zero.

The spectrum of the Hamiltonian H_{ABC} can be readily derived. The simplest approach is to consider the propagators. The propagator of A is

$$D_{AA} = \frac{-i}{q^2 + \delta^2 - i\varepsilon} \tag{2.13}$$

where $\varepsilon = 0+$; those of the B, C fields should be described by a matrix

$$\begin{pmatrix} D_{BB} & D_{BC} \\ D_{CB} & D_{CC} \end{pmatrix} \tag{2.14}$$

where the first and second subscripts denote, respectively, the initial and the final fields. Since, according to (2.3), the coupling connecting B and C is simply γm, one can easily sum up the series involving transitions $B \to C \to B$, $B \to C \to B \to C \to B$, etc. The result is

$$D_{BB} = D_{CC} = -\frac{1}{2} i \left[\frac{1}{q^2 + m^2 + i\gamma m} + \frac{1}{q^2 + m^2 - i\gamma m} \right]. \tag{2.15}$$

Similarly, by summing over transitions $B \to C$, $B \to C \to B \to C$, etc., one finds

$$D_{BC} = D_{CB} = -\frac{\gamma m}{(q^2 + m^2)^2 + \gamma^2 m^2}. \tag{2.16}$$

Thus, one sees that the degeneracy between B and C is removed by the coupling $\mathcal{L}_1$, changing the poles from $-q^2 = m^2$ to

$$-q^2 = m^2 \pm i\gamma m, \tag{2.17}$$

in complete analogy to (1.4) in electrodynamics. Identical results can also be obtained by directly diagonalizing the Hamiltonian H_{ABC}; the details will be given in the next section.

It is clear that without further interaction terms, the model is too trivial to be of any interest. We shall now generalize the model to include an additional field ψ, simulating the various lepton and hadron fields in electrodynamics. The Lagrangian for the entire system is assumed to be of the form

$$\mathcal{L} = \mathcal{L}_{ABC} + \mathcal{L}_{free}(\psi) + \mathcal{L}_{int}(\psi, A + i B) \quad (2.18)$$

where $\mathcal{L}_{ABC}$ is given by (2.1), $\mathcal{L}_{free}(\psi)$ denotes the free Hamiltonian of ψ

$$\mathcal{L}_{free}(\psi) = -\frac{1}{2}\left(\frac{\partial\psi}{\partial x_\nu}\right)^2 - \frac{1}{2}\mu^2\psi^2, \quad (2.19)$$

and $\mathcal{L}_{int}$ denotes the interaction which depends on the ABC system only through the combined field

$$\phi = A + i B \quad (2.20)$$

in analogy to (1.1). For simplicity, throughout the paper, ψ is assumed to be neutral, of zero spin, and of positive metric; i.e., in the special representation $\eta = \eta_o$, given by (2.6),

$$\psi = \psi^\dagger \quad \text{and} \quad [\eta,\psi] = 0. \quad (2.21)$$

As examples of $\mathcal{L}_{int}$, we may list

$$f\psi^2\phi, \text{ or } f\psi\phi^2, \text{ or the more general form } f\psi^m\phi^n \quad (2.22)$$

where f is real. The advantage of the model lies in the possibility of studying multiple virtual ϕ exchange processes (corresponding to multi-photon processes in the realistic case) by using simply the second order perturbation formula. Since the propagator of ϕ is, on account of (2.13), (2.15) and (2.20),

$$D_\phi = D_{AA} - D_{BB}$$

$$= -\frac{1}{2} i \left[\frac{2}{q^2 + \delta^2 - i\varepsilon} - \frac{1}{q^2 + m^2 + i\gamma m} - \frac{1}{q^2 + m^2 - i\gamma m}\right], \quad (2.2$$

which has complex poles, the appropriate Feynman integration paths for these multiple ϕ-exchange processes must be modified.

Let us consider any general Feynman graph by using, say, one of the above interaction Lagrangians in (2.22). We recall that if γ were zero, then the usual Feynman rule would hold. Each Feynman graph would then lead to a typical integral

$$\int\cdots d^3\underset{\sim}{q}''d^3\underset{\sim}{q}'d^3\underset{\sim}{q}\cdots\int_{-\infty}^{\infty}dq_o''\int_{-\infty}^{\infty}dq_o'\int_{-\infty}^{\infty}dq_o\, R(q,q',q'',\cdots)_{\gamma=0} \tag{2.24}$$

where R is a rational function of the virtual momenta q, q', q'',·· . We note that, keeping all three-momenta $\underset{\sim}{q}$, $\underset{\sim}{q}'$, $\underset{\sim}{q}''$,·· fixed, in each successive integration of the fourth-component of these virtual momenta dq_o, dq_o', dq_o'',··, the corresponding integrand continues to be a rational function of the relevant fourth-component variables q_o, q_o', q_o'',··. The pole positions of these rational functions are determined by the appropriate $\pm i\varepsilon$ factors given by the usual Feynman rule. Alternatively, we may regard (still for $\gamma=0$) all poles to be on the real axis, but the Feynman paths are slightly detoured so as to go either above or below the appropriate poles. Now, as γ increases from zero to its final value, these poles will move continuously. We shall require the contours C, C', C'',··· in each of these integrations dq_o, dq_o', dq_o'',·· to be continuously detoured in such a way that <u>none</u> of these poles ever <u>crosses</u> the corresponding contour. For $\gamma\neq0$, instead of (2.24), the integral for the same Feynman graph becomes simply

$$\int\cdots d^3\underset{\sim}{q}''d^3\underset{\sim}{q}'d^3\underset{\sim}{q}\cdots\int_{C''}dq_o''\int_{C'}dq_o'\int_{C}dq_o\, R(q,q',q''\cdots)_{\gamma\neq0}. \tag{2.25}$$

In this model, by choosing different interaction Lagrangians, one can obtain a second order Feynman graph containing as many integrations in virtual momenta as one wants. This makes it possible to check in detail the correctness of the above modified Feynman path against the usual second order perturbation expression in terms of the corresponding interaction Hamiltonian. We note that these modified Feynman paths are unambiguous if the poles do not coalesce. Nevertheless, as we shall see in the following sections (especially Section 5), there is no difficulty in applying (2.25) to cases when there is coalescence of poles, such as the double-ϕ, or multi-ϕ, exchange processes.

3. Diagonalization of H_{ABC}

The Hamiltonian H_{ABC} can be diagonalized in any given Lorentz frame by using the following Fourier expansions

$$A(x) = \sum (2\nu_k \Omega)^{-1/2} [\alpha(\underset{\sim}{k}) e^{i\underset{\sim}{k}\cdot\underset{\sim}{r}} + \alpha(\underset{\sim}{k})^\dagger e^{-i\underset{\sim}{k}\cdot\underset{\sim}{r}}], \tag{3.1}$$

$$\begin{aligned} B(x) = &\sum \frac{1}{2} (\omega_k \Omega)^{-1/2} [\beta_+(\underset{\sim}{k}) e^{i\underset{\sim}{k}\cdot\underset{\sim}{r}} + \beta_+(\underset{\sim}{k})^\dagger e^{-i\underset{\sim}{k}\cdot\underset{\sim}{r}}] \\ &- \sum \frac{1}{2} (\omega_k^* \Omega)^{-1/2} [\beta_-(\underset{\sim}{k}) e^{i\underset{\sim}{k}\cdot\underset{\sim}{r}} + \beta_-(\underset{\sim}{k})^\dagger e^{-i\underset{\sim}{k}\cdot\underset{\sim}{r}}], \end{aligned} \tag{3.2}$$

$$\begin{aligned} C(x) = &\sum \frac{1}{2} (\omega_k \Omega)^{-1/2} [\beta_+(\underset{\sim}{k}) e^{i\underset{\sim}{k}\cdot\underset{\sim}{r}} + \beta_+(\underset{\sim}{k})^\dagger e^{-i\underset{\sim}{k}\cdot\underset{\sim}{r}}] \\ &+ \sum \frac{1}{2} (\omega_k^* \Omega)^{-1/2} [\beta_-(\underset{\sim}{k}) e^{i\underset{\sim}{k}\cdot\underset{\sim}{r}} + \beta_-(\underset{\sim}{k})^\dagger e^{-i\underset{\sim}{k}\cdot\underset{\sim}{r}}] \end{aligned} \tag{3.3}$$

$$P_A(x) = -i \sum (\nu_k/2\Omega)^{1/2} [\alpha(\underset{\sim}{k}) e^{i\underset{\sim}{k}\cdot\underset{\sim}{r}} - \alpha(\underset{\sim}{k})^\dagger e^{-i\underset{\sim}{k}\cdot\underset{\sim}{r}}] \tag{3.4}$$

$$\begin{aligned} P_B(x) = &-i \sum \frac{1}{2} (\omega_k/\Omega)^{1/2} [\beta_+(\underset{\sim}{k}) e^{i\underset{\sim}{k}\cdot\underset{\sim}{r}} - \beta_+(\underset{\sim}{k})^\dagger e^{-i\underset{\sim}{k}\cdot\underset{\sim}{r}}] \\ &+ i \sum \frac{1}{2} (\omega_k^*/\Omega)^{1/2} [\beta_-(\underset{\sim}{k}) e^{i\underset{\sim}{k}\cdot\underset{\sim}{r}} - \beta_-(\underset{\sim}{k})^\dagger e^{-i\underset{\sim}{k}\cdot\underset{\sim}{r}}] \end{aligned} \tag{3.5}$$

and

$$\begin{aligned} P_C(x) = &-i \sum \frac{1}{2} (\omega_k/\Omega)^{1/2} [\beta_+(\underset{\sim}{k}) e^{i\underset{\sim}{k}\cdot\underset{\sim}{r}} - \beta_+(\underset{\sim}{k})^\dagger e^{-i\underset{\sim}{k}\cdot\underset{\sim}{r}}] \\ &- i \sum \frac{1}{2} (\omega_k^*/\Omega)^{1/2} [\beta_-(\underset{\sim}{k}) e^{i\underset{\sim}{k}\cdot\underset{\sim}{r}} - \beta_-(\underset{\sim}{k})^\dagger e^{-i\underset{\sim}{k}\cdot\underset{\sim}{r}}] \end{aligned} \tag{3.6}$$

where P_A, P_B and P_C are the conjugate momenta of the fields A, B and C, Ω is the volume of the system, the summation extends over all $\underset{\sim}{k}$,

$$\omega_k = (\underset{\sim}{k}^2 + m^2 + i\,\gamma\, m)^{1/2} \tag{3.7}$$

and $$\nu_k = (\underset{\sim}{k}^2 + \delta^2)^{1/2}. \tag{3.8}$$

The canonical commutation rules are satisfied by having

$$[\alpha(\underset{\sim}{k}),\ \alpha(\underset{\sim}{q})^\dagger] = [\beta_+(\underset{\sim}{k}),\ \beta_+(\underset{\sim}{q})^\dagger] = [\beta_-(\underset{\sim}{k}),\ \beta_-(\underset{\sim}{q})^\dagger] = \delta_{\underset{\sim}{k}\underset{\sim}{q}} \tag{3.9}$$

and all other equal-time commutators zero. Furthermore, in order to satisfy (2.4) and (2.10), we have chosen a specific representation for the metric

$$\eta = \eta_o' , \tag{3.10}$$

where $\eta_o' = \exp\{\frac{1}{2} i\pi \sum [\beta_+(\underset{\sim}{k})^\dagger - \beta_-(\underset{\sim}{k})^\dagger]\,[\beta_+(\underset{\sim}{k}) - \beta_-(\underset{\sim}{k})]\}$. (3.11).

In this representation (3.10), η satisfies

$$\eta^2 = 1, \qquad \eta^\dagger = \eta ;$$

η commutes with $\alpha(k)$, but

$$\eta\beta_-(\underset{\sim}{k})\,\eta = \beta_+(\underset{\sim}{k}) \qquad \text{and} \qquad \eta\beta_+(\underset{\sim}{k})\,\eta = \beta_-(\underset{\sim}{k}) . \tag{3.12}$$

The metric η_o' is, of course, related to η_o, Eq. (2.6), by a T-transformation (2.8). However, the transformation matrix T is <u>not</u> unitary.

Upon substituting (3.1)-(3.6) into (2.11), one finds, apart from a trivial constant additive term,

$$H_{ABC} = \sum [\nu_k\, \alpha(\underset{\sim}{k})^\dagger\, \alpha(\underset{\sim}{k}) + \omega_k \beta_+(\underset{\sim}{k})^\dagger\, \beta_+(\underset{\sim}{k}) + \omega_k^* \beta_-(\underset{\sim}{k})^\dagger\, \beta_-(\underset{\sim}{k})] \tag{3.13}$$

which, of course, agrees with our previous conclusions by using propagators.

To each eigenstate $|\, n >$ of H_{ABC} there exists a conjugate state $|\, n^c >$ such that

$$< n^c\, |\eta|\, n' > = I_{nn'} \tag{3.14}$$

where I is the unit matrix. For example, in the representation $\eta = \eta_o'$, the conjugate state for $\alpha(\underset{\sim}{k})^\dagger | \text{vac} >$ is itself, but those for $\beta_+(\underset{\sim}{k})^\dagger | \text{vac} >$ and $\beta_-(\underset{\sim}{k})^\dagger | \text{vac} >$ are, respectively, $\beta_-(\underset{\sim}{k})^\dagger | \text{vac} >$ and $\beta_+(\underset{\sim}{k})^\dagger | \text{vac} >$.

4. Single ϕ exchange processes

We consider in this section the simplest non-trivial interaction in (2.22):

$$\mathcal{L}_{int} = f\psi^2\phi \tag{4.1}$$

where f is real. For definiteness, let us first discuss the mass shift $\delta\mu$ of the ψ particle due to this interaction. [More complicated interactions will be discussed later. See Eq. (4.15).]

(i) Feynman diagram. The Feynman diagram for $\delta\mu$ is the usual one given by Fig. 1. According to (2.25), the corresponding integral is

$$\sum(p) = -i \frac{f^2}{4\pi^4} \int d^3\underset{\sim}{q} \int_C dq_o \; D_\psi(p-q) \; D_\phi(q) \tag{4.2}$$

where the ϕ-propagator D_ϕ is given by (2.23), and the free ψ-propagator is

$$D_\psi(k) = -i(k^2 + \mu^2 - i\varepsilon)^{-1}.$$

The mass shift $\delta\mu$ is related to $\sum(p)$ on the mass shell:

$$\delta\mu^2 = \sum(p) \quad \text{at } p^2 = \underset{\sim}{p}^2 - p_o^2 = -\mu^2. \tag{4.3}$$

The function $\sum(p)$ can also be used to construct the complete ψ-propagator D_ψ' with interactions. We have the usual relation

$$D_\psi'(p) = -i[p^2 + \mu^2 + \sum(p)]^{-1}. \tag{4.4}$$

The integrand in (4.2) has eight poles in the complex q_o-plane:

$$\omega,\ \omega^*,\ \nu - i\varepsilon,\ p_o + E - i\varepsilon \tag{4.5}$$

$$-\omega,\ -\omega^*,\ \nu + i\varepsilon \text{ and } p_o - E + i\varepsilon \tag{4.6}$$

where
$$\omega = (\underset{\sim}{q}^2 + m^2 + i\,\gamma\, m)^{1/2}, \tag{4.7}$$

$$\nu = (\underset{\sim}{q}^2 + \delta^2)^{1/2} \tag{4.8}$$

and
$$E = [(\underset{\sim}{p} - \underset{\sim}{q})^2 + \mu^2]^{1/2}; \tag{4.9}$$

the integration $d^3\underset{\sim}{q}$ is real, but the integration dq_o is along a path C, shown in Fig. 2.

To prove the correctness of the above expression for $\delta\mu^2$, we shall compare it directly with the result obtained by using the eigenstates of the Hamiltonian. Such a comparison can be most easily made after one carries out the q_o integration. The integral (4.2) becomes

$$\begin{aligned}\Sigma(p) = (8\pi^3)^{-1} f^2 \int d^3\underset{\sim}{q} \Bigg[&\frac{1}{\nu E}\left(\frac{1}{p_o - \nu - E + i\varepsilon} - \frac{1}{p_o + \nu + E - i\varepsilon}\right)\\ &- \frac{1}{2\omega E}\left(\frac{1}{p_o - \omega - E} - \frac{1}{p_o + \omega + E}\right)\\ &- \frac{1}{2\omega^* E}\left(\frac{1}{p_o - \omega^* - E} - \frac{1}{p_o + \omega^* + E}\right)\Bigg].\end{aligned} \tag{4.10}$$

The explicit functional form of $\Sigma(p)$ will be given later by Eq. (4.24).

<u>(ii) Hamiltonian method</u>. We start with the second order perturbation formula

$$\Delta\mathcal{E}_n = \sum_{m\neq n} (\mathcal{E}_n - \mathcal{E}_m)^{-1} < n^c \,|\eta H_{int}|\, m > < m^c \,|\eta H_{int}|\, n > \tag{4.11}$$

where $| n >$ denotes the eigenstates of

$$H_{ABC} + H_{free}(\psi),$$

$\mathcal{E}_n$ is its eigenvalue and $| n^c >$ its conjugate state, defined by (3.14). To evaluate the change in energy p_o of a ψ-particle with momentum $\underset{\sim}{p}$ (denoted simply by $\psi_{\underset{\sim}{p}}$ in the following), we observe that only the following virtual transitions are relevant:

$$\psi_{\underset{\sim}{p}} \rightarrow \psi_{\underset{\sim}{p}-\underset{\sim}{q}} \phi_{\underset{\sim}{q}} \rightarrow \psi_{\underset{\sim}{p}},$$

$$\psi_{\underset{\sim}{p}} \rightarrow \psi_{\underset{\sim}{p}} \psi_{\underset{\sim}{p}} \psi_{-\underset{\sim}{p}+\underset{\sim}{q}} \phi_{-\underset{\sim}{q}} \rightarrow \psi_{\underset{\sim}{p}} \tag{4.12}$$

and

$$\text{vacuum} \rightarrow \psi_{\underset{\sim}{p}} \psi_{-\underset{\sim}{p}+\underset{\sim}{q}} \phi_{-\underset{\sim}{q}} \rightarrow \text{vacuum}$$

where $\phi_{\underset{\sim}{q}}$ can be either

$$A_{\underset{\sim}{q}} \equiv \alpha(\underset{\sim}{q})^{\dagger} \; | \; vac > \tag{4.13}$$

or

$$B_{\underset{\sim}{q}} \equiv \beta_{\pm}(\underset{\sim}{q})^{\dagger} \; | \; vac > . \tag{4.14}$$

The energy shift δp_o is given by the difference in $\Delta\mathcal{E}_n$ between $n = \psi_{\underset{\sim}{p}}$ and the vacuum state. By using

$$H_{int} = -f\psi^2\phi$$

and the explicit representations of $A + i B$ given in section 3, it is straightforward to verify that for $p_o = (\underset{\sim}{p}^2 + \mu^2)^{1/2}$

$$\delta p_o = \frac{1}{2} p_o^{-1} \sum(p)$$

where $\sum(p)$ is given by (4.10). Thus, we establish the validity of the above modified integration path for the Feynman integral.

<u>(iii) Unitarity</u>. For the interaction (4.1), it is easy to see that its S-matrix has no non-zero $O(f)$ matrix elements. Thus, to order f^2, an explicit verification of the unitarity condition is almost trivial. For example, for the one particle state, unitarity simply means that $\delta\mu^2$ should be real. From (4.10), one sees that this is indeed the case since $\Sigma(p)$ is real at $p^2 = -\mu^2$.

To investigate unitarity for more complicated processes, we may consider a different interaction; instead of (4.1), let us assume the interaction Lagrangian $\mathcal{L}_{int}$ to be

$$\mathcal{L}_{int} = f\psi^3\phi. \tag{4.15}$$

For this new interaction, there are transitions to first order in f, such as

$$\psi_{\mathbf{p}_1} + \psi_{\mathbf{p}_2} \rightleftarrows \psi_{\mathbf{k}} + A_{\mathbf{q}}. \tag{4.16}$$

The corresponding matrix element of the S-matrix is given by

$$\langle \psi_{\mathbf{k}}, A_{\mathbf{q}} | S | \psi_{\mathbf{p}_1}, \psi_{\mathbf{p}_2} \rangle = -i(16\pi^2)^{-1}(2E_1E_2E\nu)^{-1/2}(3!)f\delta^4(p_1 + p_2 - k - q) \tag{4.17}$$

where E_1 and E_2 denote the two initial energies, and E and ν refer to the two final energies.

To order f^2, through this new interaction one has elastic scattering processes, such as

$$\psi_{\mathbf{p}_1} + \psi_{\mathbf{p}_2} \rightarrow \psi_{\mathbf{p}_1'} + \psi_{\mathbf{p}_2'}. \tag{4.18}$$

It can be readily verified that the corresponding $O(f^2)$ matrix element of the S-matrix is

$$-i(3!)^2(128\pi^2)^{-1}(E_1E_2E_1'E_2')^{-1/2}\,\delta^4(p_1 + p_2 - p_1' - p_2') \times [\Sigma(p) + \Sigma(p') + \Sigma(p'')] \tag{4.19}$$

where E'_1 and E'_2 denote the two final energies, $\sum(p)$ is given by (4.2), and the 4-momenta p, p' and p'' are, respectively,

$$p = p_1 + p_2 = p'_1 + p'_2 ,$$

$$p' = p_1 - p'_1 \qquad (4.20)$$

and

$$p'' = p_1 - p''_2 .$$

From Fig. 1, one sees that the virtual ϕ_q in $\sum(p)$ can be either $A_{\underset{\sim}{q}}$ or $B_{\underset{\sim}{q}}$, while for the physical transition (4.16) only $A_{\underset{\sim}{q}}$ can be present. Thus, in order to satisfy unitarity, only the case $\phi_{\underset{\sim}{q}} = A_{\underset{\sim}{q}}$ can contribute to the imaginary part of $\sum(p)$. From (4.9), one finds for p_o real and ≥ 0

$$\mathrm{Im} \sum(p) = -(8\pi^2)^{-1} f^2 \int (\nu E)^{-1} \, d^3\underset{\sim}{q}\, \delta(E + \nu - p_o) \qquad (4.21)$$

if

$$p_o \geq [(\delta + \mu)^2 + \underset{\sim}{p}^2]^{1/2}$$

and, as is also implied by (4.21),

$$\mathrm{Im} \sum(p) = 0, \text{ otherwise}, \qquad (4.2$$

which is exactly the condition required by unitarity.

<u>(iv) Lorentz invariance</u>. The simplest way to show that $\sum(p)$ is invariant under a Lorentz transformation is to consider first the case p_o^2 real $< \mu^2$. The contour C in (4.2) can then be transformed into an integration path along the imaginary axis from $q_o = -i\infty$ to $i\infty$. It then follows that the integral (4.2) depends only on p^2. For p_o^2 real $> \mu^2$, $\sum(p)$ is defined by its appropriate analytic continuation; consequently, $\sum(p)$ depends also only on p^2.

(v) Analyticity. The integral (4.2) can be readily evaluated. It is convenient to introduce

$$z = -p^2 = p_o^2 - \underset{\sim}{p}^2 \tag{4.23}$$

as the variable. We find

$$\begin{aligned}\Sigma(p) = -(8\pi^2)^{-1} f^2 \{ & \frac{1}{2z}(\mu^2 - M^2 - z)\ln\frac{\mu^2}{M^2} + \frac{1}{2z}(\mu^2 - M^{*2} - z)\ln\frac{\mu^2}{M^{*2}} \\ & - \frac{1}{z}(\mu^2 - \delta^2 - z)\ln\frac{\mu^2}{\delta^2} \\ & + [2M^2 - \frac{1}{2z}(\mu^2 - M^2 - z)^2]\, J(z;M,\mu) \\ & + [2M^{*2} - \frac{1}{2z}(\mu^2 - M^{*2} - z)^2]\, J(z;M^*,\mu) \\ & - [4\delta^2 - \frac{1}{z}(\mu^2 - \delta^2 - z)^2]\, J(z;\delta,\mu)\}\end{aligned} \tag{4.24}$$

in which

$$M^2 = m^2 + i\,\gamma\, m \tag{4.25}$$

and

$$J(z;\kappa,\mu) = \int_0^1 \frac{1}{\kappa^2(1-\xi) + \mu^2\xi - \xi(1-\xi)z}\, d\xi\ . \tag{4.26}$$

The latter is an elementary integral; its explicit expression is

$$J(z;\kappa,\mu) = F^{-1}\ln\left(\frac{\mu^2 + \kappa^2 - z + F}{\mu^2 + \kappa^2 - z - F}\right) \tag{4.27}$$

where

$$F = [z - (\mu+\kappa)^2]^{1/2}\,[z - (\mu-\kappa)^2]^{1/2}\ . \tag{4.28}$$

In (4.26), the denominator of its integrand vanishes at

$$z = \frac{\kappa^2}{\xi} + \frac{\mu^2}{1-\xi} \tag{4.29}$$

where κ denotes either δ, or M, or M*.

For $\kappa=\delta$, as ξ varies from 0 to 1, (4.29) describes a cut along the real axis from

$$z = (\mu+\delta)^2 \text{ to } \infty. \tag{4.30}$$

For $\kappa=M$ (or M*), (4.29) becomes a hyperbola which lies in the first (or fourth) quadrant in the complex plane $z = x + iy$; its asymptotes are

$$my - \gamma x + \gamma\mu^2 = 0$$

and

$$y = \gamma m \tag{4.31}$$

(or, $my + \gamma x - \gamma\mu^2 = 0$ and $y = -\gamma m$). Thus, according to (4.26), along the real axis $J(z;M,\mu)$ and $J(z;M^*,\mu)$ are both analytic; both functions can be analytically continued beyond (4.29). The analytic continuation of $J(z;M,\mu)$ has a branch point at $(\mu+M)^2$, and that of $J(z;M^*,\mu)$ has a branch point at $(\mu+M^*)^2$. The function $J(z;\delta,\mu)$ is everywhere analytic, except along the branch cut (4.30).

Using these explicit expressions, one can directly verify that (4.21) is valid; i.e., along the real axis, one has, for $z < (\mu+\delta)^2$,

$$\text{Im} \sum = 0$$

and, for z varying from $(\mu+\delta)^2 + i\varepsilon$ to $\infty + i\varepsilon$,

$$\text{Im} \sum = -(4\pi z)^{-1} f^2 \, [z - (\mu+\delta)^2]^{1/2} \, [z - (\mu-\delta)^2]^{1/2}. \tag{4.32}$$

5. Double ϕ exchange processes

To study double ϕ exchange processes, we assume the interaction Lagrangian $\mathcal{L}_{int}$ in (2.18) to be simply

$$\mathcal{L}_{int} = f\psi\phi^2 . \tag{5.1}$$

For definiteness, we calculate the mass shift $\delta\mu$ of the ψ field. The discussions in this section are completely parallel to those given in the previous section.

(i) Feynman diagram. The Feynman diagram for $\delta\mu$ is given by Fig. 3. Similarly to (4.2), the corresponding integral is

$$\Sigma(p) = -i \frac{f^2}{8\pi^4} \int d^3\underset{\sim}{q} \int_C dq_o \, D_\phi(p-q) D_\phi(q) \tag{5.2}$$

and

$$\delta\mu^2 = \Sigma(p) \quad \text{at} \quad p^2 = -\mu^2 . \tag{5.3}$$

From (2.23), one sees that the integrand in (5.2) has twelve poles in the complex q_o-plane:

$$\omega_q, \ \omega_q^*, \ \nu_q - i\varepsilon, \ p_o + \omega_k, \ p_o + \omega_k^*, \ p_o + \nu_k - i\varepsilon \tag{5.4}$$

$$-\omega_q, \ -\omega_q^*, \ -\nu_q + i\varepsilon, \ p_o - \omega_k, \ p_o - \omega_k^* \text{ and } p_o - \nu_k + i\varepsilon \tag{5.5}$$

where

$$\nu_q = (\underset{\sim}{q}^2 + \delta^2)^{1/2}, \qquad \nu_k = (\underset{\sim}{k}^2 + \delta^2)^{1/2}$$
$$\omega_q = (\underset{\sim}{q}^2 + m^2 + i\gamma\, m)^{1/2}, \qquad \omega_k = (\underset{\sim}{k}^2 + m^2 + i\gamma\, m)^{1/2} \tag{5.6}$$

and

$$\underset{\sim}{k} = \underset{\sim}{p} - \underset{\sim}{q}.$$

The integration $d^3\underset{\sim}{q}$ is real, but the contour[7] C for the integration dq_o is shown in Fig. 4.

After carrying out the q_o-integration, $\Sigma(p)$ becomes

$$\Sigma(p) = (16\pi^3)^{-1} f^2 \int d^3\underset{\sim}{q} \left\{ \frac{1}{\nu_q \nu_k} \left(\frac{1}{p_o - \nu_q - \nu_k + i\varepsilon} - \frac{1}{p_o + \nu_q + \nu_k - i\varepsilon} \right) \right. \tag{5.7}$$

(equation continued on next page)

$$- \frac{1}{2}\left[\frac{1}{\nu_q \omega_k}\left(\frac{1}{p_o - \nu_q - \omega_k} - \frac{1}{p_o + \nu_q + \omega_k}\right) + \text{c.c.}\right]$$

$$- \frac{1}{2}\left[\frac{1}{\omega_q \nu_k}\left(\frac{1}{p_o - \omega_q - \nu_k} - \frac{1}{p_o + \omega_q + \nu_k}\right) + \text{c.c.}\right]$$

$$+ \frac{1}{4}\left[\frac{1}{\omega_q \omega_k}\left(\frac{1}{p_o - \omega_q - \omega_k} - \frac{1}{p_o + \omega_q + \omega_k}\right) + \text{c.c.}\right] \qquad (5.7) \text{ continued}$$

$$+ \frac{1}{4}\left[\frac{1}{\omega_q \omega_k^*}\left(\frac{1}{p_o - \omega_q - \omega_k^*} - \frac{1}{p_o + \omega_q + \omega_k^*}\right) + \text{c.c.}\right]\Bigg\}$$

As we shall see, identical results can be obtained by using the appropriate eigenstates of the total Hamiltonian.

<u>(ii) Hamiltonian method</u>. To evaluate the change in energy

$$p_o = (\underset{\sim}{p}^2 + \mu^2)^{1/2} \qquad (5.8)$$

of a ψ-particle with momentum $\underset{\sim}{p}$, we note that, similarly to (4.12), only the following virtual transitions are relevant:

$$\psi_{\underset{\sim}{p}} \rightarrow \phi_{\underset{\sim}{p}-\underset{\sim}{q}}\ \phi_{\underset{\sim}{q}} \rightarrow \psi_{\underset{\sim}{p}} \qquad (5.9)$$

$$\psi_{\underset{\sim}{p}} \rightarrow \psi_{\underset{\sim}{p}}\psi_{\underset{\sim}{p}}\ \phi_{-\underset{\sim}{p}+\underset{\sim}{q}}\ \phi_{-\underset{\sim}{q}} \rightarrow \psi_{\underset{\sim}{p}}$$

and

$$\text{vacuum} \rightarrow \psi_{\underset{\sim}{p}}\ \phi_{-\underset{\sim}{p}+\underset{\sim}{q}}\ \phi_{-\underset{\sim}{q}} \rightarrow \text{vacuum.}$$

In order to apply the second order perturbation formula (4.11), it is necessary to verify that in these transitions the intermediate states are not degenerate with the initial energy p_o. For clarity, let us consider first the case

$$\mu < 2\delta \qquad (5.10)$$

so that $\qquad p_o < \nu_k + \nu_q,$

where $\underset{\sim}{k}$ denotes $(\underset{\sim}{p}-\underset{\sim}{q})$; therefore $\psi_{\underset{\sim}{p}}$ cannot decay into $A_{\underset{\sim}{k}}A_{\underset{\sim}{q}}$. Since ω_k and ω_q are both complex, the initial energy p_o must be different from either $\omega_k+\nu_q$, or $\omega_k+\omega_q$, or their complex conjugates. For the remaining intermediate-state-energy $\omega_k^*+\omega_q$, we shall take advantage of the fact that $\sum(p)$ is Lorentz invariant, a property which will be explicitly verified later. Therefore, we may take $\underset{\sim}{p}$ to be any non-zero three-vector:

$$\underset{\sim}{p} \neq 0 . \tag{5.11}$$

In order that

$$p_o = \omega_k^* + \omega_q , \tag{5.12}$$

one must have, on account of (5.6),

$$|\underset{\sim}{p}-\underset{\sim}{q}| = |\underset{\sim}{q}| \tag{5.13}$$

and

$$\text{Re } \omega_q = \frac{1}{2} p_o . \tag{5.14}$$

In the $\underset{\sim}{q}$-space, the points that satisfy these two conditions (5.13) and (5.14) must lie on a circle whose center is at $\frac{1}{2}\underset{\sim}{p}$; such a one-dimensional distribution of points clearly gives zero contribution to a three-dimensional integration,[8] such as (5.7), in which the integrand does not contain singularities worse than $(p_o-\omega_k^*-\omega_q)^{-1}$. The integration over $\underset{\sim}{q}$ is therefore absolutely convergent and well-defined; the ambiguity of the integrand over such a set of points of zero measure (a circle) in $\underset{\sim}{q}$-space does not lead to any ambiguity in the final result.

By using (4.11),

$$H_{int} = -f\psi\phi^2$$

and taking the energy difference between the $\psi_{\underset{\sim}{p}}$-state and the vacuum state, one can readily verify that the energy shift δp_o is related to the integral (5.7) by

$$\delta p_o = \frac{1}{2} p_o^{-1} \sum(p) \ . \tag{5.15}$$

This then gives a direct proof of the correctness of the above modified Feynman integration path.

We emphasize that in applying the second order perturbation formula it is important to regard the initial state

$$\underset{\sim}{p} = 0$$

as the limit $\underset{\sim}{p} \to 0$. It is easy to see that in such a limit the integral (5.7) becomes

$$\begin{aligned} &\lim_{\underset{\sim}{p}\to 0} \int d^3\underset{\sim}{q} \left\{ \cdots \left[\frac{1}{\omega_q \omega_k^*} \left(\frac{1}{p_o - \omega_q - \omega_k} - \frac{1}{p_o + \omega_q + \omega_k^*} \right) + \text{c.c.} \right] \right\} \\ &= \int d^3\underset{\sim}{q} \left\{ \cdots \left[\frac{1}{\omega_q \omega_{-q}^*} \mathcal{P} \left(\frac{1}{p_o - \omega_q - \omega_{-q}^*} - \frac{1}{p_o + \omega_q + \omega_{-q}^*} \right) + \text{c.c.} \right] \right\} \end{aligned} \tag{5.16}$$

where $\mathcal{P}$ denotes the principal value. This precaution is necessary because at $\underset{\sim}{p} = 0$, Eq. (5.13) becomes an identity.[9]

Next, we consider the case, instead of (5.10),

$$\mu \geqq 2\delta \ . \tag{5.17}$$

Therefore, the decay

$$\psi_{\underset{\sim}{p}} \to A_{\underset{\sim}{p}-\underset{\sim}{q}} A_{\underset{\sim}{q}} \tag{5.18}$$

is possible. By following the usual perturbation method for unstable particles, it can be readily shown that for $\underset{\sim}{p} \neq 0$, the energy shift δp_o is given by

$$\delta p_o = \frac{1}{2} p_o^{-1} \ \text{Re} \sum(p) \ . \tag{5.19}$$

From (5.7) one finds that the imaginary part of $\sum(p)$ at

$$p_o = (\underset{\sim}{p}^2 + \mu^2)^{1/2}$$

is

$$\mathrm{Im}\sum(p) = -(16\pi^2)^{-1} f^2 \int d^3\underset{\sim}{q} (\nu_q \nu_k)^{-1} \delta(p_o - \nu_q - \nu_k) \qquad (5.20)$$

which is related to the lifetime τ of $\psi_{\underset{\sim}{p}}$ in the laboratory system due to the decay (5.18) by

$$\tau^{-1} = -p_o^{-1} \mathrm{Im}\sum(p) \ . \qquad (5.21)$$

By using (5.16), one finds that both (5.19) and (5.21) are valid in the limit $\underset{\sim}{p} = 0$. Eqs. (5.20) and (5.21) give the explicit verification that the unitarity condition is satisfied.

(iii) Lorentz invariance. To show that $\sum(p)$ is Lorentz invariant, one may start with the integral (5.7), and verify that its value is unchanged under a Lorentz transformation. Such a direct proof is straightforward, though slightly lengthy. The details are given in the Appendix. An alternative method is to observe that for p_o^2 real but less than δ^2 and m^2, the contour C in (5.2) can be transformed to a path along the imaginary axis from $q_o = -i\infty$ to $i\infty$. One can then easily show that the integral (5.2) depends only on p^2. For p_o^2 real but larger than either δ^2 or m^2, $\sum(p)$ is related to its analytic continuation, provided $-p^2 < (M+M^*)^2$. As we shall see, across the point $-p^2 = (M+M^*)^2$, $\sum(p)$ has a non-analytic jump. It switches from one sheet of a multi-valued function to another sheet of the same multi-valued function [given by (5.22) and (5.23) below]; nevertheless, $\sum(p)$ remains a function of p^2.

(iv) Analyticity. The integral (5.2) can be readily integrated. We find

$$\sum(p) = -(32\pi^2)^{-1} f^2 \Big\{ \frac{2}{z}(M^2-\delta^2) \ln \frac{M^2}{\delta^2} + \frac{2}{z}(M^{*2}-\delta^2) \ln \frac{M^{*2}}{\delta^2}$$

$$- \frac{1}{z}(M^2 - M^{*2}) \ln \frac{M^2}{M^{*2}} \qquad (5.22)$$

$$- (8\delta^2 - 2z)\, J(z;\delta,\delta)$$

(equation continued on next page)

$$+ [4(M^2+\delta^2) - \frac{2}{z}(M^2-\delta^2)^2 - 2z]\ J(z;M,\delta)$$

$$+ [4(M^{*2}+\delta^2) - \frac{2}{z}(M^{*2}-\delta^2)^2 - 2z]\ J(z;M^*,\delta)$$

$$- (2M^2 - \frac{1}{2}z)\ J(z;M,M)$$

$$- (2M^{*2} - \frac{1}{2}z)\ J(z;M^*,M^*) \tag{5.22 continued}$$

$$- [2(M^2+M^{*2}) - \frac{1}{z}(M^2-M^{*2})^2 - z]\ J(z;M,M^*)\Big\}$$

where

$$z = -p^2 = p_0^2 - \underset{\sim}{p}^2 \ .$$

The function $J(z;\kappa,\mu)$ is given explicitly by Eq. (4.27); both κ and μ are regarded as parameters which can be either δ, or M, or M^*. Except for $J(z; M, M^*)$, which will be specially discussed later, the integral representation (4.26) holds for all other $J(z;\kappa,\mu)$ along the real axis. By using this integral representation, one finds that $J(z;M,\delta)$, $J(z;M^*,\delta)$, $J(z;M,M)$ and $J(z;M^*,M^*)$ are analytic at all real values of z; $J(z;\delta,\delta)$ is analytic for z real $< 4\delta^2$, and it has a branch cut extending from $4\delta^2$ to ∞.

We now turn to the question of $J(z;M,M^*)$. According to (4.27),

$$J(z;M,M^*) = F^{-1}\ \ln P \tag{5.23}$$

where

$$F = [z - (M+M^*)^2]^{1/2}\ [z - (M-M^*)^2]^{1/2} \tag{5.24}$$

and

$$P = \frac{(M^2+M^{*2}) - z + F}{(M^2+M^{*2}) - z - F} \ . \tag{5.25}$$

At a given z, F is a double-valued function, and $\ln P$ a multi-valued function. We may write

$$P = |P|e^{i\theta_p}$$

and

$$\ln P = i(2n\pi + \theta_p) + \ln |P|$$

where $\ln |P|$ denotes the real part and n can be any integer. Different values of n may then be regarded as labels for different sheets of complex z-plane; on each sheet, excluding the branch cuts, the function $F^{-1} \ln P$ becomes single-valued.[10] By definition, $\sum(p)$ is a single-valued function in the physical region (i.e., z real), and therefore so is $J(z;M,M^*)$. Our problem is to determine for $J(z;M,M^*)$ which sheet of $F^{-1} \ln P$ should be used at what real value of z.

We start from the Hamiltonian method; this leads immediately to the integral (5.7) in which the integration of the terms in its last square bracket results in $J(z;M,M^*)$. From (5.7), it follows that $J(z;M,M^*)$ is real at all real z if $\underset{\sim}{p} \neq 0$ and, on account of (5.16), it remains real at $\underset{\sim}{p} = 0$. By examining the zeroes of the denominators in the relevant integrands in (5.7) for the special case, say, $\underset{\sim}{p} = 0$ and $z = p_o^2$, one can readily establish that $J(z;M,M^*)$ is analytic for z real $< (M+M^*)^2$. Furthermore, as we shall see, by using (5.16) and the simple identity

$$\mathcal{P}\left(\frac{1}{x}\right) = \frac{1}{2}\left(\frac{1}{x+i\varepsilon} + \frac{1}{x-i\varepsilon}\right) \tag{5.26}$$

it can be shown that $J(z;M,M^*)$ is also analytic for z real $> (M+M^*)^2$.

It is of interest to display in detail the function $J(z;M,M^*)$ along the real axis. For definiteness, we may choose the cut for F to lie on the real axis between

$$(M-M^*)^2 \qquad \text{and} \qquad (M+M^*)^2 \ . \tag{5.27}$$

(a) For z real $< (M-M^*)^2$, we may choose F to be real and positive; $\ln P$ becomes then also real and positive, and therefore so is $J(z;M,M^*)$. In this region, the integral representation

(4.26) holds; i.e.,

$$J(z;M,M^*) = \int_0^1 \frac{d\xi}{M^2(1-\xi) + M^{*2}\xi - \xi(1-\xi)z} . \tag{5.28}$$

(b) Along the interval (5.27) from $(M-M^*)^2$ to $(M+M^*)^2$,

$$\mathrm{Re}\ F = 0 \qquad \text{and} \qquad |P| = 1 .$$

For z immediately above the real axis, one has

$$\mathrm{Im}\ F < 0 \qquad \text{and} \qquad \ln P = i\beta$$

where β is 0 at $z = (M-M^*)^2 + i\varepsilon$; it decreases continuously with increasing z according to

$$\beta = 2 \tan^{-1} \frac{[(M+M^*)^2 - z]^{1/2}\ [z - (M-M^*)^2]^{1/2}}{z - (M^2+M^{*2})} \tag{5.29}$$

and becomes

$$\beta = -2\pi \qquad \text{at} \qquad z = (M+M^*)^2 + i\varepsilon . \tag{5.30}$$

For z slightly below the real axis, one has

$$\mathrm{Im}\ F > 0 \qquad \text{and} \qquad \ln P = -i\beta .$$

Thus, $J(z;M,M^*)$ is real and positive; it is analytic along the real axis for $z < (M+M^*)^2$. The integral representation (5.28) holds for z real $< 2(M^2+M^{*2})$. Its analytic continuation (but not the integral representation itself) gives the function $J(z;M,M^*)$ for $2(M^2+M^{*2}) \leq z < (M+M^*)^2$; this analytic continuation has a branch point at $(M+M^*)^2$.

(c) For z real $> (M+M^*)^2$, F is real and negative, ln P is real and positive, and

$$\ln P = 0+ \qquad \text{at} \qquad z = (M+M^*)^2+ . \tag{5.31}$$

The function $J(z;M,M^*)$ is analytic, real and negative. The integral representation (5.28) holds again in this region. Although the integral representation can be analytically continued beyond this region from z real $> (M+M^*)^2$ to z real $> (M-M^*)^2$, but $< (M+M^*)^2$; this analytic continuation is different from the $J(z;M,M^*)$ given previously in case (b).

As already mentioned in case (b), the analytic continuation of the previous $J(z;M,M^*)$ from the region z real $< (M+M^*)^2$ to z real $> (M+M^*)^2$ encounters a branch point at $z = (M+M^*)^2$. By choosing the branch cut to be along the real axis, we may analytically continue the $J(z;M,M^*)$, given previously in case (b), to the region z real $> (M+M^*)^2$ by staying either slightly above the real axis or slightly below the real axis. Of course, neither continuation agrees with the correct $J(z;M,M^*)$ in the region z real $> (M+M^*)^2$. However, it can be readily verified that, in accordance with (5.16) and (5.26), the arithmetic mean of these two analytic continuations does give the correct $J(z;M,M^*)$, which is, as we have already shown, an analytic function in the region z real $> (M+M^*)^2$.

The function $J(z;M,M^*)$ is, therefore, analytic along the entire real axis except at the point

$$z = z_o = (M+M^*)^2 \ .$$

Across this point $z = z_o$, while $J(z;M,M^*)$ is given by the same multi-valued function F^{-1} ln P, it switches a sheet.[10] The jump ΔJ may be defined by

$$\Delta J = J(z_o-\zeta;\ M,M^*) - J(z_o+\zeta;\ M,M^*) \tag{5.32}$$

as ζ approaches zero. From (5.30) and (5.31), it follows that as $\zeta \to 0+$

$$\Delta J \to \zeta^{-1/2}\ (2\pi)\ (4MM^*)^{-1/2} \ . \tag{5.33}$$

Now, according to (5.22), the function $J(z;M,M^*)$ contributes to the physical amplitude $\sum(p)$ a term

$$\sum(z;M,M^*) \equiv -\frac{f^2}{32\pi^2 z}[z-(M+M^*)^2]\,[z-(M-M^*)^2]\,J(z;M,M^*), \qquad (5.34)$$

which is analytic and real along the entire real axis, except at the point $z = (M+M^*)^2$. Furthermore, the function $\sum(z;M,M^*)$ is continuous at $z = (M+M^*)^2$, since the jump ΔJ induces a change in $\sum(z;M,M^*)$ across the point $z = (M+M^*)^2$ by an amount

$$\Delta\sum = (8\pi)^{-1}\, f^2\, \zeta^{1/2}\, (M+M^*)^{-2}\, (MM^*)^{1/2} \qquad (5.35)$$

which becomes zero as $\zeta \to 0$. We may regard this change $\Delta\sum$ as due to a cut in $\sum(z;M,M^*)$; this cut separates the complex z-plane into two disconnected regions,[11] and it passes through the real axis at $z = (M+M^*)^2$. Physical consequences of such a cut will be discussed in section 7.

6. Multiple ϕ exchange processes

To illustrate the use of Feynman diagrams in which there is more than one closed loop, we may assume the interaction Lagrangian to be

$$\mathcal{L}_{int} = f\psi\phi^3\ . \qquad (6.1)$$

Let us consider the particular double-loop self-energy diagram given by Fig. 5. According to the general rule (2.25), the integral for this Feynman diagram is

$$-i(2\pi)^{-8}\; 3!\; f^2 \int d^3\underset{\sim}{q}\; d^3\underset{\sim}{k} \int_{C'} dq_o \int_C dk_o\; D_\phi(k)\; D_\phi(q)\; D_\phi(p-k-q)\ . \qquad (6.2)$$

By using the explicit form of D_ϕ,

$$D_\phi(q) = -i\left[\frac{1}{q^2+\delta^2-i\varepsilon} - \frac{1}{2}\left(\frac{1}{q^2+M^2}+\frac{1}{q^2+M^{*2}}\right)\right], \qquad (6.3)$$

one sees that the above integral can be decomposed into a sum of integrals, each of the form

$$\int d^3\underset{\sim}{q}\, d^3\underset{\sim}{k} \int_{C'} dq_o \int_C dk_o\, (k^2+m_1^2)^{-1}\, (q^2+m_2^2)^{-1}\, [(p-k-q)^2 + m_3^2]^{-1} \tag{6.4}$$

where m_1, m_2, m_3 can be either $\delta - i\varepsilon$, or M, or M*. We discuss first the integration in k_o. The integrand has four poles in the complex k_o plane:

$$E_1,\quad p_o - q_o + E_3\ , \tag{6.5}$$

$$-E_1 \quad \text{and} \quad p_o - q_o - E_3\ , \tag{6.6}$$

where

$$E_1 = (\underset{\sim}{k}^2+m_1^2)^{1/2}\ , \tag{6.7}$$

$$E_3 = [(\underset{\sim}{p}-\underset{\sim}{k}-\underset{\sim}{q})^2 + m_3^2]^{1/2} \tag{6.8}$$

and the real parts of E_1 and E_3 are positive. The contour C starts from $-\infty$, goes under the two poles (6.6), then above the two poles (6.5), and then to ∞. After the k_o integration, one finds

$$(6.4) = \frac{1}{2}\pi i \int d^3\underset{\sim}{q}\, d^3\underset{\sim}{k} \int_{C'} dq_o \frac{1}{E_1 E_3 (q^2+m_2^2)} \times \left[\frac{1}{q_o - p_o + E_1 + E_3} - \frac{1}{q_o - p_o - E_1 - E_3}\right] \tag{6.9}$$

in which the integrand has four poles in the complex q_o-plane:

$$E_2,\quad p_o + E_1 + E_3, \tag{6.10}$$

$$-E_2 \quad \text{and} \quad p_o - E_1 - E_3\ , \tag{6.11}$$

where

$$E_2 = (\underset{\sim}{q}^2+m_2^2)^{1/2} \tag{6.12}$$

and Re $E_2 > 0$. The contour C' starts from $-\infty$, goes under the two poles (6.11), then above the two poles (6.10), and then to ∞. After the k_o-integration, the integral (6.4) becomes

$$\frac{1}{2}\pi^2 \int d^3\underset{\sim}{q}\, d^3\underset{\sim}{k} \frac{1}{E_1E_2E_3}\left[\frac{1}{p_o - (E_1+E_2+E_3)} - \frac{1}{p_o + (E_1+E_2+E_3)}\right] . \tag{6.13}$$

Just as in the previous two sections, one may use the Hamiltonian to directly evaluate the energy shift δp_o of a ψ-particle with an arbitrary momentum $\underset{\sim}{p}$. To compare with the Feynman integral (6.2), the relevant virtual transitions are

$$\begin{aligned} &\psi_{\underset{\sim}{p}} \rightarrow \phi_{\underset{\sim}{k}}\phi_{\underset{\sim}{q}}\phi_{\underset{\sim}{p}-\underset{\sim}{k}-\underset{\sim}{q}} \rightarrow \psi_{\underset{\sim}{p}} \\ &\psi_{\underset{\sim}{p}} \rightarrow \psi_{\underset{\sim}{p}}\psi_{\underset{\sim}{p}}\phi_{-\underset{\sim}{k}}\phi_{-\underset{\sim}{q}}\phi_{-\underset{\sim}{p}+\underset{\sim}{k}+\underset{\sim}{q}} \rightarrow \psi_{\underset{\sim}{p}} \\ \text{and}\quad &\text{vacuum} \rightarrow \psi_{\underset{\sim}{p}}\phi_{-\underset{\sim}{k}}\phi_{-\underset{\sim}{q}}\phi_{-\underset{\sim}{p}+\underset{\sim}{k}+\underset{\sim}{q}} \rightarrow \text{vacuum} . \end{aligned} \tag{6.14}$$

It can then be readily verified that the second order perturbation formula (4.11) leads to the same result as that given by the Feynman integral (6.2). By following the same discussion given in section 5, one can show that the unitarity condition is satisfied. The actual proof of unitarity is in fact simpler in the present case.[9] It is clear that, in a similar way, all the above considerations can be extended to the general case

$$\mathcal{L}_{int} = f\psi^m\phi^n .$$

However, further details will not be given in this paper; instead, we will discuss in the next section some possible physical consequences of the theory.

7. Causality

For definiteness, let us take a model interaction Lagrangian

$$\mathcal{L}_{int} = f\psi^2\phi^2 \tag{7.1}$$

and discuss the particular Feynman graph given in Fig. 6. Such a graph would contribute to the S-wave elastic scattering

$$\psi + \psi \rightarrow \psi + \psi \tag{7.2}$$

a phase shift δ given by

$$e^{2i\delta} = 1 - i(8\pi)^{-1} \, v \sum(p) \tag{7.3}$$

where v is the velocity of the ψ-particle in the C.M. system, $\sum(p)$ is, to order f^2, given by (5.22). In this section, we are interested in the motion of wave packets for such a scattering at an energy near $(M+M^*)$. The S-state wave function of the wave packet at large distance in the C.M. system can be written as (after neglecting the mass μ of the ψ-particle)

$$\chi(r,t) = \chi^{in}(r,t) + \chi^{out}(r,t) \ , \tag{7.4}$$

where

$$\chi^{in}(r,t) = r^{-1} \int_0^\infty C_k \, dk \, e^{-ik(r+t)} \ , \tag{7.5}$$

$$\chi^{out}(r,t) = r^{-1} \int_0^\infty C_k \, dk \, e^{2i\delta} \, e^{ik(r-t)} \ ; \tag{7.6}$$

r denotes the distance of one of the ψ-particles from their center of mass and $|C_k|^2$ describes the momentum distribution of the wave packet. This momentum distribution $|C_k|^2$ is assumed to have a peak at

$$k_o = \frac{1}{2} (M+M^*) \tag{7.7}$$

and a width $(\delta k) << k_o$; i.e.

$$|C_k|^2 << 1 \qquad \text{if} \qquad |k-k_o| << (\delta k). \tag{7.8}$$

It is normalized to unity

$$\int |C_k|^2 \, dk = 1 \ , \tag{7.9}$$

and therefore

$$|C_k|^2 \sim (\delta k)^{-1} \qquad \text{if} \qquad |k-k_o| < (\delta k). \tag{7.10}$$

Furthermore, we may assume the phase angle of C_k to be either zero or a slowly varying function of k; for example, C_k can be simply

$$(2/\pi)^{1/2} (\delta k)^{3/2} [(k-k_o)^2 + (\delta k)^2]^{-1} . \tag{7.11}$$

As discussed in section 5, at the total energy (M+M*), the scattering amplitude $\sum$ switches from one sheet of a multi-valued function to another sheet of the same function; its jump $\Delta\sum$ is given by (5.35). We may therefore separate $\chi^{out}(r,t)$ into two parts:

$$\chi^{out}(r,t) = \chi^{out}_{\Delta\sum} + \chi^{out}_{o} \tag{7.12}$$

where, after setting v = 1,

$$\chi^{out}_{\Delta\sum} = -i(8\pi)^{-1} \int C_k \, dk \, r^{-1} e^{ik(r-t)} \Delta\sum \tag{7.13}$$

and χ^{out}_o denotes the remainder. In general, if, e.g., C_k is of the form (7.11), χ^{out}_o approaches zero exponentially[12] as $(r-t) \to \infty$. As we shall see, $\chi^{out}_{\Delta\sum}$ exhibits quite a different asymptotic behavior.

For k near $k_o = \frac{1}{2}$ (M+M*), the square of the total energy in the C.M. system is

$$z \cong (2k)^2 \cong (M+M^*)^2 + 4(k-k_o)(M+M^*) .$$

One finds, by using (5.35),

$$\chi^{out}_{\Delta\sum} \cong -i(32\pi^2)^{-1} f^2 (M+M^*)^{-3/2} (MM^*)^{1/2} \times r^{-1} \int_o^\infty (k-k_o)^{1/2} dk \, C_k \, e^{ik(r-t)} . \tag{7.14}$$

The asymptotic form of $\chi^{out}_{\Delta\Sigma}$ can be most simply derived by converting this integral to an integration parallel to the imaginary axis from $k=k_o$ to $k_o+i\infty$, plus other less dominant terms depending on the poles, or cuts, of C_k. By using (7.11), or (7.10), and approximating $MM^* \cong k_o^2$, one finds, apart from a constant multiplicative phase factor, for

$$(r-t) >> (\delta k)^{-1} , \tag{7.15}$$

$$\chi^{out}_{\Delta\Sigma} \sim (128\pi^2)^{-1} f^2 k_o^{-1/2} (\delta k)^{-1/2} r^{-1} (r-t)^{-3/2} e^{ik_o(r-t)} . \tag{7.16}$$

Thus, at a fixed t, the intensity $|\chi^{out}_{\Delta\Sigma}|^2$ is proportional to r^{-5} as $r \to \infty$, and at a fixed r it is proportional to t^{-3} as $t \to -\infty$, showing a clear "time-advancement" effect. Such a precursory signal does not violate relativistic invariance. This is because for the time interval earlier than any given time, say $-|t_o|$, where t_o can be arbitrarily large, the initial wave packet has necessarily a tail extending to all distances. Such a tail must exist even in the ordinary theory of massive bosons (with a positive definite metric), since the energy spectrum cannot extend over the entire real axis. While in the ordinary theory such a tail and its subsequent effects can always be made to decrease exponentially with r at large distances, this is no longer the case in the present theory. From (7.16), one sees that the shape of the wave packet can be completely controlled in the present theory only at $t = -\infty$. Since for any realistic experiment the initial time is never $-\infty$, this naturally imposes restrictions on the types of initial conditions that can be realized in all experiments. As we shall see, such restrictions are <u>not</u> inconsistent with our present knowledge on experiments.

In order to see whether such effects may be experimentally detected, we shall apply the above results to, say, the colliding beam reaction

$$e^+e^- \to 2\Gamma \to \mu^+\mu^- \tag{7.17}$$

at a total energy $(M+M^*)$, where M denotes the position of the heavy photon pole Γ given by (1.4), and 2Γ represents the relevant virtual state. [All our subsequent discussions are, of course, also applicable to other similar processes such as e^+e^-, or $\bar{p}p \to 2\Gamma \to e^+e^-$, or $\pi^+\pi^-$, or ...]. Similarly to (7.12), we may decompose the outgoing wave packet χ^{out} of $\mu^+\mu^-$ into two parts: a $\chi^{out}_{\Delta\sum}$ due to a similar jump $\Delta\sum$ as that given by (5.35), and a remaining part χ^{out}_o due to (essentially)

$$e^+e^- \to \gamma \quad (\text{or } \Gamma) \quad \to \mu^+\mu^- . \tag{7.18}$$

For simplicity, we will discuss only the intensity of $\mu^+\mu^-$ averaged over all directions. Let us consider at a given macroscopic time $t \gg (\delta k)^{-1}$ and γ^{-1}, the detection of $\mu^+\mu^-$ at different positions r with a space resolution

$$\tau > (\delta k)^{-1} . \tag{7.19}$$

The overall precursory effect of $\chi^{out}_{\Delta\sum}$ may be represented by

$$P(\tau,\delta k) = \frac{\int_{t+\tau}^{\infty} \left| \chi^{out}_{\Delta\sum} \right|^2 r^2\, dr}{\int_{o}^{\infty} \left| \chi^{out}_{o} \right|^2 r^2\, dr} . \tag{7.20}$$

Alternatively, we may consider the detection of $\mu^+\mu^-$ at a given position r, but over different time intervals with a time resolution τ. The relative fraction of $\mu^+\mu^-$ detected during the early period $t < r - \tau$ is given by

$$\frac{\int_{-\infty}^{r-\tau} \left| \chi^{out}_{\Delta\sum} \right|^2 dt}{\int_{-\infty}^{\infty} \left| \chi^{out}_{o} \right|^2 dt} . \tag{7.21}$$

Both $r^2 \, |\chi_{\Delta\Sigma}^{out}|^2$ and $r^2 |\chi_o^{out}|^2$ are functions of (r-t) only; furthermore, we may neglect $r^2 \, |\chi_o^{out}|^2$ for $t >> r$, since it decreases exponentially in $|r-t|$. Thus,

$$(7.20) = (7.21) \ .$$

For any practical arrangement, there would always be inaccuracies in both t and r, the parameter τ therefore represents a combination of time and space resolutions.

Since the power dependence of $\chi_{\Delta\Sigma}^{out}$, given by (7.16), can be traced to a simple phase space consideration, we expect a similar form to hold for the more complicated process (7.17). Thus, we find, as an order of magnitude estimation and for $\tau >$, or $\sim$, $(\delta k)^{-1}$

$$P(\tau,\delta k) \sim \alpha^2 \frac{1}{k_o(\delta k)} \cdot \frac{1}{\tau^2} \tag{7.22}$$

where the factor α^2 is due to the rates of (7.17) and (7.18) being proportional to α^4 and α^2 respectively. Assuming (arbitrarily) $k_o = \frac{1}{2}(M+M^*) \sim 10$ Gev, $(\delta k) \sim 10$ Mev and $\tau \sim 10^{-10}$ sec. (as can be realized in some present day high energy experiments), we find[13]

$$P \sim 10^{-31} \tag{7.23}$$

which is, of course, too small to be detected at present. Nevertheless, it is hoped that there may still be ways to subject such theoretical speculations to direct experimental tests. Perhaps, the space-time resolution τ can be eventually improved to atomic distances $\sim 10^{-8}$ cm, or 3×10^{-19} sec, in which case we would have $P \sim 10^{-14}$; this may be further helped by enhancement factors, such as Lorentz contraction or other possible coherence effects. Only through direct experimental tests can we obtain a well-founded basis for our theoretical concept of causality relations.

I wish to thank the members of the CERN Theoretical Division for their kind hospitality, and to thank many of my colleagues, especially J. S. Bell, B. Ferretti, V. Glaser, J. Prentki, G. C. Wick, W. Thirring, L. Van Hove and Bruno Zumino, for discussions.

Appendix

To show that the integral (5.7) is Lorentz invariant, we shall start from a $p_\mu = (\underset{\sim}{p}, ip_o)$ where $\underset{\sim}{p}$ is an arbitrary non-zero vector parallel to the z-axis and

$$p_o = (\underset{\sim}{p}^2 + \mu^2)^{1/2} . \tag{A.1}$$

In the following, we need only consider Lorentz transformations in the (z,t) plane; i.e., it transforms $p_\mu \to p'_\mu$ where

$$p'_1 = p_1 = 0, \; p'_2 = p_2 = 0$$

$$p'_3 = (1-u^2)^{-1/2} (p_3 - up_o) \tag{A.2}$$

and

$$p'_o = (1-u^2)^{-1/2} (p_o - up_3) .$$

By using the identity

$$\frac{1}{2E_q E'_k}\left[\frac{1}{p_o - E_q - E'_k} - \frac{1}{p_o + E_q + E'_k}\right]$$
$$= \frac{1}{E_q}\left[\frac{1}{(p_o - E_q)^2 - E'^2_k}\right] + \frac{1}{E'_k}\left[\frac{1}{(p_o + E'_k)^2 - E^2_q}\right] , \tag{A.3}$$

we may express (5.7) as

$$\int dq_1 \, dq_2 \, \{ \cdots \}$$

in which $\{ \cdots \}$ denotes a sum over terms, each of the form of a constant, multiplied by either

$$\int_{-\infty}^{\infty} \frac{dq_3}{E_q} \frac{1}{(p_o - E_q)^2 - E_k'^2} \tag{A.4}$$

or

$$\int_{-\infty}^{\infty} \frac{dk_3}{E_k'} \frac{1}{(p_o + E_k')^2 - E_q^2} \tag{A.5}$$

where

$$E_q = (\underset{\sim}{q}^2 + \lambda^2)^{1/2}$$

$$E_k' = (\underset{\sim}{k}^2 + \lambda'^2)^{1/2}$$

$$\underset{\sim}{k} = \underset{\sim}{p} - \underset{\sim}{q}$$

and λ, λ' can be either δ, or M, or M*. The integral (A.4) can also be written as

$$I(p) = \int_{-\infty}^{\infty} \frac{dq_3}{q_o} \frac{1}{\mu^2 + 2(p \cdot q) + \lambda^2 - \lambda'^2} \tag{A.6}$$

where

$$p \cdot q = \underset{\sim}{p} \cdot \underset{\sim}{q} - p_o q_o$$

and

$$q_o = E_q \quad .$$

Similarly, the integral (A.5) can be cast into the same form (A.6), provided one replaces $q_\mu = (\underset{\sim}{q}, iE_q)$, λ and λ' by $k_\mu = (\underset{\sim}{k}, -iE_k')$, λ' and λ respectively. We observe that under the Lorentz transformation (A.2), $q_\mu \to q_\mu'$ where

$$q_1' = q_1, \qquad q_2' = q_2$$

$$q_3' = (1-u^2)^{-1/2} (q_3 - uq_o) \tag{A.7}$$

and

$$q_o' = (1-u^2)^{-1/2} (q_o - uq_3);$$

the integration path from $q_3 = -\infty$ to $q_3 = +\infty$ is mapped into a contour Γ in the complex q_3'-plane. Therefore, the integral (A.6) satisfies the equality

$$I(p) = \int_\Gamma \frac{dq_3'}{q_o'} \frac{1}{\mu^2 + 2(p'\cdot q') + \lambda^2 - \lambda'^2} \tag{A.8}$$

where except for $\lambda = \lambda' = \delta$, Γ is different from the real axis. It is sufficient to consider only infinitesimal Lorentz transformations, in which case Γ is infinitesimally near the real axis. In (A.8), one can transform Γ back to the real axis, provided the denominator

$$\mu^2 + 2(p'\cdot q') + \lambda^2 - \lambda'^2 \neq 0 \tag{A.9}$$

on the real axis (in the complex q_3' plane). For $(\lambda,\lambda') = (M,\delta)$, (M^*,δ), (M,M) and (M^*,M^*) one can easily verify that (A.9) holds. For $(\lambda,\lambda') = (M,M^*)$, a violation of (A.9) implies, for $p_3' = |\underset{\sim}{p}'| \neq 0$,

$$q_3' = \frac{1}{2} p_3' \tag{A.10}$$

and

$$\text{Re } q_o' = \text{Re } (\underset{\sim}{q}'^2+M^2)^{1/2} = \frac{1}{2} p_o' . \tag{A.11}$$

which are the same two conditions given by Eqs. (5.13) and (5.14). As already shown in section 5, these two conditions can be satisfied only by a set of $\underset{\sim}{q}'$ vectors of measure zero (all on a circle); such vectors give no contribution to $\sum(p)$. Thus, in (A.8) one can always deform Γ back to the real axis in the complex q_3'-plane. Therefore

$$I(p) = I(p'), \tag{A.12}$$

provided both $\underset{\sim}{p}$ and $\underset{\sim}{p}'$ are $\neq 0$. By regarding $\underset{\sim}{p} = 0$ as the limiting point of $\underset{\sim}{p} \to 0$, we establish, then, the Lorentz invariant property of $\sum(p)$.

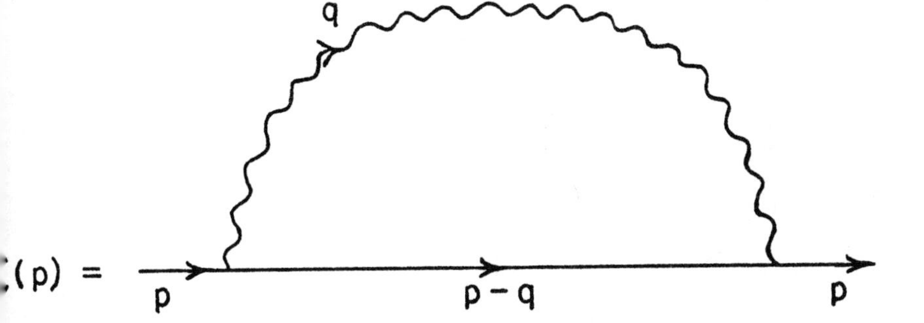

FIGURE 1

Feynman graph for $\sum(p)$ defined by Eq. (4.2). The wavy and solid lines denote, respectively, the ϕ and ψ propagators.

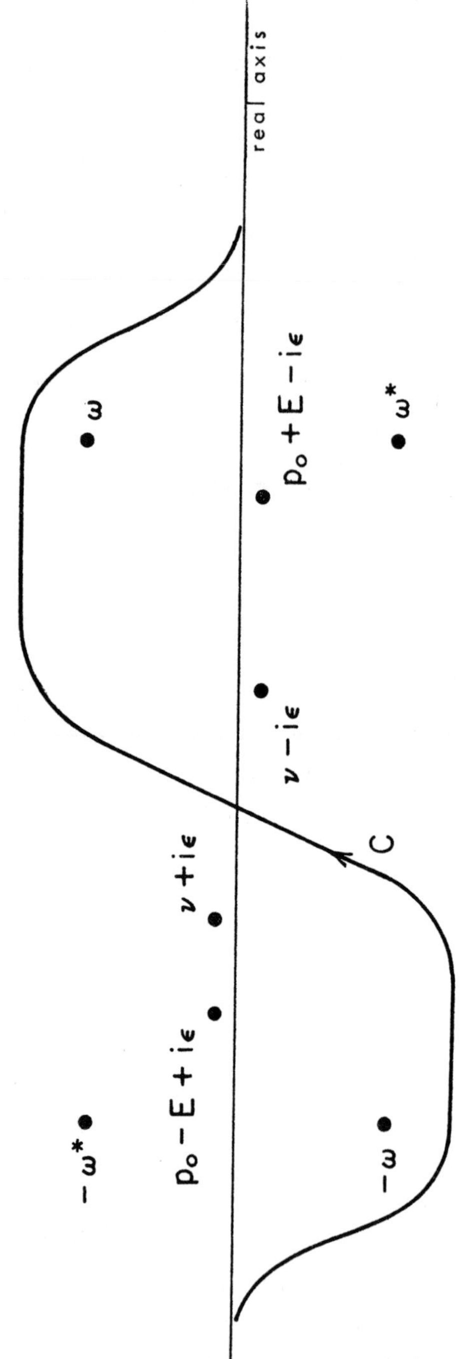

FIGURE 2

Contour C for the integral (4.2

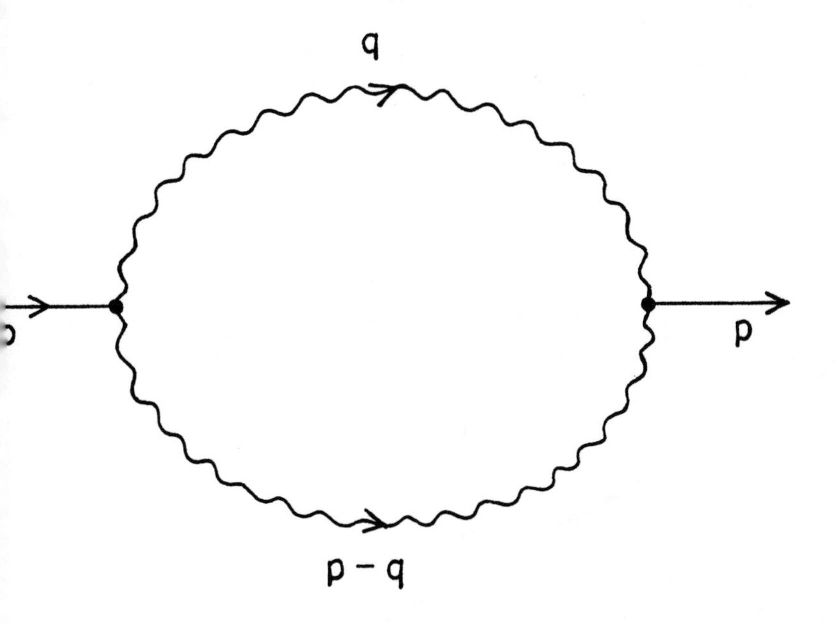

FIGURE 3

Feynman graph for $\Sigma(p)$ defined by Eq. (5.2).

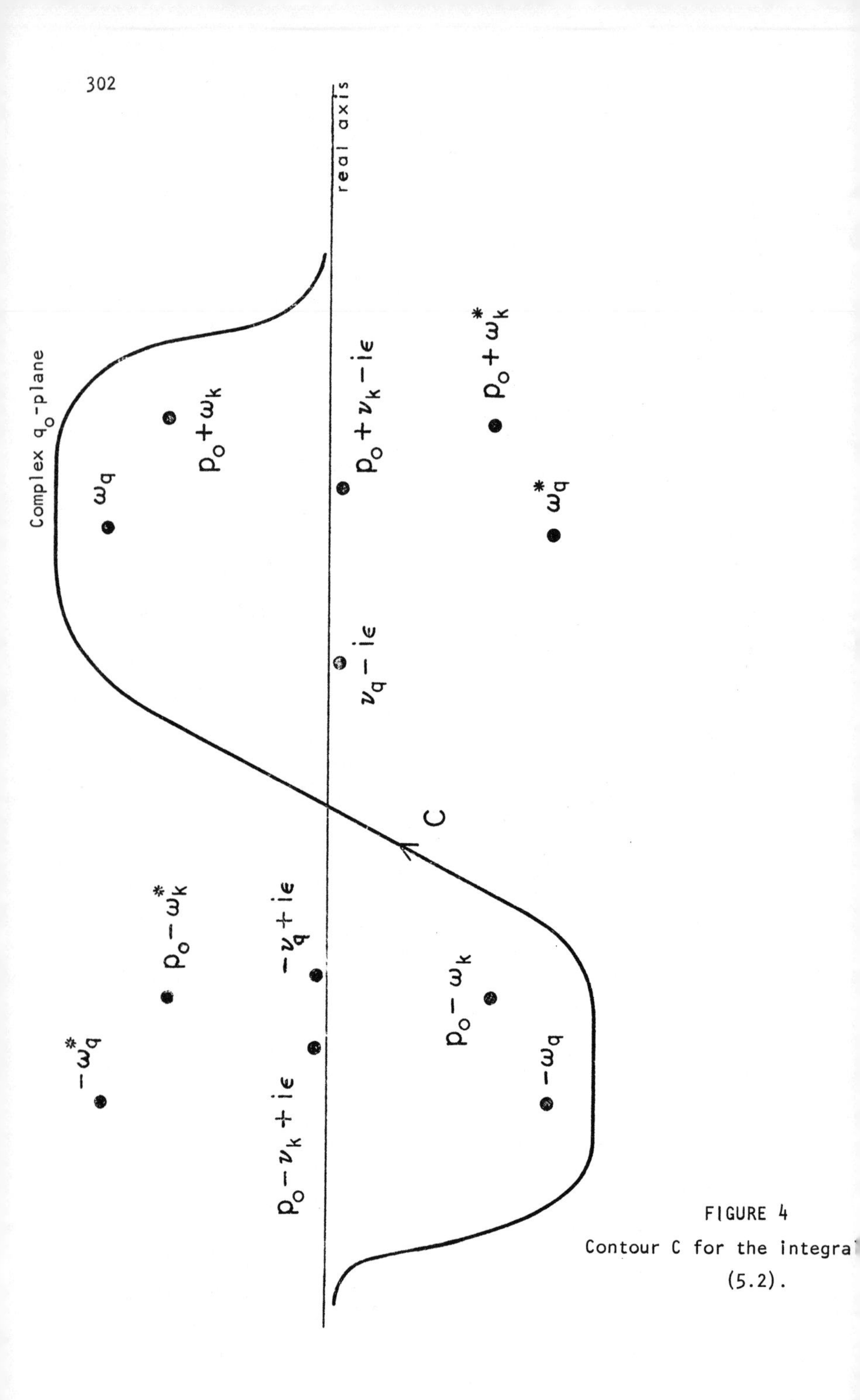

FIGURE 4
Contour C for the integral (5.2).

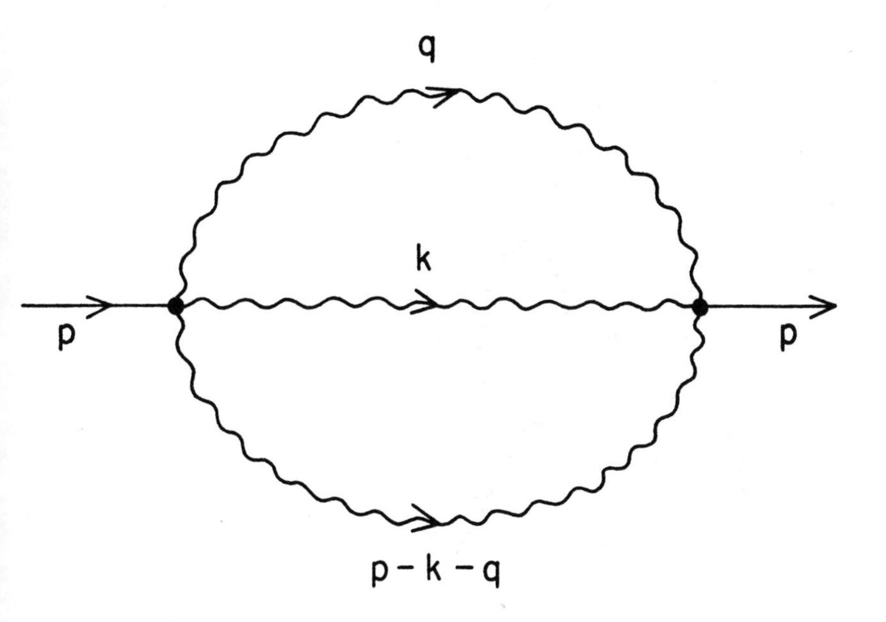

FIGURE 5

A self-energy graph for the interaction (6.1)

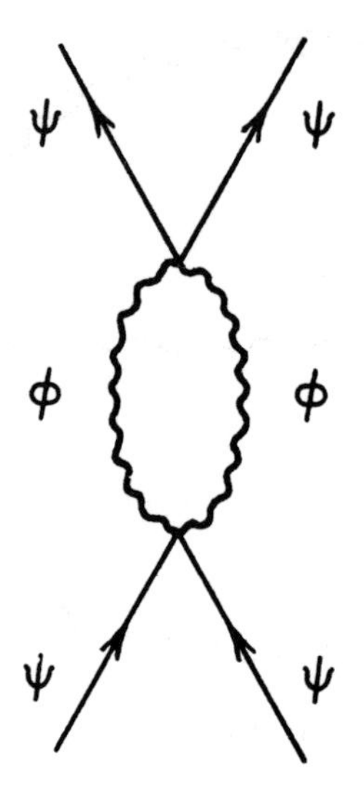

FIGURE 6

A $\psi\psi$ scattering graph for the interaction (7.1).

References

1. T. D. Lee and G. C. Wick, Nuclear Physics B9, 209 (1969).

2. Divergence difficulties, of course, has a long history in physics. It already existed in the classical work on the theory of the electron by Lorentz. For solutions to such difficulties in classical electrodynamics see G. Wentzel, Z. Phys. 86, 479 (1933); 87, 726 (1934); and P.A.M. Dirac, Proc. Roy. Soc. London 167, 148 (1938). For a summary of the present divergence difficulties, see, e.g., Ref. 3.

3. T. D. Lee, Proceedings of the 1969 Topical Conference on Weak Interactions (CERN). Further details will be given in a separate publication.

4. If one modifies only the photon propagator, then there remains a charge renormalization in the theory. However, all physical observables, such as electromagnetic mass shifts of hadrons and radiative corrections to weak decays are finite.

5. The free field case of this model is, apart from some trivial differences, the same one considered by Froissart and Glaser some years ago. See M. Froissart, Suppl. Nuovo Cimento 14, 197 (1959).

6. We recall that under the transformation $\eta \to T^{\dagger} \eta T$ the operator O for a physical observable transforms as $O \to T^{-1} OT$, so that $\eta^{-1} O^{\dagger} \eta$ transforms in the same way as O. Thus, (2.4) and (2.5) are independent of the representation of η, but (2.7) holds only in the representation $\eta = (-1)^{N_B}$.

7. In Fig. 4, the contour C is shown only for the case when the poles do not coalesce. Coalescence of poles implies that, in the usual perturbation formula using the Hamiltonian, the relevant intermediate state is degenerate with the initial state. This situation will be analyzed in detail in Section 5(ii). As we shall see, the resulting Feynman integral $\sum(p)$ remains well defined, and it satisfies both unitarity and Lorentz invariance.

8. To show this, it is only necessary to study the q-integration near the singularities of $(p_o - \omega_k^* - \omega_q)^{-1}$. Let us use the cylindrical coordinates: $\underset{\sim}{q} = (\rho \cos \phi, \rho \sin \phi, z)$ where the z axis is parallel to $\underset{\sim}{p}$. The singularities of $(p_o - \omega_k^* - \omega_q)^{-1}$ lie on the circle

$$z = \frac{1}{2} |\underset{\sim}{p}| \qquad \text{and} \qquad \rho = \rho_o$$

where ρ_o is its radius, determined by

$$p_o = \frac{1}{2} \omega_o$$

and

$$\omega_o = (\rho_o^2 + \frac{1}{4} \underset{\sim}{p}^2 + m^2 + i\gamma m)^{1/2}.$$

In the neighborhood of this circle, it is convenient to write

$$\rho = \rho_o + r \cos \theta$$

and

$$z = \frac{1}{2} |\underset{\sim}{p}| + r \sin \theta.$$

The region $r \leq \varepsilon << 1$ denotes, then, a doughnut which contains all singularities of $(p_o - \omega_k^* - \omega_q)^{-1}$; its cross-section is a small circle of radius ε. Neglecting $O(\varepsilon^2)$, one finds, inside this doughnut region,

$$p_o - \omega_k^* - \omega_q = r(a \cos \theta + i\, b \sin \theta)$$

where a and b are two real constants, different from zero, and given explicitly by

$$a = -\rho_o \left(\frac{1}{\omega_o} + \frac{1}{\omega_o^*}\right) \qquad \text{and} \qquad b = i\,\frac{1}{2} \left(\frac{1}{\omega_o} - \frac{1}{\omega_o^*}\right) \cdot |\underset{\sim}{p}| .$$

Since $d^3\underset{\sim}{q} = r\, \rho_o\, dr\, d\theta\, d\phi$, the $\underset{\sim}{q}$-integration of $(p_o - \omega_k^* - \omega_q)^{-1}$ over this doughnut region is simply

$$2\pi\, \varepsilon\, \rho_o \int_0^{2\pi} d\theta\, (a \cos \theta + i\, b \sin \theta)^{-1},$$

which goes to zero as $\varepsilon \to 0$.

9. We note that this particular circumstance occurs only if the intermediate (virtual) state consists of two particles of masses M and M*. No special precaution is needed if the intermediate state consists of three or more particles. Take, e.g., the case of three particles of energies ω_k, ω^*_q and ν_{p-k-q}. A similar requirement as (5.12) would be $p_0 = \omega_k + \omega^*_q + \nu_{p-k-q}$ which always imposes two separate conditions (a) $|\underset{\sim}{k}| = |\underset{\sim}{q}|$, so that $\mathrm{Im}\,(\omega_k + \omega^*_q) = 0$, and (b) $\mathrm{Re}\,\omega_k = \frac{1}{2}(p_o - \nu_{p-k-q})$, independently of whether $\underset{\sim}{p} = 0$ or not.

10. For definiteness, let us choose the cut for F to be along (5.27). The different sheets can then be defined by assigning $\theta_p = 0$ and F real > 0 for z real < $(M-M^*)^2$, and by choosing the two cuts for $F^{-1} \ln P$ to be along the real axis between $(M-M^*)^2$ and $(M+M^*)^2$ and between $(M+M^*)^2$ to ∞ for all sheets $n \neq 0$; for the sheet $n = 0$, there is only one cut from $(M+M^*)^2$ to ∞. Since as $F \to -F$, one has, according to (5.25), $P \to P^{-1}$; the function $F^{-1} \ln P$ switches sheets from $n \to -n$. The double-valueness of F, therefore, does not increase the total number of sheets for $F^{-1} \ln P$.

An interesting feature is that, through the process of analytic continuation, all sheets of even n are connected with each other, and all sheets of odd n are also connected with each other. However, there is no connection between an even n sheet and an odd n sheet. This can be most easily seen by noting that as $z \to \infty$, $F \to -z$, $P \to z^2 (MM^*)^{-2}$ and, therefore, as the phase of z changes from 0 to 2π, $\ln P$ increases like $\ln z^2$ by an amount $i4\pi$. Hence, the sheet n is connected to the sheets (n+2) and (n-2) through the cut from $(M+M^*)^2$ to ∞; through the other cut from $(M-M^*)^2$ to $(M+M^*)^2$, the sheet n is connected to the sheet -n.

Along the real axis (i.e., the physical region), for $z < (M+M^*)^2$, $J(z;M,M^*)$ is given by the value of $F^{-1} \ln P$ on the sheet $n = 0$; for $z > (M+M^*)^2$, $J(z;M,M^*)$ is given by the value of $F^{-1} \ln P$ on the sheet $n = 1$ taken at points slightly above the real axis, which is of course the same as F^{-1} ln P on the sheet n = -1 but taken at points slightly below the real axis.

11. Since the completion of this work, Dr. R. E. Cutkosky has kindly informed me that he, Landshoff, Olive and Polkinghorne have examined a similar case like the double ϕ exchange process studied in this section, and they have independently arrived at a similar conclusion, though through a different route of reasoning.

12. After subtracting out $\Delta\sum$, the remaining amplitude still has, according to (5.22), complex branch points at $-q^2 = (M+\delta)^2$, $(M^*+\delta)^2$, $4M^2$ and $4M^{*2}$; such complex branch points, like the complex poles discussed in reference 1, contribute, among others, a factor $e^{-\lambda(r-t)}$ to the asymptotic form of χ_o^{out} as $(r-t) \to \infty$, where λ is the magnitude of the nearest distance of these complex singularities to the real axis.

13. The momentum resolution δk of a high energy beam usually refers to that of an incoherent mixture of initial wave packets; each may have a coherent momentum width δk_c, much smaller than δk, though in general still much larger compared to τ^{-1}. Such a wave packet can be represented by, e.g., instead of (7.11),

$$c_k = (2/\pi)^{1/2} (\delta k_c)^{3/2} [(k-k_o-\Delta)^2 + (\delta k_c)^2]^{-1}$$

where Δ has a probability distribution $I(\Delta)$. The momentum resolution δk of the beam refers to the width of $I(\Delta)$. For a given Δ, the overall precursory effect is, according to (7.20)

$$P_\Delta(\tau,\delta k_c) \sim \frac{\alpha^2 (\delta k_c)^3}{k_o\ [\Delta^2+(\delta k_c)^2]^2\ \tau^2}$$

where the Δ dependence comes only from the numerator in (7.20). The average value $\int d\Delta\ I(\Delta)\ P_\Delta\ (\tau,\delta k_c)$ gives (7.22) which depends only on the experimental resolution δk.

Received 4/14/69

Forward Dispersion Relations and Microcausality

by

Reinhard Oehme

The University of Chicago

A brief review is given of the structure of forward dispersion relations and of their derivation on the basis of microcausality, spectral conditions, and other basic notions of relativistic field theory. After a short discussion of the global aspects of local commutativity, the effects of possible violations of microcausality are studied. Modified dispersion formulae are derived under the assumption of exponentially decreasing bounds for the matrix elements of commutators in space-like directions. Simple mathematical models are considered in order to study qualitatively the implications of possible acausal singularities.

1. Introduction

It is now almost fifteen years since dispersion relations for pion nucleon scattering have been introduced. In 1955, the dispersion relations for forward $\pi^{\pm}$ p-scattering[1] were found to be in agreement with experimental data up to laboratory energies of about 200 MeV.[2] In later years, there was occasional excitement about possible discrepancies with experiments at higher energies, but, on closer examination, one found that it were the data which required correction and not the dispersion formulae. By now, thanks to the efforts of Lindenbaum and his group,[3] we have precision experiments up to and beyond energies of 20 BeV, and still the relations seem to be in agreement with the data. With a new generation of big machines becoming operational or being under construction, it may be of some interest to reconsider the background of these dispersion formulae, and to speculate about possible modifications in case there should be a deviation from the micro-causal Lorentz structure of field theory.*

In the following, we briefly review the structure of for-N-dispersion relations and their derivation on the basis of microcausality, spectral conditions, and other assumptions. Then, after a short discussion of local commutativity in relativistic quantum field theories, we present some exploratory studies of the effects of specific violations of microcausality. We derive modified dispersion relations, and we discuss their implications qualitatively on the basis of simple, mathematical models.

2. The dispersion formulae

Dispersion relations for the forward scattering of light have a long history, and their connection with causality is well understood, at least in classical electrodynamics.[6] In relativistic field theory, dispersion formulae for forward Compton scattering[7] can be obtained essentially on the basis of the assumed local commutativity of the electro-magnetic field operators and their currents. Real problems appear, however, if we want to write dispersion relations for the scattering of massive particles at each other.[1,7,8] The main obstacles to a straightforward generalization of the photon relations are due to the finite mass of the projectile and the possibility of it being a charged particle. Historically, in 1955, we circumvented the first problem by using heuristic methods, but we solved the second one.[1] The important, but today almost trivial, sounding step was the recognition that the analytic function of

---

*We do not consider in this paper the more intuitively physical concept of macroscopic causality or phenomenological causality, as Wigner[4] prefers to call it. There is a considerable recent literature[5] about macroscopic causality in S-matrix theory. A violation of microcausality does not necessarily imply a deviation from phenomenological causality, but this is a rather involved question, because there are many difficulties in formulating the latter concept.

the energy variable which has, for example, the proton-proton scattering amplitude as a boundary value at the positive real axis, has a boundary value at the negative energy axis which corresponds to the complex conjugate of the antiproton-proton scattering amplitude. With these crossing rules, and reasonable assumptions about the high energy limits of total cross sections, we obtained the forward dispersion relations for $\pi^{\pm}p$-scattering.[1] If $T_{\pi^{\pm}p}(\omega)$ is the forward scattering amplitude in the laboratory system, and if we define

$$T^{(\pm)}(\omega) \equiv \frac{1}{2}\left[T_{\pi^-p}(\omega) \pm T_{\pi^+p}(\omega)\right], \tag{2.1}$$

we can write these relations in the familiar form

$$\begin{aligned} T^{(+)}(\omega) &= T^{(+)}(\mu) + \frac{f^2}{m} \frac{\omega^2 - \mu^2}{\left(\omega^2 - \frac{\mu^4}{4m^2}\right)\left(1 - \frac{\mu^2}{4m^2}\right)} \\ &+ \frac{2(\omega^2 - \mu^2)}{\pi} \int_{\mu}^{\infty} d\omega' \frac{\mathrm{Im}\, T^{(-)}(\omega')\, \omega'}{(\omega'^2 - \omega^2)(\omega'^2 - \mu^2)}, \end{aligned} \tag{2.2}$$

and

$$\begin{aligned} T^{(-)}(\omega) &= 2f^2 \frac{\omega}{\omega^2 - \frac{\mu^4}{4m^2}} + \\ &+ \frac{2\omega}{\pi} \int_{\mu}^{\infty} d\omega' \frac{\mathrm{Im}\, T^{(-)}(\omega')}{\omega'^2 - \omega^2}, \end{aligned} \tag{2.3}$$

with

$$\mathrm{Im}\, T_{\pi^{\pm}p}(\omega) = \frac{\sqrt{\omega^2 - \mu^2}}{4\pi} \sigma^{tot}_{\pi^{\pm}p}(\omega), \tag{2.4}$$

and

$$f^2 = \frac{\mu^2}{4m^2} \frac{g^2_{\pi NN}}{4\pi} \approx 0.08.$$

(μ = pion mass, m = nucleon mass.)

The asymptotic bound, which is required for the validity of the once subtracted dispersion relations (2.2), can actually be derived from local field theory.[9] In particular the unitarity condition plays an important role in this proof. Some additional assumptions and the familiar Phragmén-Lindelöf Theorems also give the asymptotic bound required in Eq. (2.3).[10]

In order to compare our relations with experiments, we take the real part for real $\omega \geq \mu$, in which case we obtain on the right hand side principle value integrals involving total cross-sections. The present situation of the comparison is shown in Figs. 1 and 2, at least at higher energies.* Of course, one can test also several other versions of Eqs. (2.2) and (2.3), and in particular one can make additional subtractions in order to de-emphasize uncertainties due the extrapolation of $\sigma_{\pi p}^{tot}$ to $\omega \to \infty$. For these tests we refer to Lindenbaum's articles.[11] We mention here only the interesting sum rule obtained from Eq. (2.3) for $\omega = \mu$:

$$\frac{1}{3}(a_1 - a_3)\frac{m+\mu}{m\mu} = \frac{2f^2}{\mu^2\left(1 - \frac{\mu^2}{4m^2}\right)} + \frac{1}{4\pi^2}\int_{\mu}^{\infty} d\omega' \, \frac{\sigma_{\pi^- p}^{tot}(\omega') - \sigma_{\pi^+ p}^{tot}(\omega')}{\sqrt{\omega'^2 - \mu^2}} . \tag{2.5}$$

Here a_1 and a_2 are the scattering lengths in the states with isotopic spin $I = \frac{1}{2}$ and $\frac{3}{2}$ respectively. The sum rule (2.5) was first evaluated in 1955,[3] and its immediate success was most encouraging, because until then only the p-wave πN- amplitudes could be handled theoretically with reasonable success.[14]

Quite apart from their direct use in comparison with experiments, the early success of dispersion relations, and their extension to non-forward scattering,[15] was essential in initiating the use of "analyticity" in elementary particle theory, a

---

*For more recent comparisons at lower energies see, e.g., Ref. 12.

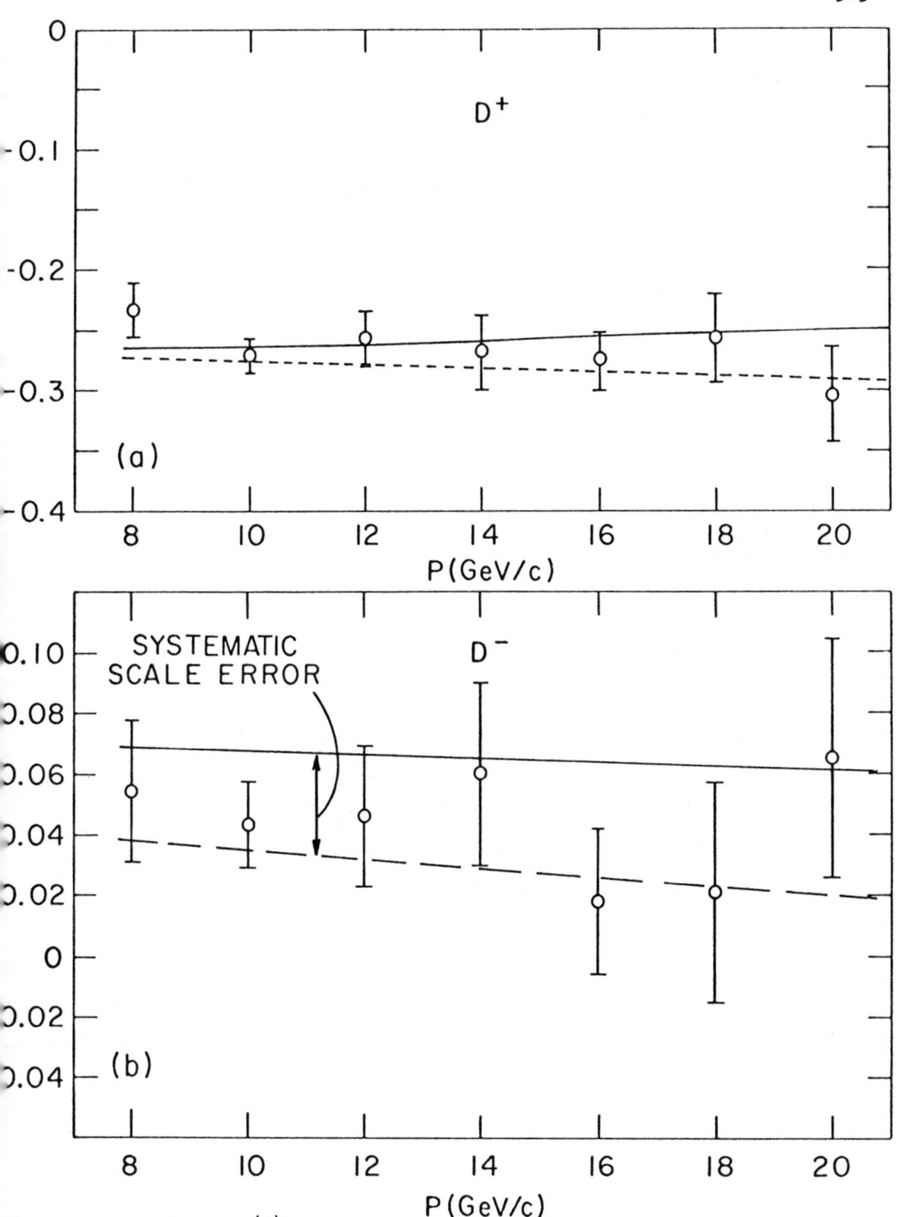

Figs. 1,2: $D^{\pm} \equiv \mathrm{Re}T^{(\pm)}$ as a function of the momentum in the laboratory system. The solid curves are the predictions of the dispersion relations. The dashed curves are the result of a displacement by 0.04, the systematic error which is predominantly of a scale nature. [Taken from the article by S. J. Lindenbaum in Proc. Int. Symp. on Cont. Phys. (Int. Atom. Energy Agency, Vienna, 1969); reprinted with the permission of the publishers.]

development which provided a general framework for the formulation of many new ideas.[16]

3. Derivation of forward dispersion relations

With the apparent agreement of the dispersion formulae with experiments, we may ask what we have learned from this result. Under what conditions would the dispersion relations not be true, or would they be modified in such a way as to give a clear contradiction with experiments. In particular, what would be the effect of specific violations of the causal Lawrence-structure, and, if this violation can be characterized by a "fundamental length", what are the upper limits for this length in view of the present situation? We will see that these are difficult questions in principle, because they require actually a consistent and explicit formulation of the S-matrix within the framework of non-local theories, and at the time of writing such a formalism does not exist.

In order to see the problems involved, it is helpful to recall the essential points in the derivation of forward dispersion relations. From the basic notions of relativistic field theory,[17] we obtain a representation of the amplitude as a Fourier-transform of a space-time function (or distribution) which has support only in the future light cone.[18] Taking, for instance, $\pi^o\pi^o$ - scattering as an example, we obtain for the invariant function

$$F = 16\pi \sqrt{s}\, T_{c,m}.$$

The representation

$$F(K,P) = \frac{i}{(2\pi)^3} \int d^4 x\, e^{-iK\cdot x}\, \theta(x_o) \langle p'|[j(\tfrac{x}{2}), j(-\tfrac{x}{2})]|p\rangle, \tag{3.1}$$

where

$$K = \tfrac{1}{2}(k+k'),\ P = \tfrac{1}{2}(p+p'),\ k + p = k'+p',$$

$$k^2 = k'^2 = -\mu^2,\ p^2 = p'^2 = -m^2,\ s = -(k+p)^2, \text{ and}$$

$$j(x) \equiv (\ \ - \mu^2)\ \phi(x).$$

Here $\phi(x)$ is the Heisenberg field. For every given S-matrix, which is unitary in the Hilbert space of physical states, we can introduce interpolating Heisenberg fields in terms of specific extrapolations of the S-matrix elements off the mass shell.[19] Generally, these fields do <u>not</u> satisfy the local communtativity condition*

$$[\phi(x),\ \phi(x')] = 0 \text{ for } -(x-x')^2 < 0, \tag{3.2}$$

but here we make the explicit assumption that Eq. (3.2) is satisfied. We specialize Eq. (3.1) to forward scattering, and go into the laboratory system where $K = \{\hat{e}\sqrt{\omega^2 - \mu^2},\ i\omega\}$ and $p = \{0, i\mu\}$. Then we find:

$$F(\omega) = \int d^4 x\ e^{i\omega x_o - i\sqrt{\omega^2 - \mu^2}\ \hat{e}\cdot\vec{x}}\ \theta(x_o)\ \chi(x_o, |\vec{x}|) \tag{3.3}$$

with

$$\chi(x_o, |\vec{x}|) \equiv \frac{i}{(2\pi)^3} \langle p|[j(\tfrac{x}{2}),\ j(-\tfrac{x}{2})]|p\rangle. \tag{3.4}$$

being a real function, which is odd under $x_o \to -x_o$ for our case of neutral currents. The Eq. (3.3) defines the amplitude $F(\omega)$ for real ω with $|\omega| \geq \mu$. Actually, $F(\omega)$ is only defined up to a polynomial in ω, because $\chi(x_o|\vec{x}|)$ is a distribution on the light cone, and the arbitrariness in the definition of $\theta(x_o)\ \chi(x_o, |\vec{x}|)$ is reflected in the unspecified coefficients of the polynomial.

If the field $\phi(x)$ satisfies the local communtativity, we find that $\chi(x_o, |\vec{x}|)$ vanishes for $\vec{x}^2 - x_o^2 > 0$, and hence the integral in Eq. (3.3) has support only in the future cone $x_o \geq |\vec{x}|$.

---

*For Fermion fields, we have, of course, anticommutators in place of commutators.

However, because of

$$|\mathrm{Im}\ \sqrt{\omega^2 - \mu^2}| \geq |\mathrm{Im}\ \omega| \text{ for } |\mathrm{Im}\ \omega| > 0,$$

the representation (3.3) is still restricted to real ω with $|\omega| \geq \mu$. Let us now write

$$F(\omega) = \int_0^\infty dr\ F(\omega,r), \tag{3.5}$$

with

$$F(\omega,r) = s(\omega,r) \int_0^\infty dx_o\ e^{i\omega x_o}\ \chi(x_o,r), \tag{3.6}$$

where

$$s(\omega,r) \equiv 4\pi\ \frac{r\ \sin(r\ \sqrt{\omega^2 - \mu^2})}{\sqrt{\omega^2 - \mu^2}}\ . \tag{3.7}$$

Then it follows from

$$\chi(x_o,r) = 0 \text{ for } x_o^2 - r^2 < 0 \tag{3.8}$$

that $F(\omega,r)$ is analytic in the upper half-plane and polynomially bounded. Except for subtractions, which may be necessary, we can thus write a Hilbert representation for $F(\omega,r)$:

$$F(\omega,r) = \frac{1}{\pi} \int_{-\infty}^{+\infty} d\omega'\ \frac{\mathrm{Im}\ F(\omega',r)}{\omega' - \omega}\ . \tag{3.9}$$

We must now evaluate the absorptive part of $F(\omega,r)$, which is given by

$$\begin{aligned} \mathrm{Im}\ F(\omega,r) &= s(\omega,r)\ \frac{-i}{2} \int_{-\infty}^{+\infty} dx_o\ e^{i\omega x_o}\ \chi(x_o,r) \\ &= \pi\ s(\omega,r) \sum_{|n\rangle} \langle p|j(0)|n\rangle \langle n|j(0)|p\rangle\ N_n^2 \\ &\quad \times \{\delta(\omega + \mu - p_{no}) - \delta(\omega - \mu + p_{no})\}, \end{aligned} \tag{3.10}$$

with $N_n(p_n)$ being a normalization factor. Since $p_{no} \geq 2\mu$ in the case of $\pi\pi$-scattering, we find that

$$\text{Im } F(\omega, r) = 0 \text{ for } |\omega| < \mu. \tag{3.11}$$

With Eq. (3.11) we may write

$$F(\omega, r) = \frac{1}{\pi} \left(\int_{-\infty}^{-\mu} + \int_{\mu}^{\infty} \right) d\omega' \, \frac{\text{Im } F(\omega', r)}{\omega' - \omega} . \tag{3.12}$$

Specializing ω to the region $|\omega| \geq \mu$ on the real axis, we can now perform the r-integration and obtain

$$F(\omega) = \frac{1}{\pi} \left(\int_{-\infty}^{-\mu} + \int_{\mu}^{\infty} \right) d\omega' \, \frac{\text{Im } F(\omega')}{\omega' - \omega} . \tag{3.13}$$

This is the desired dispersion relation (without possible subtractions), which now can be extended analytically into the cut ω-plane.

The proof of forward dispersion relations,[20] which we have sketched above, ignores many fine points, but it gives the essence of the implications of local commutativity for the validity of these formulae. This proof was first obtained by the author in the early summer of 1956.*

Our proof is easily generalized to cases with single particle intermediate states, like πN-scattering, where there is a single nucleon state below the threshold $\omega = \mu$ ($\mu = m_\pi$).[21,22,23] In order to obtain a gap relation corresponding to Eq. (3.11), we can project out the single nucleon terms with factors like $[(k \pm p)^2 + m^2]$.[22] For example, in the case of the amplitude $T^{(-)}(\omega)$ defined in Eq. (2.1), we consider in place of $T^{(-)}(\omega, r)$ the function

$$\left(\omega^2 - \frac{\mu^4}{4m^2}\right) T^{(-)}(\omega, r), \tag{3.14}$$

---

* See in particular the appendix of Ref. 20. Although the title of this paper refers to static nucleons, the fully relativistic case is discussed in the appendix.

where $\omega = \mp\mu^2/2m$ corresponds to $(k \pm p)^2 = -m^2$. This is an analytic function in the upper half of the complex ω-plane, and its imaginary part on the real ω-axis has a gap for $|\omega| < \mu$. In the place of Eq. (3.12), we can then write a dispersion relation which, however, should contain at least one subtraction. Making the subtraction at $\omega = \mu^2/2m$, and performing the ω-integration, we obtain the dispersion formula

$$T^{(-)}(\omega) = C \frac{\omega}{\omega^2 - \frac{\mu^4}{4m^2}} + \frac{2\omega}{\pi} \int_{\mu}^{\infty} d\omega' \frac{\mathrm{Im}\, T^{(-)}(\omega')}{\omega'^2 - \mu^2}, \quad (3.15)$$

where we have used the symmetry relation

$$T^{(-)}(\omega) = -T^{(-)}(-\omega). \quad (3.16)$$

This relation can be inferred from the Fourier representation of $T^{(-)}(\omega)$, which is similar to Eq. (3.1), but with the appropriate charge structure.[1] In Eq. (3.15), the constant C is an unspecified number which can be identified with $2f^2$, the πNN-vertex on the mass shell. Actually, this identification can be proven rigorously by applying to the pion-nucleon vertex function the same methods we have used above for the scattering amplitude.[21-23]

4. Local commutativity

As given in the previous chapter, the derivation of forward dispersion relations shows the relevance of the support properties in space-time for the analytic structure of the amplitude $F(\omega)$. However, we may want to see more directly how the dispersion formulae are changed if we relax some of the basic assumptions (or axioms) of local, relativistic field theory. Already in 1955, before we had a detailed proof of ordinary dispersion relations,[20] and before axiomatic field theory was properly formulated,[24] we have been speculating about possible modifications of these relations in case local commutativity is not valid in the neighborhood of the light cone.[25] By now, there are some very interesting theorems concerning the global character of local commutativity. These theorems must be discussed briefly before we venture into space-like regions.

Suppose we consider a relativistic field theory without local commutativity, but satisfying all other axioms, in particular Lorentz covariance and the spectral conditions. Such a theory can be formulated without difficulties, for instance, in terms of Wightman functions.[24] Using this formalism, we may be interested in exploring the implications of limited locality. For instance, we may want to postulate that, in the case of boson fields, the field operators $\phi(x)$ commute only for space-like separations with $\vec{x}^2 - x_o^2 > \ell^2$, where ℓ is a small length. The interesting fact is that this is not possible. One can prove that the vanishing of the commutator

$$[\phi(x), \phi(x')]$$

for some space-like neighbourhood implies that it must vanish for all space-like points.[26] Hence we are back to the case of exact locality.

As we may expect, this theorem about the perseverance of local commutativity can be extended to cases where the commutator vanishes sufficiently fast as the space-like distance $(\vec{x} - \vec{x}')^2 - (x_o - x_o')^2$ goes to infinity. It can be shown[27] that an asymptotic bound like

$$\exp\ [-\alpha|(x-x')^2|^{\beta/2}], \qquad \alpha > 0,$$

with $\beta>1$ is sufficient to imply local commutativity for all space-like points. The proof of these statements rests mainly on the analytic properties of Wightman functions in <u>space-time</u>, which follow from the spectral conditions in their most general form (positive energy states only) and from Lorentz covariance.

Although these theorems are very interesting, it is perhaps good to remember that possible deviations from the microcausal Lorentz-scheme may well not be restricted to the locality axiom alone. It is not implausible to expect a simultaneous violation of other axioms like, for example, the exact Lorentz covariance in small dimensions, which could be so as to prevent acausal regions from spreading if seen from rapidly moving observers. Another interesting possibility is that a violation of microcausality may be connected with an indefinite metric in the Hilbert space of state vectors.[28] In these and other cases, the general assumptions leading to the theorem on the global character of local commutativity are no more complete, and the theorem is presumably not valid. We may then feel free to assume that the commutator in the representation (3.4) vanishes in some restricted space-like region.

We have mentioned before that a relativistic quantum field theory without local commutativity can be easily formulated in terms of Wightman functions. However, in its pure form, this theory makes no direct connection with experiments, because it does not define the S-matrix. Recent work in axiomatic field theory has made it possible, to some extent, to obtain from the Wightman formalism the Fourier representation of scattering amplitudes in terms of retarded products.[29] However, it is not at all clear whether this can be done if local commutativity is relaxed, for example, by assuming an exponential bound like

$$\langle\Psi|[\phi(\tfrac{x}{2}),\ \phi(-\tfrac{x}{2})]|\Psi\rangle \sim \exp\{-\alpha\sqrt{\vec{x}^2 - x_o^2}\}$$

for $\vec{x}^2 - x_o^2 \to \infty$, $\alpha > 0$. Certainly, the manifestly covariant form of retarded products is lost in this case, because the time ordering is not defined in an invariant fashion for space-like points in the R_4.

5. Modified dispersion formulae

In the following we want to explore the effect of specific violations of microcausality in a very simple-minded and model-dependent way. We take the representation (3.3) for the forward scattering amplitude and make assumptions about the support of the matrix element of the commutator in space-like regions. If the function $\chi(x_o,r)$ is non-zero for $r^2 - x_o^2 > 0$, then the restriction of the x_o integration to $x_o \geq 0$ is no more a manifestly invariant statement. Unless we want to introduce an external time-like vector, we must replace x_o by $-p\cdot x(-p^2)^{-1/2}$ which equals x_o only in the laboratory frame. But since it is the laboratory system in which the dispersion relations are evaluated and compared with experiments, we restrict ourselves in the following to this frame of reference.

Let us consider first the possibility that the function $\chi(x_o,r)$, $r \equiv |\vec{x}|$, is exponentially bounded in the space-like region by exp $(-ar)$, where $\ell = a^{-1}$ is a small length, which we certainly assume to be smaller than the Compton wave length μ^{-1} of the projectile. With this bound, the acausal region where $\chi(x_o,r)$ has a sizable support is essentially restricted to the neighborhood of the vertex of the light cone. Later, we will consider bounds like exp $(-a\sqrt{r^2 - x_o^2})$ and others.

We write the function $F(\omega,r)$, defined in Eq. (3.6), in the form

$$F(\omega,r) = F_c(\omega,r) + F_a(\omega,r), \tag{5.1}$$

with

$$F_c(\omega,r) = s(\omega,r) \int_r^\infty dx_o\, e^{i\omega x_o}\, \chi(x_o,r), \tag{5.2}$$

and

$$F_a(\omega,r) = s(\omega,r) \int_0^r dx_o\, e^{i\omega x_o}\, \chi(x_o,r). \tag{5.3}$$

In Eq. (5.3), we have retained the support condition $x_o \geq 0$, because this guarantees that the imaginary part of $F(\omega,r)$ [and

$F(\omega)]$ along the real axis is identical with the absorptive part, which is given by Eq. (3.10) as a sum over a complete set of intermediate states.* We find

$$\text{Im } F(\omega,r) = s(\omega,r) \; \frac{-i}{2} \int_{-\infty}^{+\infty} dx_o \; e^{i\omega x_o} \; \chi(x_o,r), \tag{5.4}$$

which vanishes again for ω in the gap $-\mu < \omega < +\mu$. In general, we must assume that $\text{Im } F_c(\omega,r)$ and $\text{Im } F_a(\omega,r)$ do not vanish separately in this interval.

The functions $F_c(\omega,r)$ and $F_a(\omega,r)$ are both regular in the upper half ω-plane; F_c is polynomially bounded for $\omega_i \to \infty$ ($\omega_i \equiv \text{Im } \omega$), but F_a can have an r-dependent essential singularity there. In the following, we assume sufficient boundedness at infinity whenever there is no exponential increase. Then we can write unsubtracted Hilbert representations; they can be easily generalized.

For $F_c(\omega,r)$, we have the representation

$$F_c(\omega,r) = \frac{1}{\pi} \int_{-\infty}^{+\infty} d\omega' \; \frac{\text{Im } F_c(\omega',r)}{\omega' - \omega}, \tag{5.5}$$

and for $F_a(\omega,r)$, we write

$$F_a(\omega,r) = \frac{1}{2\pi i} \int_{-\infty}^{+\infty} d\omega' \; \frac{F_a(\omega',r)}{\omega' - \omega} - \frac{1}{2\pi i} \int_{-\infty}^{+\infty} d\nu \; \frac{F_a(\nu+ia_o,r)}{\nu + ia_o - \omega} . \tag{5.6}$$

---

*It may well be that we should relax this condition. It is conceivable that a possible violation of microcausality is related to deviations from the usual unitarity connections, in particular at very high energies. Formally, the inclusion of a limited region with $x_o < 0$ does not change the modified dispersion relations derived in this section (for more details, see Sec. 7).

We also note that for propagators (two point functions) the spectral condition and Lorentz covariance already imply the support conditions in space-time which are required by microcausality.

In the upper half ω-plane, the function $F_a(\omega,r)$ is exponentially bounded for $r \to \infty$ provided

$$\mathrm{Im}\, \sqrt{\omega^2 - \mu^2} < a$$

We assume that $a > \mu$ (generally we expect $a >> \mu$), and we replace this inequality by

$$\mathrm{Im}\, \omega < a_o \equiv \sqrt{a^2 - \mu^2}. \tag{5.7}$$

Taking ω real and $|\omega| \geq \mu$, adding Eqs. (5.5) and (5.6) together and taking the real part, we obtain the relation

$$\begin{aligned} \mathrm{Re}\, F(\omega,r) &= \frac{P}{\pi} \left(\int_{-\infty}^{-\mu} + \int_{\mu}^{\infty} \right) d\omega' \, \frac{\mathrm{Im}\, F(\omega',r)}{\omega' - \omega} \\ &\quad - \mathrm{Re} \left\{ \frac{1}{2\pi i} \int_{-\infty}^{+\infty} d\nu \, \frac{F_a(\nu + ia_o, r)}{\nu + ia_o - \omega} \right\} , \end{aligned} \tag{5.8}$$

where we can now perform the r-integration without difficulties on both sides of the equation. The result is a representation for $\mathrm{Re}\, F(\omega)$, which may be written in the form

$$\begin{aligned} F(\omega) &= \frac{1}{\pi} \left(\int_{-\infty}^{-\mu} + \int_{\mu}^{\infty} \right) d\omega' \, \frac{\mathrm{Im}\, F(\omega')}{\omega' - \omega} - \\ &\quad - \frac{1}{2\pi i} \int_{-\infty}^{+\infty} d\nu \, \frac{F_a(\nu + ia_o)}{\nu + ia_o - \omega} + \frac{1}{2\pi i} \int_{-\infty}^{+\infty} d\nu \, \frac{F_a^*(\nu + ia_o)}{\nu - ia_o - \omega} , \end{aligned} \tag{5.9}$$

and which can now be extended to complex values of ω. Hence we find that the scattering amplitude $F(\omega)$ is the boundary value of an analytic function which is regular in the cut ω-plane, at least for $|\mathrm{Im}\, \omega| < a_o$, and which is a real analytic function, i.e., $F^*(\omega^*) = F(\omega)$. In Eq. (5.9) we can replace the acausal part $F_a(\nu+ia_o)$ in the last two integrals by $F(\nu+ia_o)$ and $F^*(\nu+ia_o) = F(\nu-ia_o)$ respectively. Taking both integrals together, we could also write their contribution in the form

$$\frac{1}{\pi} \int_{-\infty}^{+\infty} d\nu \, \frac{a_o \, \mathrm{Re}\, F(\nu+ia_o) + (\omega-\nu)\, \mathrm{Im}\, F(\nu+ia_o)}{(\nu - \omega)^2 + a_o^2} . \tag{5.10}$$

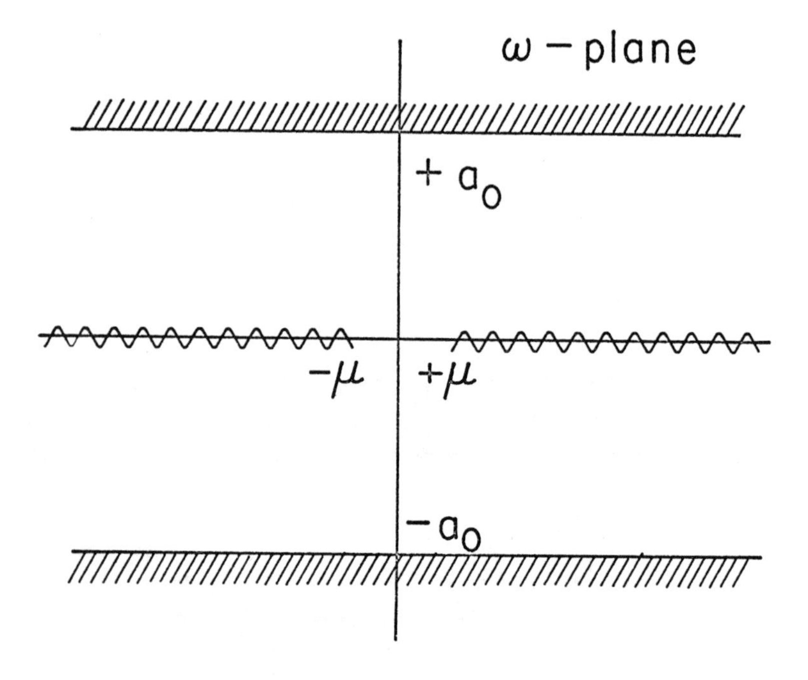

Fig.3

Region of regularity for an acausal amplitude with exponential space-like bound of range $a = (a_o^2 + \mu^2)^{1/2}$

The integrals over the lines $\mathrm{Im}\,\omega = \pm a_o$ in Eqs. (5.9) or (5.10) do not mean, of course, that these lines are necessarily branch lines or even natural boundaries. Given our assumptions, all we can say is that there are no singularities in the domain $|\mathrm{Im}\,\omega| < a_o$ which are not also present in the micro-causal case. The possible singularities in the region $|\mathrm{Im}\,\omega| \geq a_o$ depend upon the details of the function $\chi(x_o,r)$, and if we assume a particular form for this function, we generally will be able to replace the integral (5.10) by the particular contributions of these singularities. Some specific cases are discussed in the following chapter.

In our previous discussions, we have assumed that $\chi(x_o,r)$ behaves essentially like exp (-ar) for $r \to \infty$, $r^2 - x_o^2 > 0$. Instead of restricting the relevant acausal region to a domain near the vertex of the light cone, we may consider the possibility that $\chi(x_o,r)$ has a sizable space-like support in a region $r^2 - x_o^2 < \ell^2$ near the light cone. Let us assume that, except for finite powers of $\sqrt{r^2 - x_o^2}$, the function $\chi(x_o,r)$ is bounded by

$$\exp\left[-a\sqrt{r^2 - x_o^2}\right] \qquad (5.11)$$

for $r^2 - x_o^2 \to \infty$. With this bound, we can essentially obtain the same modified dispersion formulae (5.9) as before. There is, however, some difficulty with our simple method of proof, but it is only due to the finite mass of the projectile, and therefore, it can perhaps be circumvented by using more elaborate methods.* The difficulty is related to the behavior of $F_a(\omega,r)$ for $r \to \infty$ and $\omega_i \leq a$. From Eq. (5.3) and the bound (5.11), we find for $\omega = \nu + ia$ an exponential term of the form

$$\exp\left[\left(\mathrm{Im}\sqrt{(\nu + ia)^2 - \mu^2} - a\right)r\right]$$

---

*We have constructed simple models where $\chi(x_o,r)$ has a behavior corresponding to Eq. (5.11), and we find that the singularities of the amplitude $F(\omega)$ are consistent with the modified dispersion relation (5.9).

which, for $a >> \mu$ and $\nu = 0$, corresponds to an exponential increase like

$$\exp\,[\frac{\mu^2}{2a}\,r].$$

Evidently, there is no difficulty for $\mu = 0$. With a finite mass, we may supplement the space-like bound (5.11) by a long range factor $\exp\,[-\frac{\mu^2}{2a}\,r]$ or $\exp\,[-\mu r)$ in order to preserve our simple proof. A damping factor like this may anyhow be required in order to prevent difficulties with macroscopic causality because, with the bound (5.11), the undamped acausal region is in principle spread out over macroscopic distances, although it becomes very thin.

We have not explored here the much more elaborate methods of proof which use analytic continuations in mass variables.[22,23,30] A priori, they involve replacement of the root $\sqrt{\omega^2 - \mu^2}$ in Eq. (5.2) and (5.3) by $\sqrt{\omega^2 - \beta}$. With $\beta < 0$, one can derive the modified dispersion relations (5.9) without difficulties, but the essential problem is the analytic continuation in β from $\beta < 0$ to $\beta = \mu^2$, which may be rather difficult to prove in non-local theories.

6. Nonlocal models

In order to visualize the kind of singularities one may expect if the matrix element of the commutator vanishes exponentially for large, space-like distances, we may consider a simple mathematical model. Suppose $\chi(x_o,r)$ is such that the function $F_a(\omega,r)$ defined in Eq. (5.3) is of the form

$$F_a(\omega,r) = s(\omega,r)\int_a^\infty d\alpha\;\rho(\omega,\alpha)\;\frac{e^{-\alpha r}}{4\pi r}\,, \tag{6.1}$$

where $\rho(\omega,\alpha)$ is an analytic function of ω for $\mathrm{Im}\,\omega > 0$; it may be a distribution in α.* With this Ansatz, we find

---

*Note that $\rho(\omega,\alpha)$ generally is not real for real values of ω. It must, however, be of such a form that $\mathrm{Im}\,F_a + \mathrm{Im}\,F_c = \mathrm{Im}\,F$ is the absorptive part given in Eqs. (5.4) and (3.10).

$$F_a(\omega) = \int_a^\infty d\alpha \frac{\rho(\omega,\alpha)}{\omega^2 + \alpha^2 - \mu^2} , \tag{6.2}$$

i.e., a function which is analytic in the upper half of the ω-plane except for the imaginary axis above $\mathrm{Im}\,\omega = a_o \equiv \sqrt{a^2 - \mu^2}$.

In order to write the modified dispersion relations (5.9) for the model described above, it is convenient to consider separately the two symmetry cases with

$$F^{(\pm)}(\omega) = \pm F^{(\pm)}(-\omega). \tag{6.3}$$

Ignoring the subtractions, which can be easily supplied, we obtain then the formulae

$$\begin{aligned} F^{(+)}(\omega) = \frac{2}{\pi} \int_\mu^\infty d\omega' \frac{\omega' \,\mathrm{Im}\, F^{(+)}(\omega' + io)}{\omega'^2 - \omega^2} \\ - \frac{2}{\pi} \int_{a_o}^\infty d\lambda \frac{\lambda \,\mathrm{Im}\, F^{(+)}(i\lambda + o)}{\lambda^2 + \omega^2} , \end{aligned} \tag{6.4}$$

and

$$\begin{aligned} F^{(-)}(\omega) = \frac{2\omega}{\pi} \int_\mu^\infty d\omega' \frac{\mathrm{Im}\, F^{(-)}(\omega')}{\omega'^2 - \omega^2} + \\ + \frac{2\omega}{\pi} \int_{a_o}^\infty d\lambda \frac{\mathrm{Re}\, F^{(-)}(i\lambda + o)}{\lambda^2 + \omega^2} . \end{aligned} \tag{6.5}$$

We can easily obtain a more general pattern of singularities for $\mathrm{Im}\,\omega \geq a_o$, for example by including oscillatory terms in r. Because of the symmetry relations (6.3) and the reality condition $F^*(\omega^*) = F(\omega)$, the singularities appear generally in all four quadrants in a symmetric way. For example, if $F^{(-)}(\omega,r)$ has a pole at $\omega = a_r + ia_i$, there must be three other poles giving a total contribution of the form

$$2\omega R \left\{ \frac{1}{(a_r + ia_i)^2 - \omega^2} + \frac{1}{(a_r - ia_i)^2 - \omega^2} \right\} , \tag{6.6}$$

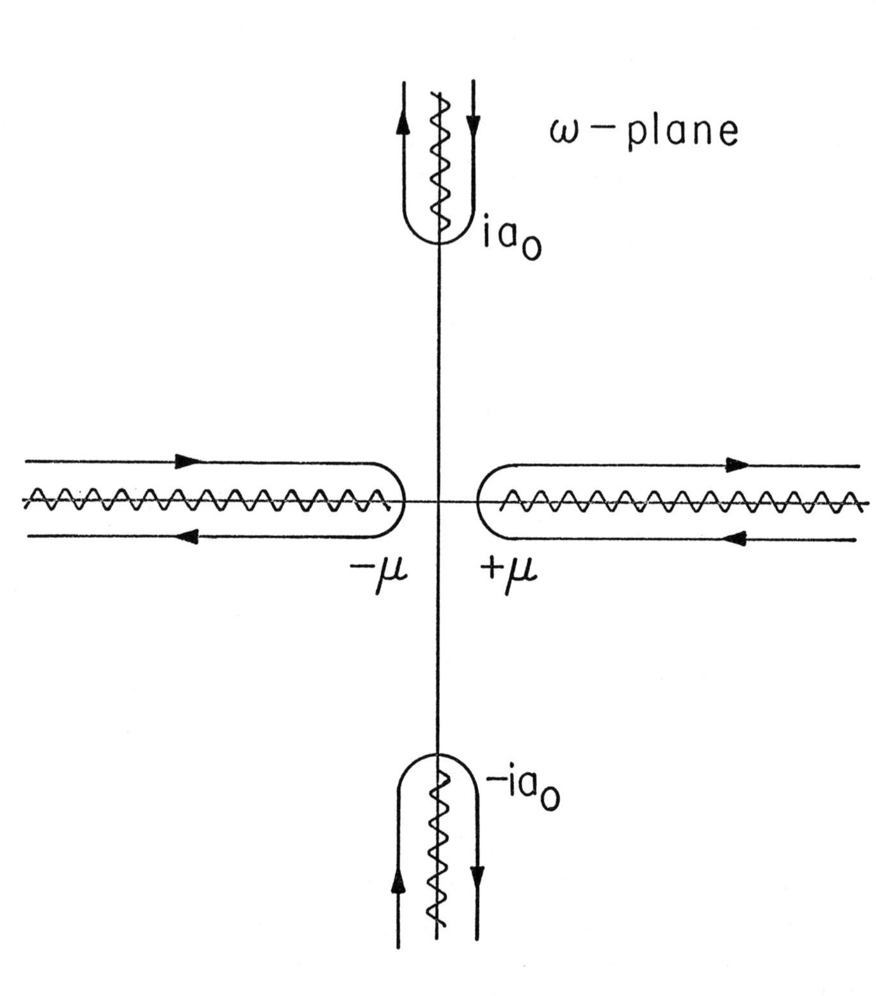

Fig.4

Integration contours which lead to the modified dispersion formulae (6.4) and (6.5) for a special nonlocal model.

where R is a real constant.

If we are willing to consider more comprehensive deviations from the micro-causal and Lorentz covariant structure of field theory, we may also study models where the matrix element $\chi(x_o, r)$ of the commutator vanishes exactly outside some space-like region. In this case, the character of the singularities in the upper half of the ω-plane changes, and it may be of some interest to consider an example. Suppose we take the model (6.1) and cut off the support in r for $r > R$. Then we obtain for $F_a(\omega)$

$$F_a(\omega) = \int_a^\infty d\alpha \frac{\rho(\omega,\alpha)}{\omega^2 - \mu^2 + \alpha^2} \left\{1 - e^{-\alpha R}\left(\cos R\sqrt{\omega^2 - \mu^2} + \alpha \frac{\sin R\sqrt{\omega^2 - \mu^2}}{\sqrt{\omega^2 - \mu^2}}\right)\right\} \tag{6.7}$$

which is a regular function for $\text{Im}\,\omega > 0$ except for an essential singularity at infinity. As before, we can write a modified dispersion formula corresponding to Eq. (5.9), where a_o can be chosen very large but finite. Although there is an important difference in the character of the singularities, for sufficiently large values of R, the functions (6.7) and (6.2) are not very different for real, physical values of ω. The presence of the essential singularity produces oscillation of the amplitude as a function of ω, but these are strongly damped if $R \gg a^{-1}$.

On the other hand, we may cut off the space-like support of $\chi(x_o, r)$ very near the vertex of the light cone. Suppose we take $R = a^{-1} = \ell$, with $\ell \ll \mu^{-1}$. Then the function $F_a(\omega)$ is of the form

$$F_a(\omega) = \int_{\ell^{-1}}^\infty d\alpha \frac{\rho(\omega,\alpha)}{\omega^2 - \mu^2 + \alpha^2} \left\{1 - e^{-\alpha\ell}\left(\cos \ell\sqrt{\omega^2 - \mu^2} + \ell\alpha \frac{\sin \ell\sqrt{\omega^2 - \mu^2}}{\ell\sqrt{\omega^2 - \mu^2}}\right)\right\} \tag{6.8}$$

For real $\omega > \mu$ with $\omega\ell \ll 1$, the second term in the bracket becomes approximately equal to $1 - (1+\ell\alpha)\exp[-\alpha\ell]$, and with

$$\bar{\rho}(\omega,\alpha) = \rho(\omega,\alpha)\ \{1 - e^{-\alpha\ell}(1 + \ell\alpha)\},$$

the function $F_a(\omega)$ is given by

$$F_a(\omega) \approx \int_{\ell^{-1}}^{\infty} d\alpha\ \frac{\bar{\rho}(\omega,\alpha)}{\alpha^2}\ , \tag{6.9}$$

which vanishes for $\ell \to 0$. We see that Eqs. (6.9) and (6.2) are essentially the same except for a redefinition of $\rho(\omega,\alpha)$. For our present purpose, the relevant point is only that the exponential model and the related cut-off model have essentially equivalent physical amplitudes for $\omega\ell << 1$. For $\omega\ell \gtrsim 1$, these models can be different. In particular, Eq. (6.8) gives rise to oscillations of $F(\omega)$, which may be small compared to the full amplitude since they appear only in an additive term.

The function $F_a(\omega)$ in Eq. (6.8) increases exponentially for $\mathrm{Im}\ \omega \to \infty$, but not faster than exp $(-i\omega\ell)$. Therefore the product

$$e^{i\omega\ell}F_a(\omega) \tag{6.10}$$

is bounded (polynomially bounded, in general). Consequently also the expression

$$e^{i\omega\ell}F(\omega) \tag{6.11}$$

is bounded for $\mathrm{Im}\ \omega \to \infty$, and we can write a dispersion representation for this function:[25]

$$\begin{aligned} e^{i\omega\ell}F(\omega) &= \frac{1}{2\pi i}\int_{-\infty}^{+\infty} d\omega'\ \frac{e^{i\omega'\ell}F(\omega')}{\omega' - \omega} \\ &= \frac{1}{\pi}\int_{-\infty}^{+\infty} d\omega'\ \frac{\cos\omega'\ell\ \mathrm{Im}\ F(\omega') + \sin\omega'\ell\ \mathrm{Re}\ F(\omega')}{\omega' - \omega}\ , \end{aligned} \tag{6.12}$$

where $\mathrm{Im}\ \omega \geq 0$. In this form we see also that the usual dispersion formula is approximately valid for $\omega \ell \ll 1$.*

The models we have discussed in this section can, of course, be easily generalized. In particular, we may consider an exponential fall-off or a cut-off depending upon $\sqrt{r^2 - x_0^2}$ instead of r. Here we do not want to consider these details. Models like Eq. (6.1) may be rather unrealistic, and therefore we want to refrain from numerical estimates of the possible influence of acausal terms in the modified dispersion relations. In the case of modified dispersion formulae like Eqs. (6.4) and (6.5), we must recall that an essentially constant acausal term can be effectively removed by writing a subtracted dispersion relation, and therefore it may not influence the comparison with experiments. For $|\omega| \ll \ell_0$, the second integrals in these equations are essentially constants.

7. Concluding remarks

We have explored in this article some modifications of dispersion relations which follow from possible violations of microcausality and, in certain cases, of other basic notions. However, we have restricted our considerations to scattering amplitudes, and we have made specific assumptions only about the particular matrix elements of the commutators which appear in the Fourier representations of these amplitudes. What really should be done is a consistent formulation of a nonlocal field theory with exponential bounds for all matrix elements of commutators in space-like regions. This theory should contain explicit expressions for S-matrix elements in terms of Fourier transforms of space-time functions, which then may make it possible to derive some analytic properties of scattering amplitudes on the basis of the fundamental assumptions of the nonlocal theory. Such a general axiomatic approach has not been attempted here, and hence there is no proof that it would lead to the modified dispersion relations discussed in

---

*Modified dispersion relations of the type (6.12) for $T^{(+)}(\omega)$ have been compared with the data by S. J. Lindenbaum, see Ref. 11.

Section 5. On the other hand, there is certainly no guarantee that possible violations of microcausality will occur in complete isolation, and it may well be that modifications of other fundamental assumptions, like the definiteness of the metric in Hilbert space, or even Lorentz covariance in small dimensions, have to be considered simultaneously.

In all our discussions we have maintained the support condition $x_o \geq 0$ for the Fourier representation of the scattering amplitude in the laboratory system. As we have pointed out before, this makes it possible to retain the unitarity connection between $\text{Im}\, F(\omega)$ and the set of intermediate states. However, if we are willing to ignore some aspects of this connection and include a limited region with $x_o < 0$, we find that our modified dispersion relation in Section 5 are not changed, at least as long as the acausal domain for negative values of x_o is finite $(x_o \geq -\ell)$ or exponentially bounded for $x_o \to -\infty$ $(\sim \exp[-\alpha|x_o|],\ a = \ell^{-1})$.[31] There are, of course, changes in the specific acausal singularities which we find for $|\text{Im}\, \omega| > a$. As in the case of the r-distribution, a fixed cut-off generates essential singularities, and an exponential one gives poles and branch points.

It is perhaps of interest to observe that, with the restriction of the support in the representation (3.4) to $x_o \geq 0$, the amplitude

$$F(k_o,\vec{k}) = \int d^4x\, e^{ik_o x_o - i\vec{k}\cdot\vec{x}}\, \theta(x_o)\, \chi(x_o,|\vec{x}|) \tag{7.1}$$

is an analytic function of k_o in the upper half of the complex k_o-plane for fixed values of $\vec{k}$. Consequently we have, except for possible subtractions, the Low equation

$$F(k_o,\vec{k}) = \sum_{|n>} |\langle p|j(0)|n\rangle|^2\, N_n^2 \times \left\{ \frac{\delta(\vec{k} - \vec{p}_n)}{p_{no} - p_o - k_o - i\varepsilon} + \frac{\delta(\vec{k} + \vec{p}_n)}{p_{no} - p_o + k_o + i\varepsilon} \right\}, \tag{7.2}$$

with $p = \{\vec{p}, ip_o\} = \{o, i\mu\}$. In view of the modified dispersion relations which we have derived in Section 5, we see here explicitly the essential difference between actual dispersion relations for physical amplitudes on the mass shell and the relations of the type (7.2).

The acausal models we have considered generally contain a characteristic length ℓ which describes the size of the space-time region where there is a larger deviation from the microcausal structure. Since we have been working in the laboratory system, this length is defined relative to the laboratory energy ω. Essentially, our models are such that, for properly subtracted dispersion formulae, the modifications due to acausal singularities are negligible as long as $\omega\ell << 1$. The actual value of ℓ, which could be inferred by comparing the modified dispersion relation with the experimental data, is certainly model dependent, and therefore we do not try to give a number. In view of the models we have considered, it appears to be plausible that good agreement of the modified dispersion formulae with experimental data up to a laboratory ω_{max} generally would require that a possible length ℓ is smaller than $\ell_o(\text{lab}) \sim \omega_{max}^{-1}$.

We have mentioned in this paper only forward dispersion relations, mainly for pion nucleon scattering, because these formulae do not contain continuous unphysical contributions. Corresponding forward relations for nucleon-nucleon scattering[32] necessarily involve integrations over nucleon-antinucleon channels, and there are continuous contributions which can only be expressed indirectly in terms of physical quantities. This complicates the comparison with experiments, but does not make it impossible. The nucleon-nucleon forward relations have not yet been proven rigorously, mainly because it is difficult to incorporate the relevant unitarity relations in the $N\bar{N}$-channel.[23]

Dispersion relations for nonforward πN-scattering at fixed momentum transfers also contain an unphysical continuum. These relations can be proven in a limited range of momentum transfers, and the proof requires a much more detailed use of microcausality and spectral conditions than the derivation of forward relations.[22,23,30] It is an interesting open problem to find the modifications of these nonforward dispersion relations in the presence of violations of microcausality and of other basic notions of field theory.

Finally, we mention that dispersion relations for vertex functions[33] can in several cases be derived[23] by the simple method described in Section 3, and it would be of interest to study their modifications due to possible violations of microcausality. We hope to return to this problem elsewhere.

Acknowledgements

It is a pleasure to thank Professor S. J. Lindenbaum for many discussions, and in particular for his kind permission to reproduce two of his figures. I am also indebted to Professor R. Haag, Professor H. Lehmann and Professor K. Symanzik for helpful conversations.

References

1. M. L. Goldberger, H. Miyazawa and R. Oehme, Phys. Rev. 96, 986 (1955).
2. H. L. Anderson, W. C. Davidon and U. E. Kruse, Phys. Rev. 100, 339 (1955).
3. K. J. Foley, R. S. Jones, S. J. Lindenbaum, W. A. Love, S. Ozaki, E. D. Platner, C. A. Quarles and E. H. Willen, Phys. Rev. Letters 19, 193, 330 (1967).
4. E. P. Wigner, in Dispersion Relations and Their Connection with Causality, ed. by E. P. Wigner (Academic Press, New York and London, 1964), p. 40; W. Schützer and J. Tiomno, Phys. Rev. 83, 249 (1951); N. G. Van Kampen, Phys. Rev. 89, 1062 (1953); G. Wanders, Nuovo Cimento 14, 168 (1959).
5. See, for example, D. I. Blokhintsev, I.C.T.P. Report 67/36; H. P. Stapp, Proc. XIII Intern. Conf. on High Energy Physics (Univ. of Calif. Press, Berkeley, 1967), p. 19. These papers contain many further references.

6. Sellmeier, Poggendorf Ann. 143, 271 (1871); R. Kronig, J. Opt. Soc. Am. 12, 547 (1926); H. A. Kramers, Physica 12, 543 (1946); Atti del Congresso Internazionale de Fisici, Como (Nicolo Zanichelli, Bologna, (1927) p. 545; J. S. Toll, The Dispersion Relation for Light and its Application to Problems Involving Electron Pairs, Thesis, Princeton, 1952; Phys. Rev. 104, 1760 (1956); these papers contain further references.
7. M. Gell-Mann, M. L. Goldberger and W. E. Thirring, Phys. Rev. 95, 1612 (1954).
8. R. Karplus and M. A. Ruderman, Phys. Rev. 98, 771 (1955); M. L. Goldberger, Phys. Rev. 99, 979 (1955); R. Oehme, Phys. Rev. 100, 1503 (1955).
9. M. Froissart, Phys. Rev. 123, 1053 (1961); A. Martin, Nuovo Cimento 42, 930 (1966).
10. I. Ya. Pomeranchuk, Zhur. Eksp. i Teoret. Fiz. 34, 725 (1958); N. N. Meiman, Zhur. Eksp. i Teoret. Fiz. 43, 2277 (1962); L. Van Hove, Phys. Letters 5, 252 (1963); A. A. Logunov, Nguyen Van Hien, I. T. Todorov and O. A. Krustalev, Phys. Letters 7, 69 (1963); A. Martin, Nuovo Cimento 39, 704 (1965).
11. S. J. Lindenbaum, "Asymptotic Energies," paper presented at the International Symposium on Contemporary Physics, Trieste, June, 1968; (Report BNL 12811).
12. G. Höhler, G. Ebel and J. Giesecke, Z. für Physik 180, 430 (1964); J. Hamilton and W. S. Woolcock, Rev. Mod. Phys. 35, 737 (1963).
13. R. Oehme, unpublished letter, July 1955; for a recent evaluation see G. Höhler, J. Baacke and R. Strauss, Phys. Letters 21, 223 (1966).
14. G. F. Chew, Phys. Rev. 95, 1669 (1954); F. E. Low, Phys. Rev. 97, 1392 (1955); G. F. Chew and F. E. Low, Phys. Rev. 101, 1570 (1956); G. C. Wick, Rev. Mod. Phys. 27, 339 (1955).
15. R. Oehme, Phys. Rev. 100, 1503 (1955); 102, 1174 (1956); M. L. Goldberger, Y. Nambu and R. Oehme, reported in: Proceedings of the Sixth Annual Rochester Conference (Interscience, New York, 1956) p. 1-7. This report contains references to related work

of Gell-Mann and Polkinghorne, Symanzik, Salam, Capps and Takeda, Fradkin and others; A. Salam and W. Gilbert, Nuovo Cimento 3, 607 (1956); G. F. Chew, M. L. Goldberger, F. E. Low and Y. Nambu, Phys. Rev. 106, 1337 (1957); S. Mandelstam, Phys. Rev. 112, 1344 (1958).

16. For a survey and for references see, for example, G. F. Chew, S-Matrix Theory of Strong Interactions (Benjamin, New York, 1961); R. Oehme, "Complex Angular Momentum in Elementary Particle Scattering," in Strong Interactions and High Energy Physics, ed. by R. G. Moorhouse (Oliver and Boyd, Edinburgh and London, 1964) p. 129; M. L. Goldberger and K. M. Watson, Collision Theory (J. Wiley and Sons, New York, 1964).
17. See, for example, R. Jost, The General Theory of Quantized Fields (Ann. Math. Soc., Providence, 1965).
18. H. Lehmann, K. Symanzik and W. Zimmermann, Nuovo Cimento 1, 205 (1955); N. N. Bogoliubov and D. V. Shirkov, Introduction to the Theory of Quantized Fields (Interscience, New York, 1959); H. Umezawa and S. Kamefuchi, Progr. Theor. Phys. 6, 543 (1951); M. L. Goldberger, Phys. Rev. 97, 508 (1955); F. E. Low, Phys. Rev. 97, 1392 (1955); Y. Nambu, Phys. Rev. 98, 803 (1955).
19. H. Lehmann, K. Symanzik and W. Zimmermann, Nuovo Cimento 6, 319 (1957).
20. R. Oehme, Nuovo Cimento 10, 1316 (1956).
21. K. Symanzik, Phys. Rev. 100, 743 (1957).
22. N. N. Bogoliubov, B. V. Medvedev and M. K. Polivanov, Voprossy Teorii Dispersionnykh Sootnoshenii (Fizmatgiz, Moscow, 1958).
23. H. J. Bremermann, R. Oehme and J. G. Taylor, Phys. Rev. 109, 2178 (1958).
24. R. Haag, Kgl. Danske Videnskab. Selskab, Mat. Fys. Medd. 29, 12 (1955); A. S. Wightman, Phys. Rev. 101, 860 (1956); A. S. Wightmann and J. Garding, Ark. Fys. 28, 129 (1964).
25. R. Oehme, Phys. Rev. 100, 1503 (1955).

26. R. Jost and O. Steinmann, see A. S. Wightmann, J. Indian Math. Soc. 24, 625 (1960); R. F. Streater and A. S. Wightman, PCT, Spin and Statistics, and all that. (Benjamin, New York, 1964) p. 134.
27. H. J. Borchers and K. Pohlmeyer, Commun. Math. Phys. 8, 269 (1968); K. Pohlmeyer, Commun. Math. Phys. 7, 80 (1968).
28. T. D. Lee and G. C. Wick, Nuclear Physics B9, 209 (1969), this paper contains further references.
29. O. Steinmann, Commun. Math. Phys. 10, 245 (1968); K. Hepp, Commun. Math. Phys. 1, 95 (1965).
30. H. Lehmann, Suppl. Nuovo Cimento 14, 153 (1958), Nuovo Cimento 10, 579 (1958); R. Omnes, in Relations de dispersion et particules elementaires (Herman, Paris, 1960) p. 319; M. Froissart, in Dispersion Relations and their Connection with Causality, ed. by E. P. Wigner (Academic Press, New York, 1964) p. 1; K. Hepp, Helv. Phys. Acta 37, 639 (1964).
31. D. I. Blokhintsev and G. I. Kolerov, Nuovo Cimento 34, 163 (1964).
32. M. L. Goldberger, Y. Nambu and R. Oehme, Ann. Phys. (New York) 2, 226 (1956).
33. P. Federbush, M. Goldberger and S. B. Treiman, Phys. Rev. 112, 642 (1958); R. Oehme in Werner Heisenberg and die Physik unserer Zeit, ed. by F. Bopp (Vieweg, Braunschweig, 1961) p. 240, this paper contains further references.

Received 5/2/69

The Spatial Extent of Magnetic Monopoles

by

C. J. Goebel

University of Wisconsin

Dirac[1] showed that if magnetic charge exists, it must occur in units $e' = e^{-1/2}$ $(h = c = 1)$; this follows merely from quantization of angular momentum. Schwinger[2] argued that the unit should be $e' = e^{-1}$, which amounts to requiring integral quantization of the field angular momentum; we shall accept this. So a particle which carries a magnetic charge, a monopole, is coupled very strongly to the electromagnetic field; its coupling to a photon is characterized by the number

$$e'^2 = e^{-2} = 137. \tag{1}$$

This is an order of magnitude stronger than the pion-nucleon coupling, $g^2/4\pi = 15$; it is so strong that many calculations that have been attempted, such as ionization energy loss of a monopole in matter, and monopole pair production cross sections, are very doubtful.

One's first reaction is that strong coupling theory (we mean the well known static model strong coupling theory) ought to be applicable; unfortunately it is not. For instance, consider the photon-monopole elastic amplitude, Compton scattering. The theory of this is identical (up to vacuum polarization effects) to ordinary Compton scattering, if we exchange $\vec{E}$ and $\vec{H}$ fields of the radiation field, and substitute e' for e. It is thus a rigorous result that the amplitude at zero frequency, the scattering length, is given by the Thompson formula

$$f(\omega=0) = -e'^2/M \equiv -r_m \quad (\text{"classical magnetic radius"}) \tag{2}$$

where M is the mass of the monopole. The difficulty is that this is a large repulsive scattering length, which loosely speaking requires[3] that the range of interaction between photon and monopole be at least $r_m = 137M^{-1}$. [Because this is large compared to the Compton radius M^{-1}, relativistic effects do not help, unlike the case of electrically charged particles, where the classical electronic radius r_e is dynamically irrelevant because $r_e = e\, m^{-1} << m^{-1}$.]

In contrast, in meson models in which the strong coupling theory applies, the meson-nucleon scattering is strong and attractive, which induces a bound state (isobar) pole which saturates the amplitude; the corresponding phenomenon in the present case would be the binding of a ghost.

This is well known in the classical theory, where the magnetic dipole amplitude for an electromagnetic wave scattering on a point monopole is

$$f_M(\omega) = -[r_m^{-1} + 2/3\ i\omega]^{-1} . \tag{3}$$

This has a non-causal "ghost" pole at $\omega = i\ 3/2\ r_m^{-1} = i\ 3/2\ M/137$; it represents Dirac's pre-acceleration.[4] In order for the classical theory to be causal, the monopole's charge must be spread out, with a radius of at least r_m, so that the form factor singularities replace the ghost pole. [Put more dynamically, the condition for the absence of runaway motion, or pre-acceleration, is that the bare mass m_o be non-negative, which requires that the electromagnetic self mass $\sim e'^2 \langle \frac{1}{r} \rangle$ must be no larger than the total mass M, which means $\langle \frac{1}{r} \rangle \lesssim M/e'^2 = 1/r_m$.)

We argue[5] that the same must be true in the quantum theory of the monopole: The radius of a monopole must be of the order of r_m.[6,7] We start with a rigorous formula, the dispersion relation for the reciprocal of the (laboratory) forward Compton amplitude, $f^{-1}(\omega)$, which implies the sum rule

$$r_m^{-1} = \frac{M}{137} = \frac{2}{\pi} \int_0^\infty d\omega\ R(\omega) \tag{4}$$

where

$$R(\omega) = \frac{\sigma_{tot}}{4\pi|f|^2}. \tag{5}$$

The l.h.s. of (4) is small (on the scale of M) and so this puts a limit on R. For instance, R could not be of order 1 over a frequency range larger than $0 < \omega < M/137$. Thus, on the scale of M, R must quickly fall, which means that higher partial waves must quickly come in, corresponding to a range of interaction of order r_m. One argues this by writing

$$R = \frac{\sigma_{tot}}{\sigma_{e\ell}} \cdot \frac{\sigma_{e\ell}}{4\pi|f|^2} \geq \frac{\sigma_{e\ell}}{4\pi|f|^2} = \frac{\sigma_{e\ell}/4\pi}{(d\sigma_{e\ell}/d\Omega)_{\theta=0}} \approx \frac{1}{4(\Delta\theta)^2} \tag{6}$$

where $\Delta\theta$ is the width of the forward peak of $d\sigma_{e\ell}/d\Omega$, $\Delta\theta \approx ka$, where a is the range of the force. Thus (4) puts a lower limit on the range a. For instance, if $d\sigma_{e\ell}/d\Omega$ were of the form $d\sigma_{e\ell} \sim (1+a^2k^2)^{-2}$, (4) yields $a \geq \frac{1}{2} r_m$.

However, in the present case the effect of soft photons complicates the argument a bit. The elastic Compton cross section vanishes because, except for forward scattering, additional soft pions are always emitted by the recoil of the monopole. Hence it makes no sense to discuss the magnitude of R by writing it as $R = \frac{\sigma_{tot}}{\sigma_{e\ell}} \cdot \frac{\sigma_{e\ell}}{4\pi|f|^2}$, because the two factors are infinite and vanishing, respectively (and the vanishing has nothing to do with the range of the phonon-monopole interaction, but rather with the distance an almost-real monopole can travel). Instead, one should use the concept of quasielastic scattering, defined in the usual way as scattering in which the incident photon loses less than an energy Δ. Then $\sigma_{q-e\ell}/4\pi|f|^2$ will properly reflect the size of the monopole.

It might be noted that soft photon emission by a monopole is actually not a strong effect, despite the largeness of e', if the monopole has size r_m. The spectrum of soft photons emitted by a non-relativistic monopole in a process in which the velocity change of the monopole is ΔV is of the order

$$\frac{dE_{rad}}{d\omega} \sim e'^2 (\Delta V)^2 . \tag{7}$$

If this spectrum is cut off at $\omega \approx r_m^{-1}$, "because the monopole has the size r_m", the total radiative soft photon energy is of order

$$E_{rad} \sim e'^2 (\Delta V)^2 \; r_m^{-1} = (\Delta V)^2 M .$$

Further, if the monopole size is r_m means that the momentum transfer to the monopole is never larger than r_m^{-1}, then $\Delta V \lesssim r_m^{-1}/M = 1/137$, and so

$$E_{rad} \lesssim (1/137)^2 M.$$

More specifically, in Compton scattering, we have under these "size" assumptions,

$$\omega_{Max} \sim \min\,(k,\; r_m^{-1})$$
$$(\Delta V)_{Max} \sim \min\,\left(\frac{k}{M},\; \frac{r_m^{-1}}{M}\right), \tag{8}$$

where k is the incident photon momentum, with the result

$$\left(\frac{E_{rad}}{k}\right)_{Max} \sim \min\,\frac{e'^2 k^2}{M^2}\,, \qquad \frac{e'^2}{r_m^{\,3} M^2 k} \leq 1/e'^2 \tag{9}$$

which means that the fraction of the incident photon energy that is radiated off as soft photons is at most $\approx$ 1%. (It would be large only in the "intermediate coupling case", $e^2 \approx e'^2 \approx 1$.)

We have argued that a monopole must be as big as its "classical radius" $r_m = 137\; M^{-1}$. We now sketch how this can come about self-consistently, in a model in which there are only monopoles and photons. The monopole size r_m would result from the mediation of neutral mesons with masses of the order of r_m^{-1}. [For instance, such a vector C = -1 meson would produce a one-photon form factor; a scalar C = +1 meson would contribute to a two-photon form factor.] But just such mesons result as bound states of a monopole-antimonopole pair. When the monopole and antimonopole are at the same position, their magnetic charge distributions cancel; for small separations r, the magnetostatic energy is of order Mr^2/r_m^2. Hence the energy of the system is

$$\simeq 2M_0 + p^2/M + cMr^2/r_m^2, \quad c = O(1), \tag{10}$$

where M_0 is the classical bare mass of each monopole.

The level spacing of the harmonic oscillator (10) is

$$\omega = c' r_m^{-1}, \quad c' = O(1). \tag{11}$$

(One verifies that in low lying states of (10) the monopoles are nonrelativistic and localized in the harmonic region of the potential, $\langle p^2/M^2\rangle \sim \langle r^2/r_m^2\rangle \sim \frac{\omega}{M} \sim \frac{1}{137} \ll 1$). Thus if M_0 is small, the bound states of the monopole pair system have masses of order r_m^{-1}, as advertized.

A more specific model results from assuming that the form factor of the monopole is dominated by a vector meson of mass μ; the monopole-antimonopole interaction is then $-\left(e'^2/q^2\right)\left(\frac{1}{1+q^2/\mu^2}\right)^2$, which in coordinate space is

$$-\frac{e'^2}{r}\left[1 - \left(1 + \frac{\mu r}{2}\right) e^{-\mu r}\right]. \tag{12}$$

The energy of the pair system for small r is then

$$H \simeq \left(2M - \frac{e'^2\mu}{2}\right) + p^2/M + \frac{1}{2}\,\frac{e'^2\mu}{6}\,\mu^2 r^2 \tag{13}$$

which is of the form (10), with $M - M_0$ identified as $e'^2\mu/4$; this is indeed just the classical self energy of the assumed charge distribution. According to (13), the low lying states have energies of order μ if $\mu \approx 4M/e'^2$.

Of course these mesons (monopole-pair bound states) are unstable to decay into photons. But if their decay is not so rapid as to wash them out as states altogether, one would expect their decay to have little effect on their role of producing the structure of the monopole.

We conclude by considering monopoles in the real world. If, as we have argued, a monopole of mass M has a size r_m which is produced by the mediation of neutral mesons of mass $\approx r_m^{-1} = M/137$, these neutral mesons should be observable through their couplings to photons. No such scalar or vector mesons are known, below the σ and ρ_0, mass $\approx$ 750 MeV. This lower limit on r_m^{-1} puts a lower limit on the monopole mass of the order 137 x .75 $\approx$ 100 BeV, or if one takes Eqs. (12) and (13) seriously,

$$M > \frac{e'^2}{4} m_\rho \approx 25 \text{ BeV}. \tag{14}$$

If one chooses to _identify_ the σ and ρ_0 with the monopole "structure mesons", one gets M $\approx$ 25 BeV as an _estimate_ of the monopole mass. The identification of hadrons with bound states of monopole pairs, or more generally of any assortment of monopoles whose total magnetic charge adds up to zero, is attractive[8] but has at least one difficulty: It is hard to disguise the fact that such states are composed of large magnetic charges e'; their magnetic polarizabilities are of order m^{-3} (m = typical hadron mass) rather than the usual e^2m^{-3}.

References

1. P.A.M. Dirac, Proc. Roy. Soc. (London) A133, 60 (1931); Phys. Rev. 74, 817 (1948).
2. J. Schwinger, Phys. Rev. 144, 1087 (1966).
3. C. Goebel and G. Shaw, Phys. Letters 27B, 291 (1968); see footnote ***.
4. P.A.M. Dirac, Proc. Roy. Soc. (London) A167. 148 (1938).

5. The argument of this paragraph was given in Ref. 3.
6. This is the same estimate of monopole size made by L.I. Schiff, Phys. Rev. 160, 1257 (1967) [also Phys. Rev. Letters 17, 714 (1966)], based on equating its mass to its classical magnetostatic self energy. He did not explain why a bare mass, traditionally arbitrary (even infinite) in field theory, would not upset this argument. He did argue that a monopole might obey classical, not quantum, dynamics, and the sum rule, Eq. (4), does show that the classic causality argument on the monopole size applies. But Schiff's deduction of classical behavior was based on an argument that an extended monopole must continuously acquire angular momentum as a point charge approaches; this is surely faulty, because the argument would imply the same about any extended charged system, such as an H^- ion, as a point magnetic charge approaches.
7. If the usual electromagnetic potentials are used, monopoles must have strings attached[1] (string = singular line of the vector potential). It seems rather awkward to treat extended monopole this way, although in principle it should be possible [cf. G. Wentzel, Prog. Theoret. Phys. (Kyoto) Suppl. Nos. 37 and 38, 163 (1966), and Schiff, Ref. 6]. Perhaps it would be simpler to interchange E and H, and attach strings to the electric charges (if any) of the system.
8. It amounts to the identification of quarks with monopoles. I thank G. Zweig for a discussion on this matter.

Received 5/14/69

A New Representation for the Plasma Conductivity Tensor in a Magnetic Field

by

Burton D. Fried and Clyde L. Hedrick
University of California, Los Angeles

1. Introduction

Conventional expressions for the conductivity tensor $\underset{\sim}{\sigma}$, or the dielectric tensor

$$\underset{\sim}{\varepsilon} = 1 + 4\pi i \underset{\sim}{\sigma}/\omega,$$

of a collisionless plasma in a magnetic field involve infinite sums of Bessel function terms, each associated with an harmonic of the cyclotron frequency, Ω. For many purposes, it is convenient to express these sums in closed form; Berk and Rosenbluth[1] have shown how the series may be summed in the particular case of electrostatic waves, and Majumdar[2] has examined the special case of propagation perpendicular to the magnetic field, assuming the unperturbed velocity distribution function is Maxwellian. We give here general expressions for the elements of $\underset{\sim}{\sigma}$ in which no Bessel function expansions are used, thus eliminating the necessity for subsequent resumming. The closed form expressions given here are valid for any particle distribution functions, since the integrations over v_z and $v_\perp$ are not carried out explicitly. The compact nature of our closed form representation facilitates the extension to more complicated problems (inhomogeneous plasmas, nonlinear interaction of waves). It also is well suited to the derivation of asymptotic expressions when the frequency is large (compared to cyclotron frequency) or the wavelength is small (compared to cyclotron radius). The conventional expressions, involving sums over harmonics, can be recovered at any point by making use of a standard identity for Bessel functions.

2. General expression for conductivity tensor

We start from the linearized, relativistic Vlasov equation, for a species with charge q and rest mass m, Fourier and Laplace transformed from $(\underline{x},t)$ to $(\underline{k},\omega)$:

$$\begin{aligned} \partial f/\partial\phi + i(\omega - \underline{k}\cdot\underline{v})f/\Omega_z &= \gamma(\underline{E} + \underline{v} \times \underline{B}/c) \cdot \nabla_u f_o/B_o \\ &\equiv g(\underline{k},\underline{v},\omega) \end{aligned} \tag{1}$$

where B_o is the z component of external magnetic field; $\Omega_z = qB_o/m\gamma c = \bar{\Omega}_z/\gamma$ is the (signed) cyclotron frequency; $\underline{u} = \gamma\underline{v}$; $\gamma = (1-v^2/c^2)^{-1/2}$, and ϕ is the azimuth angle in velocity space ($v_x = v_\perp\cos\phi$; $v_y = v_\perp\sin\phi$). Choosing the x-z plane to contain $\underline{k}$ and $\underline{B}_o$, we can write (1) in the form

$$(\partial/\partial\phi)\,\{f\exp[-iH(\phi)]\} = g\exp[-iH(\phi)] \tag{2}$$

where

$$\begin{aligned} H(\phi) &= \lambda\sin\phi - \nu\phi \\ \lambda &= k_x v_\perp/\Omega_z \\ \nu &= \tilde{\omega}/\Omega_z \\ \tilde{\omega} &= \omega - k_z v_z \end{aligned} \tag{3}$$

It is conventional to integrate this from $q\infty$ to ϕ, taking advantage of the fact that $\exp[-iH(\phi)] \to 0$ as $\phi \to q\infty$. (As usual, we make the initial-value-problem choice, Im $\omega > 0$, whenever this is required to resolve questions of convergence.) A Bessel expansion of $\exp(-iH)$ then leads to the usual infinite sums.

Instead of following this procedure, we integrate (2) from ϕ to $(\phi + 2\pi)$. Since

$$f(\phi + 2\pi) = f(\phi)$$

and

$$H(\phi + 2\pi) = H(\phi) - 2\pi\nu$$

we find

$$f(\phi) = (e^{2\pi i\nu} - 1)^{-1} \int_{\phi}^{\phi+2\pi} d\phi' \; g(\phi') \exp\{-i[H(\phi') - H(\phi)]\}$$

or

$$f(\phi) = P(\nu) \int_{0}^{2\pi} d\phi' \; g(\phi' + \phi) \exp\{-i\lambda[\sin(\phi'+\phi) - \sin\phi] + i\nu\phi'\} \tag{4}$$

where

$$P(\nu) = (e^{2\pi i\nu} - 1)^{-1} = (-1/2)(1 + i \cot \pi\nu). \tag{5}$$

We note that while $P(\nu)$ is singular at every Doppler shifted harmonic of the cyclotron frequency, the integral it multiplies is a well behaved function of both ν and λ. One essential advantage of (4), and of the expression for $\underset{\sim}{\sigma}$ which follows from it, is this collection of all of the singularities into a single factor.

Concerning f_o, we assume only that it satisfies the unperturbed Vlasov equation, i.e., that it is a function of v_z and $v_\perp = (v_x^2+v_y^2)^{1/2}$, but not ϕ. Using the Maxwell equation $\underline{B} = c(\underline{k} \times \underline{E})/\omega$, we can write the first factor of the integrand in (4) as

$$\begin{aligned} g(\phi'+\phi) = \gamma(B_o)^{-1} \{ &E_x \cos(\phi'+\phi)[\partial f_o/\partial u_\perp + k_z G/\omega] \\ &+ E_y \sin(\phi'+\phi)[\partial f_o/\partial u_\perp + k_z G/\omega] \\ &+ E_z [\partial f_o/\partial u_z - k_x G \cos(\phi'+\phi)/\omega]\} \end{aligned} \tag{6}$$

where

$$G = v_\perp \, \partial f_o/\partial u_z - v_z \, \partial f_o/\partial u_\perp \tag{7}$$

measures the anisotropy of the distribution function, vanishing if $f_o = f_o(v^2)$.

The conductivity tensor $\underset{\sim}{\sigma}$ is defined, as usual, by

$$\underline{j} = \sum nq \int d\,\underline{u}\,\underline{v}\,f = \underset{\sim}{\sigma} \cdot \underline{E}, \tag{8}$$

the summation being over species. We thus have

$$\underset{\sim}{\sigma} = \sum_{\text{species}} (\omega_p{}^2/4\pi\bar{\Omega}_z)\,\tilde{\underset{\sim}{\sigma}} \tag{9}$$

$$\tilde{\underset{\sim}{\sigma}} = \int d\underline{v}\,P(\nu)\gamma \int_0^{2\pi} d\phi'\,\exp(i\nu\phi')\,\exp\{-i[\lambda\sin(\phi'+\phi) - \lambda\sin\phi]\,\underset{\sim}{S}(\lambda,\nu,\phi,\phi'). \tag{10}$$

The tensor

$$\underset{\sim}{S} = \begin{pmatrix} \underset{\sim}{S} & \underline{S}_{\cdot z} \\ \underline{S}_{z\cdot} & S_{zz} \end{pmatrix} \tag{11}$$

is specified by

$$\underset{\sim}{S} = v_\perp \left(\frac{\partial f_o}{\partial u_\perp} + \frac{k_z G}{\omega}\right) \begin{Bmatrix} \cos\phi\,\cos(\phi'+\phi) & \cos\phi\,\sin(\phi'+\phi) \\ \sin\phi\,\cos(\phi'+\phi) & \sin\phi\,\sin(\phi'+\phi) \end{Bmatrix}$$

$$S_{\cdot z} = v_\perp \left(\frac{\partial f_o}{\partial u_z} - \frac{k_x G}{\omega}\cos(\phi'+\phi)\right)\ [\cos\phi,\ \sin\phi]$$

$$S_{z\cdot} = \left(\frac{\partial f_o}{\partial u_\perp} + \frac{k_z G}{\omega}\right) v_z\ [\cos(\phi'+\phi),\ \sin(\phi'+\phi)] \tag{12}$$

$$S_{zz} = v_z \left[\frac{\partial f_o}{\partial u_z} - \frac{k_x G}{\omega}\ \cos(\phi'+\phi)\right]$$

Of the four integrations in (10)--v_z, $v_\perp$, ϕ and ϕ'--the first two require specification of f_o, while the last two can be done once and for all. For any function of ϕ and ϕ', say $F(\phi,\phi')$, we define a function of (λ,ν) by

$$\langle F\rangle = (2\pi)^{-1} \int_0^{2\pi} d\phi' \int_0^{2\pi} d\phi\ F(\phi,\phi') \exp\{-i[\lambda \sin(\phi'+\phi) - \lambda \sin\phi - \nu\phi']\} \quad (13)$$

On computing $\langle \sin\phi\rangle$, $\langle \cos\phi\rangle$ and the other similar expressions involved in $\langle \underset{\sim}{S}\rangle$, we find that all can be expressed[3] in terms of a single function of λ,ν:

$$A(\lambda,\nu) = \langle 1\rangle = \int_0^{2\pi} e^{i\nu\phi}\ J_o[2\lambda \sin(\phi/2)]\ d\phi \quad (14)$$

together with the derivatives $\partial A/\partial\lambda$ and $\partial^2 A/\partial\lambda^2$. The expressions for the elements of $\langle \underset{\sim}{S}\rangle$ can be written most compactly by using in place of A the slightly different function

$$\begin{aligned} C(\lambda,\nu) &= A(\lambda,\nu) + i/\nu P \\ &= \int_0^{2\pi} e^{i\nu\phi}\ [J_o(2\lambda \sin(\phi/2)) - 1] d\phi \end{aligned} \quad (15)$$

and, in place of $\partial^2 A/\partial\lambda^2$ or $\partial^2 C/\partial\lambda^2$, the function

$$\begin{aligned} D(\lambda,\nu) &= \frac{1}{2}\ [A(\lambda,\nu + 1) + A(\lambda,\nu - 1)] \\ &= \int_0^{2\pi} d\phi\ e^{i\nu\phi} \cos\phi\ J_o[2\lambda \sin(\phi/2)]. \end{aligned} \quad (16)$$

We note that (14) can also be written as

$$A(\lambda,\nu) = 2\pi\ e^{i\pi\nu}\ J_\nu(\lambda)\ J_{-\nu}(\lambda).$$

The details of this are given in Appendix I.

In terms of these functions, the elements of $\langle \underset{\sim}{S}\rangle$ are then

$$\langle \underset{\sim}{S}\rangle = \left(\frac{\partial f_o}{\partial u_\perp} + \frac{k_z G}{\omega}\right) v_\perp \begin{Bmatrix} (\nu/\lambda)^2 C & (i\nu/2\lambda) C' \\ -(i\nu/2\lambda) C' & D - (\nu/\lambda)^2 C \end{Bmatrix} \quad (17)$$

$$\langle S_{\cdot z}\rangle = \left(\frac{\partial f_o}{\partial u_z} - \frac{\nu}{\lambda}\frac{k_x G}{\omega}\right) v_\perp \, [\nu C/\lambda, \ -iC'/2] \tag{18}$$

$$\langle S_{z\cdot}\rangle = \left(\frac{\partial f_o}{\partial u_\perp} + \frac{k_z G}{\omega}\right) v_z \, [\nu C/\lambda, \ iC'/2] \tag{19}$$

$$\langle S_{zz}\rangle = \left(A\,\frac{\partial f_o}{\partial u_z} - \frac{\nu C}{\lambda}\frac{k_x G}{\omega}\right) v_z \tag{20}$$

where a prime denotes differentiation with respect to λ, $C'(\lambda,\nu) = (\partial/\partial\lambda)\, C(\lambda,\nu)$. In each of these expressions, the first factors are essentially the same, since

$$\begin{aligned}\left(\frac{\partial f_o}{\partial u_z} - \frac{\nu}{\lambda}\frac{k_x G}{\omega}\right) v_\perp &= \left(\frac{\partial f_o}{\partial u_\perp} + \frac{k_z G}{\omega}\right) v_z \\ &= v_\perp \frac{\partial f_o}{\partial u_z} - \frac{\hat{\omega}}{\omega} G\end{aligned} \tag{21}$$

independent of f_o. Defining

$$R = v_\perp \frac{\partial f_o}{\partial u_z} - \frac{\tilde{\omega}}{\omega} G \tag{22}$$

we have

$$\langle \underset{\sim}{S}\rangle = R(v_\perp/v_z) \begin{Bmatrix} (\nu/\lambda)^2 C & (i\nu/2\lambda)C' \\ -(i\nu/2\lambda)C' & D-(\nu/\lambda)^2 C \end{Bmatrix} \tag{23}$$

$$\langle S_{\cdot z}\rangle = R[\nu C/\lambda, \ -iC'/2] \tag{24}$$

$$\langle S_{z\cdot}\rangle = R[\nu C/\lambda, \ iC'/2] \tag{25}$$

$$\langle S_{zz}\rangle = RAv_z/v_\perp - i(v_z/v_\perp)(\Omega_z/\omega)(G/P). \tag{26}$$

The well-known symmetry relations

$$S_{xy} = -S_{yx}, \ S_{xz} = S_{zx}, \ S_{yz} = -S_{zy} \tag{27}$$

are immediately apparent.

Since (10) can be written

$$\tilde{\underset{\sim}{\sigma}} = \int d\underline{u}\ P(\nu)\ \langle \underset{\sim}{S} \rangle \tag{28}$$

we need only multiply the elements of $\langle \underset{\sim}{S} \rangle$ as given above by $P(\nu)$ and carry out the integration over v_z and $v_\perp$ (or, equivalently, over λ and ν), to obtain $\tilde{\underset{\sim}{\sigma}}$. A sum over species, as given by (9), then yields the conductivity tensor, $\underset{\sim}{\sigma}$. The result, of course, depends upon the choice of f_o.

We note that the usual form of an expansion in cyclotron harmonics follows immediately if in (14) we use the identity

$$J_o(2\lambda\ \sin\phi/2) = \sum_n J_n^{\,2}(\lambda)\ \exp(-in\phi). \tag{29}$$

The ϕ integration can then be done, leaving a sum of resonant denominators,

$$A(\lambda,\nu) = \sum_n J_n^{\,2}(\lambda)\ [i(\nu-n)\ P(\nu)]^{-1}. \tag{30}$$

The singular factor, $P(\nu)$, in (28) is cancelled exactly, i.e., a product over poles at $\nu = n$ is replaced by an infinite sum of such poles. In the limit of small λ, the expansion (30) is useful, since it is usually sufficient to retain only a few terms. However, if $\lambda \gtrsim 1$, the closed form expression (14) is likely to be more convenient.

Although the dispersion relation which follows from using $\underset{\sim}{\sigma}$ together with Maxwell's equations,

$$\mathrm{Det}\ \{(1-n^2)\delta_{rs} + n_r n_s + 4\pi i\sigma_{rs}/\omega\} = 0; \qquad \underline{n} = \underline{k}c/\omega$$

covers all cases, it is useful to give explicitly the scalar susceptibility in the case of electrostatic waves ($\nabla \times \underline{E} = 0$). Computing

$$\rho = \sum nq \int f\ d\underline{u},$$

using (4) for f, we find

$$\rho = \chi E \qquad \chi = \sum_{\text{species}} (\omega_p^{\,2}/4\pi\bar{\Omega}_z)\,\tilde{\chi}$$

$$\tilde{\chi} = \int d\underline{u}\; P(\nu)\gamma[k_z A\; \partial f_o/\partial u_z + k_x C(\nu/\lambda)\; \partial f_o/\partial u_\perp].$$

The expressions for $<\underset{\sim}{S}>$ given in (23) through (26) represent the simplest closed form expressions for $\underset{\sim}{\sigma}$ which we have been able to find with f_o still unspecified. In some situations it is useful to approximate f_o by a delta function,

$$f_o = \delta(v_z - V_z)\; \delta(v_\perp^2 - V_\perp^2)/\pi,$$

in which case the integrations in (28) are trivial and we need only multiply $<\underset{\sim}{S}>$ by $P(\nu)$, setting $\nu = (\omega - k_z V_z)/\Omega_z$ and $\lambda = k_x V_\perp/\Omega_z$. A more conventional choice for f_o, such as a Maxwellian, with possibly unequal parallel and perpendicular temperatures, leads, of course, to the usual expressions if the Bessel function expansion (29) is used. For any choice of f_o, a power series expansion of the J_o term in (14) through (16) provides a convenient way of deriving the small λ approximation to $\underset{\sim}{\sigma}$.

We have found this formalism to be of greatest value in deriving asymptotic expressions for $\underset{\sim}{\sigma}$ (when λ and/or ν is large); in treating the case of inhomogeneous plasmas; and in studying nonlinear phenomena (wave-wave coupling). In each of these cases some extension is, of course, required, but the compact, closed form of these expressions, as contrasted with the conventional Bessel sums, considerably facilitates the analysis. For example, in the inhomogeneous case, with the density, n, a function of x, f_o must be considered a function of $v_y + \Omega_z x$ (i.e., the canonical momentum, P_y/m). If the quantity k_x is interpreted as an operator, $-i(\partial/\partial x)$, the same algebraic steps involved in the derivation of (4) can be carried out, with due regard for the order of the operators involved (and with a slight generalization to a non-zero k_y). An expansion in $\nabla n/n$,

$$f_o(x + v_y/\Omega_z) = f_o(x) + f_o'(x)\; v_y/\Omega + \dots ,$$

reduces the problem to the one treated here, with an addition factor of $v_y = v_\perp \sin(\phi + \phi')$ inside the < > for each new term in the expansion.

Computation of the resulting products of three trigonometric functions can be simplified using the techniques given in the Appendix. A detailed exposition of this is contained in a forthcoming paper.

Appendix

1. Matrix elements of $\langle \underset{\sim}{S} \rangle$.

The calculation of $\langle \underset{\sim}{S} \rangle$, with $\underset{\sim}{S}$ given by (12) and $\langle\ \rangle$ defined by (13), is greatly simplified by the following Lemma. If f and g are periodic functions with period 2π, then

$$\langle f(\phi)\ g(\phi'+\phi)\rangle = \langle f(-\phi-\phi')\ g(-\phi)\rangle \tag{A.1}$$

(The proof follows directly from the definition of $\langle\ \rangle$.)

As a first application of (1) we have

$$\begin{aligned} \langle\cos\phi\rangle &= \langle\cos(\phi'+\phi)\rangle \\ \langle\sin\phi\rangle &= -\langle\sin(\phi'+\phi)\rangle \end{aligned} \tag{A.2}$$

From

$$A(\lambda,\nu) \equiv \langle 1\rangle = (2\pi)^{-1}\int_0^{2\pi} d\phi' \int_0^{2\pi} d\phi\ \exp\{-i\lambda[\sin(\phi'+\phi) - \sin\phi] + i\nu\phi'\}$$

we have

$$A' = \partial A/\partial\lambda = -i\ \langle\sin(\phi'+\phi) - \sin\phi\rangle = 2i\ \langle\sin\phi\rangle \tag{A.3}$$

which expresses $\langle\sin\phi\rangle$ in terms of A'. (We shall henceforth use a prime to denote differentiation with respect to λ.) Also

$$A'' = 2\ \langle\sin\phi\ \sin(\phi'+\phi) - \sin^2\phi\rangle, \tag{A.4}$$

a result we shall use presently.

If $F(\phi,\phi')$ is periodic, with period 2π, in both ϕ and ϕ', then integration by parts gives

$$\langle\partial F/\partial\phi'\rangle = (2\pi)^{-1}\int_0^{2\pi} d\phi F(\phi,0)/P(\nu) + i\langle F[\lambda\cos(\phi'+\phi) - \nu]\rangle \tag{A.5}$$

Setting F = 1, sinϕ, cosϕ, and cos(ϕ+ϕ') in (A.5) give, respectively,

$$\langle\cos(\phi'+\phi)\rangle = [\nu\langle 1\rangle + i/P(\nu)]/\lambda = \nu C/\lambda \quad \text{(A.6)}$$

$$\langle\sin\phi \cos(\phi'+\phi)\rangle = (\nu/\lambda) \langle\sin\phi\rangle = (\nu/2i\lambda) A' \quad \text{(A.7)}$$

$$\langle\cos\phi \cos(\phi'+\phi)\rangle = (\nu/\lambda) \langle\cos\phi\rangle = (\nu/\lambda)^2 C \quad \text{(A.8)}$$

$$\langle\cos^2\phi\rangle = \langle\cos\phi \cos(\phi'+\phi)\rangle - i \langle\sin\phi\rangle/\lambda \quad \text{(A.9}$$

and this combines with (A.4) to yield

$$\langle 1\rangle = \langle\sin\phi \sin(\phi'+\phi)\rangle - \frac{1}{2} A'' + \langle\cos\phi \cos(\phi'+\phi)\rangle - i \langle\sin\phi\rangle/\lambda \quad \text{(A.1}$$

or

$$\langle\sin\phi \sin(\phi'+\phi)\rangle = \frac{1}{2} A'' + A + A'/2\lambda - (\nu/\lambda)^2 C. \quad \text{(A.11}$$

Using the expression (14) for A, which will be derived presently, and the Bessel equation for J_o, we easily find that

$$A'' + A'/\lambda + 2A = 2D,$$

where D is defined by (16), and so (A.11) can be simplified:

$$\langle\sin\phi \sin(\phi'+\phi)\rangle = D - (\nu/\lambda)^2 C. \quad \text{(A.12}$$

The final matrix element follows from (A.1) and (A.7):

$$\langle\cos\phi \sin(\phi'+\phi)\rangle = -(\nu/2i\lambda) A'.$$

Collecting our results, we have for the nine matrix elements required for $\langle \underset{\sim}{S} \rangle$:

$$<\cos\phi> \; = \; <\cos(\phi'+\phi)> \; = \; \nu C/\lambda$$

$$<\sin\phi> \; = \; - <\sin(\phi'+\phi)> \; = \; C'/2i$$

$$<\cos\phi \, \cos(\phi'+\phi)> \; = \; (\nu/\lambda)^2 \, C \tag{A.13}$$

$$<\cos\phi \, \sin(\phi'+\phi)> \; = \; - <\sin\phi \, \cos(\phi'+\phi)> \; = \; - (\nu/2i\lambda) \, C'$$

$$<\sin\phi \, \sin(\phi'+\phi)> \; = \; D - (\nu/\lambda)^2 \, C$$

$$<1> \; = \; A$$

2. Other representations for $A(\lambda,\nu)$.

The expression (14) for A is a special case of an identity satisfied by

$$A_p(\lambda,\nu) \equiv <\exp[ip(\phi'+\phi)]> \tag{A.14}$$

where p is an integer. From the definition (13) of < >, we have

$$2\pi \, A_p(\lambda,\nu) = \int_0^{2\pi} d\phi' \int_0^{2\pi} \exp\{-i\lambda[\sin(\phi'+\phi)-\sin\phi] + i\nu\phi' + ip(\phi'+\phi)\}$$

which, with the change of variables from ϕ to

$$t = \phi + \phi'/2 - \pi/2$$

gives

$$2\pi \, A_p = i^P \int_0^{2\pi} d\phi' \, \exp[i(\nu+p/2)\phi'] \int_0^{2\pi} dt \, \exp[2i\lambda \, \sin(\phi'/2) \, \sin t + ipt] \tag{A.15}$$

where the periodicity of the integrand in t has allowed us to change the limits of integration from $[(\phi'-\pi)/2, \; (\phi'+3\pi)/2]$ to $(0,2\pi)$. Since the second integral in (A.15) is just Bessel's representation of

$J_{-p}(2\lambda \sin\phi'/2)$ we have the result

$$A_p = (i)^{-p} \int_0^{2\pi} d\phi' \exp[i(\nu+p/])\phi'] J_p[2\lambda \sin(\phi'/2)]. \tag{A.16}$$

In the case $p = 0$ this just gives (14), since $A_o = A$.

References

1. H. Berk and M. N. Rosenbluth, private communication.
2. S. K. Majumdar, Indian J. of Phys. 39, 511 (1965).
3. These details are included in the Appendix.

Received 3/20/69

Potential Problems Involving the Space Outside a Circular Disk

by

Otto Laporte and George W. Ford
The University of Michigan

In this paper a method will be explained by means of which it is possible to construct a Laplace-equation Green's function for an infinite space external to a circular disk on which either the function or its normal derivative should vanish. The process can be modified to apply to a space exclusive of an infinite plane screen with a circular aperture in it.

A variety of problems can be attacked with this technique; it offers a starting point for the problem of the diffraction around disks and through apertures in the case of long wave lengths using the method of Kleinman and Ar[1] and it may be used for problems of inviscid hydrodynamics which are currently gaining in interest for problems concerned with superfluid helium.

The method, at first thought to be new, is however in part already anticipated in a paper by Beltrami.[2] Since it does not seem to be generally known and since we believe to have added an essential generalization, it will be explained in some detail below.

The potentials to be discussed in the next sections are all axially symmetric and will be described by cylindrical coordinates z, and $\rho^2 = x^2 + y^2$.

Expressions will be derived for a potential which either assumes a given function $G(\rho)$ on a disk $z = 0$, $\rho < a$, or for a potential whose normal derivative assumes a given function on the disk. In the first, the Dirichlet type of problem, which is patently the problem of electrostatics, we shall arrive at a symmetric potential, i.e. at a potential which is an even function of z, while for the hydrodynamic problem one is led to an anti-symmetric potential, i.e. one which is an odd function of z. In each case there is what one might call a prototype solution from which the general solution is obtained by superposition.

The even prototype is

$$\phi_e = \int_0^\infty d\lambda J_o(\rho\lambda) e^{-\lambda|z|} \frac{J_{1/2}(\lambda a)}{\sqrt{\lambda}} \quad . \qquad (1)$$

This remarkable solution of Laplace's equation which for $z = 0$ becomes discontinuous was found long ago by Weber. In this case it and its derivative have the values:

$$\left.\begin{aligned} \rho<a: \quad \phi_e &= \sqrt{\frac{\pi}{2a}} \\ \rho>a: \quad \phi_e &= \sqrt{\frac{2}{\pi a}} \sin^{-1}\frac{a}{\rho} \end{aligned}\right\} \qquad (2)$$

This potential is clearly an even function of z; it has a discontinuous derivative for $\rho = a$ but is completely continuous on the plane $z = 0$ for $\rho \gtrless a$. The normal derivative, which is proportional to the surface charge density is the well known Weber's Discontinuous Factor

$$\left.\begin{aligned} \mp\left(\frac{\partial\phi_e}{\partial z}\right)_o = \int_0^\infty \sqrt{\lambda} d\lambda J_o(\rho\lambda) J_{1/2}(a\lambda) &= \sqrt{\frac{2}{\pi}} \frac{1}{\sqrt{a^2-\rho^2}}, \quad \rho < a \\ &= 0 \qquad\qquad\qquad \rho > a \end{aligned}\right\} \qquad (3)$$

It seems plausible to suppose that by superimposing the solutions corresponding to an infinite number of disks of different radii with appropriately chosen potentials one could construct a potential which assumes a given function at $z = 0$, and for $\rho < a$, while for $z = 0$ and $\rho > a$ its normal derivative vanishes.

The odd prototype is the integral

$$\phi_\sigma = \frac{z}{|z|} \int_0^\infty d\lambda J_o(\rho\lambda) e^{-\lambda|z|} \frac{J_{3/2}(\lambda a)}{\sqrt{\lambda}} \qquad (4)$$

which has properties which are quite the opposite of ϕ_e: For $z = 0$

and

$$\left.\begin{array}{ll} \rho<a: \phi_o(0,\rho) & = \sqrt{\dfrac{2}{\pi a}}\sqrt{1-\dfrac{\rho^2}{a^2}} \\ \rho>a: & = 0 \end{array}\right\} \quad (5)$$

while for

$$\left.\begin{array}{ll} \rho<a: \left(\dfrac{\partial\phi_o}{\partial z}\right)_o & = \sqrt{\dfrac{\pi}{2a^3}} \\ \rho>a: \left(\dfrac{\partial\phi_o}{\partial z}\right)_o & = -\sqrt{\dfrac{2a^3}{\pi}}\ \dfrac{1}{\rho^3}\left(\dfrac{\rho}{a}\right)^2 \{(1-\dfrac{a^2}{\rho^2})^{-1/2} - \dfrac{\sin^{-1}a/\rho}{a/\rho}\} \end{array}\right\} \quad (6)$$

This is clearly a dipole layer of moment proportional to $\sqrt{a^2 - \rho^2}$ while the normal derivative suffers a jump which is constant.*

We have called the solutions (1) and (4) the prototype because they belong to the simplest mixed boundary value problems. Integral (1) is the expression for the potential of a conducting circular disk at potential $\sqrt{\pi/2a}$. The integral (4) is used in the (ideal inviscid) hydrodynamic problem of a disk in a constant infinite flow. The velocity potential

$$V_\infty \left(z + \sqrt{\frac{2a^3}{\pi}}\ \mathrm{sgn}(z) \int_0^\infty d\lambda e^{-\lambda|z|} J_o(\rho\lambda)\ \frac{J_{3/2}(a\lambda)}{\sqrt{\lambda}}\right) \quad (7)$$

for this case has characteristically the factor sgn(z) which takes care of the fact that the potential has opposite values on the two sides of the disk. Only in this way has the potential a continuous first derivative for $z = 0$ and $\rho > a$. At the edge the velocity shows the same square root singularity as in the familiar plane case.

The integrals (1) and (4) will now be generalized. In the even case a potential will be constructed which assumes a given function ρ on the disk and in the odd case one whose normal derivative assumes a given function on the disk.

---

*Both integrals (1) and (4) can be evaluated in terms of elementary functions even for $z \neq 0$. But since these expressions are quite lengthy and will not be made use of here they are omitted. For $z = 0$ the integrals all lead to hypergeometric functions which in the present case can be summed.

Even Function. We pose the following boundary value problem: A solution of Laplace's equation is to be constructed such that on the circular disk for

$$z = 0, \text{ and } \rho < a \qquad \phi(\rho,0) = -G(\rho)$$

where $G(\rho)$ is a given function due to some singularity or source system which has to be axially symmetric. Furthermore, there has to be a suitable boundary condition at infinity. The condition to be imposed there depends upon the nature of the source system--whether it is a single source or multipole system--as well as the state of the disk--whether it is grounded or charged. We make the ANSATZ

$$\phi = \int_0^\infty d\lambda J_o(\rho\lambda)e^{-\lambda|z|}A(\lambda) \qquad (8$$

and seek to determine $A(\lambda)$ so that*

$$\int_0^\infty d\lambda J_o(\rho\lambda)A(\lambda) = -G(\rho).$$

We carry out a linear transformation of the following kind:

$$\int_0^\rho \frac{\eta d\eta}{\sqrt{\rho^2-\eta^2}} G(\eta) = -\int_0^\infty d\lambda A(\lambda) \int_0^\rho \frac{\eta d\eta}{\sqrt{\rho^2-\eta^2}} J_o(\eta\lambda).$$

On the right hand side the order of the λ and η integrations has been reversed. The η integral turns out to be performable since it is the Hankel inverse of the integral (3). One has therefore

$$\int_0^\rho \frac{\eta d\eta}{\sqrt{\rho^2-\eta^2}} G(\eta) = -\sqrt{\frac{\pi}{2}}\sqrt{\rho}\int_0^\infty d\lambda A(\lambda)\frac{J_{1/2}(\lambda\rho)}{\sqrt{\lambda}}.$$

Up to now $A(\lambda)$ was left general. Generalizing (1), let it now be specialized to be:

$$A(\lambda) = \sqrt{\frac{\pi}{2}}\sqrt{\lambda}\int_0^\infty \sqrt{\xi}\, d\xi J_{-1/2}(\lambda\xi)f(\xi). \qquad (9)$$

*Perhaps it is not superfluous to state that it would be quite wrong to use the ordinary Hankel inversion on this equation, for it would mean that one would have to make some kind of assumption as to $G(\rho)$ for $\rho>a$

Upon introducing this into the above and reversing the order of integration the λ integral turns out to be the Dirichlet Discontinuous Factor[3] and it is therefore possible to change the upper limit of the ξ integral from a to ρ:

$$\int_0^\rho \frac{\eta d\eta}{\sqrt{\rho^2-\eta^2}} G(\eta) = -\frac{\pi}{2}\int_0^\rho d\xi f(\xi)$$

so that

$$f(\rho) = -\frac{d}{d\rho}\int_0^\rho \frac{\eta d\eta}{\sqrt{\rho^2-\eta^2}} G(\eta) \tag{10}$$

Substituting (9) and (10) into (8) we have the result:

$$\phi(\rho,z) = -\sqrt{\frac{2}{\pi}}\int_0^\infty \sqrt{\lambda}d\lambda J_o(\rho\lambda)e^{-\lambda|z|}\int_0^a \sqrt{\xi}d\xi J_{-1/2}(\lambda\xi)\frac{d}{d\xi}\int_0^\xi \frac{\eta d\eta}{\sqrt{\xi^2-\eta^2}} G(\eta) \tag{11}$$

If one puts $G(\eta)$ equal to a constant, the even prototype integral (1) arises. The potential (11) shares with (1) the important property that $[\partial\phi(\rho,z)/\partial z]_{z=0} = 0$ for all G's for which the integrals converge. (To see this one may interchange the ξ and λ integrations. Then, before putting $z = 0$, it is necessary to perform one integration by parts to avoid divergence.)

As an example we shall calculate the interaction of a point charge situated on the axis $z = 0$ and at a distance $z_o > 0$ with a conducting disk at $z = 0$, $\rho < a$.

$$\phi = \phi_{prim} + \phi_{sec};\ \phi_{prim} = \frac{1}{\sqrt{\rho^2+|z-z_o|^2}} \tag{12}$$

the boundary conditions are

$$\left.\begin{aligned} &z = 0,\ \rho < a: \quad \phi = 0 \\ &z \text{ and } \rho \to \infty: \quad \phi \sim \frac{\text{const}}{\sqrt{\rho^2+z^2}} \end{aligned}\right\} \tag{13}$$

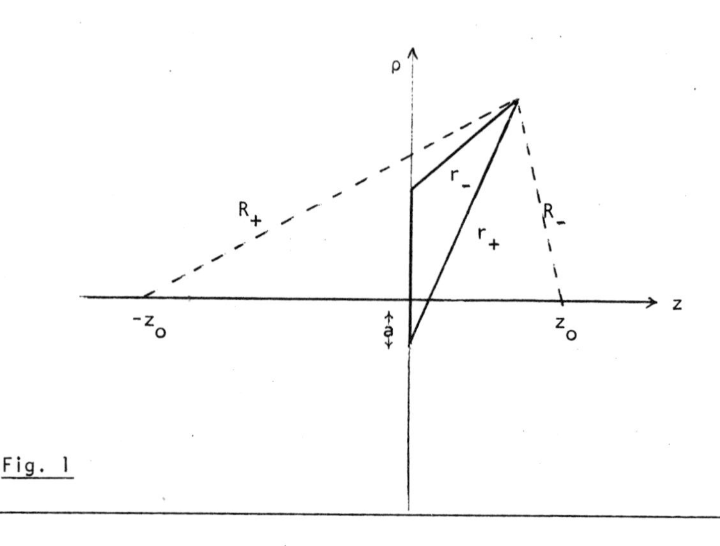

Fig. 1

The somewhat lengthy calculation leads to the following expression, in which both prolate and oblate ellipsoidal coordinates occur. With

$$r_{\pm}^2 = z^2 + |\rho \pm a|^2, \quad R_{\pm}^2 = |z \pm z_o|^2 + \rho^2; \quad 2\Lambda = r_+ + r_-$$

we obtain:

$$\pi\phi_{sec} = -\left(\frac{1}{R_+} + \frac{1}{R_-}\right) tg^{-1}\left(\frac{R_+ + R_-}{2z} \frac{a}{\sqrt{\Lambda^2 - a^2}}\right) - \left(\frac{1}{R_+} - \frac{1}{R_-}\right) tg^{-1}\left(\frac{R_+ - R_-}{2z} \frac{a}{\sqrt{\Lambda^2 - a^2}}\right) \tag{14}$$

and in particular for $z = 0$

$$\begin{aligned} \rho < a: \quad \phi_{sec} &= -\frac{1}{\sqrt{\rho^2 + z_o^2}} \\ \rho > a: \quad \phi_{sec} &= -\frac{2}{\pi(\rho^2 + z_o^2)} \, tg^{-1}\left(\frac{a}{z_o} \frac{\sqrt{\rho^2 + z_o^2}}{\sqrt{\rho^2 - a^2}}\right) \end{aligned} \tag{15}$$

The transition from the second to the first formula is continuous (although the derivatives are not). One also sees that for ρ and $z \to \infty$

$$\lim \phi_{sec} = -\left(\frac{2}{\pi} \operatorname{tg}^{-1} \frac{a}{z_o}\right) \frac{1}{r} .$$

The constant of the boundary condition (12) at infinity is therefore

$$1 - \frac{2}{\pi} \operatorname{tg}^{-1} \frac{a}{z_o} .$$

If the plate were kept insulated one would have to add a potential of the form of (1), multiplied with $(2/\pi a)^{1/2} \operatorname{tg}^{-1} a/z_o$ so as to destroy the second term above.

Odd Function. Now the problem is to construct a solution of Laplace's equation such that on the circular disk $z = 0$, $\rho < a$:

$$\frac{\partial}{\partial n} (\phi_{prim} + \phi_{sec}) = 0 \tag{16}$$

where ϕ_{prim} is due to some axially symmetric source system. In accordance with (7) we write analogously with (8):

$$\phi_{sec} = \operatorname{sgn}(z) \int_o^\infty d\lambda J_o(\rho\lambda) e^{-\lambda|z|} \overline{A}(\lambda) \tag{17}$$

and abbreviate

$$\left(\frac{\partial \phi_{prim}}{\partial z}\right)_o = \overline{G}(\rho) .$$

We now have on the positive side

$$\begin{aligned} z = +0: \quad \frac{\partial}{\partial n} (\phi_{prim} + \phi_{sec}) &= \overline{G}(\rho) + \frac{\partial}{\partial z} \left[\int \ldots e^{-\lambda z}\right]_o \\ &= \overline{G}(\rho) - \int \lambda d\lambda J_o \overline{A} = 0 ; \end{aligned} \tag{18}$$

on the negative side

$$z = -0: \frac{\partial}{\partial n} (\phi_{prim} + \phi_{sec}) = -\overline{G}(\rho) - \frac{\partial}{\partial z}(-) \int \ldots e^{+\lambda z} = -\overline{G}(\rho) + \int \lambda d\lambda J_o \overline{A} = 0 .$$

Thus, thanks to the odd character of ϕ_{sec} we have the same boundary condition. As before we carry out a linear transformation which however is now:

$$\int_0^\rho \eta d\eta \sqrt{\rho^2-\eta^2}\, \overline{G}(\eta) = \int_0^\infty \lambda d\lambda \overline{A}(\lambda) \int_0^\rho \eta d\eta \sqrt{\rho^2-\eta^2}\, J_o(\eta\lambda).$$

The order of integration was reversed. The η integral can again be performed since it is the Hankel transform of the integral in (4). The result is:

$$\int_0^\rho \eta d\eta \sqrt{\rho^2-\eta^2}\, \overline{G}(\eta) = \sqrt{\frac{\pi}{2}}\, \rho^{3/2} \int_0^\infty \lambda d\lambda \frac{J_{3/2}(\rho\lambda)}{\lambda^{3/2}}\, \overline{A}(\lambda).$$

We now choose $\overline{A}(\lambda)$ to be somewhat different from (9), namely:

$$\overline{A}(\lambda) = \sqrt{\frac{\pi}{2}}\, \sqrt{\lambda} \int_0^a \sqrt{\xi} d\xi J_{1/2}(\lambda\xi) \overline{f}(\xi) \qquad (19$$

and reverse the order of integration again:

$$\int_0^\rho \eta d\eta \sqrt{\rho^2-\eta^2}\, \overline{G}(\eta) = \frac{\pi}{2}\, \rho^{3/2} \int_0^a \sqrt{\xi} d\xi f(\xi) \int_0^\infty d\lambda J_{1/2}(\xi\lambda) J_{3/2}(\rho\lambda).$$

This time the λ integral is also a discontinuous function; we find

$$\int_0^\infty d\lambda J_{1/2}(\xi\lambda) J_{3/2}(\rho\lambda) = \sqrt{\frac{\xi}{\rho^3}} \text{ for } \xi < \rho; \; = 0 \text{ for } \xi > \rho$$

according to Nielsen.[4] Using this we have the following form for $f(\rho)$:

$$f(\rho) = -\frac{2}{\pi} \frac{1}{\rho} \frac{d}{d\rho} \int_0^\rho \eta d\eta \sqrt{\rho^2-\eta^2}\, \overline{G}(\rho).$$

The final result in the odd case is therefore:

$$\phi_{sec} = -\sqrt{\frac{2}{\pi}} \operatorname{sgn}(z) \int_0^\infty \sqrt{\lambda} d\lambda J_o(\rho\lambda) e^{-\lambda|z|} \int_0^a \sqrt{\xi} d\xi J_{1/2}(\lambda\xi) \frac{1}{\xi} \frac{d}{d\xi} \int_0^\xi \eta d\eta \sqrt{\xi^2-\eta^2}\, \overline{G}(\eta). \qquad (20$$

As before, upon equating $\overline{G}$ to a constant, the odd prototype potential (4) results. The odd ϕ_{sec} of eq. (20) retains many properties of its prototype, the most characteristic one being its vanishing for $z = 0$

and $\rho > a$. This can be seen immediately after reversing the order of integration. By virtue of (3) the λ integral vanishes for all resonable $\overline{G}(\rho)$.

The function (11) and the z derivative of (20) are therefore potentials which assume given functions of ρ on the disk. But they differ as to their behavior in the z plane for $\rho > a$. Therefore it is necessary to complete the formulation of the boundary value problem to be done by having a condition at infinity.

If one substitutes for $\overline{G}$ the normal derivative of a simple source

$$\overline{G}(\rho) = -\frac{z_o}{\sqrt{z_o^2+\rho^2}^3}$$

the resulting $f(\rho)$ becomes:

$$f(\rho) = -\frac{2}{\pi}\frac{\rho}{\rho^2+z_o^2}.$$

This, then, would be a Neumann Green's function for the case where the source is on the z axis. The resulting expression for the potential involves like (14) only elementary functions, but it is too lengthy to be reproduced here. For $z = 0$, i.e., on the plane, the formulae are rather simple and of some interest:

$$\rho < a: \quad \phi_{sec}(\rho,0) = \pm \frac{1}{\sqrt{z_o^2+\rho^2}} \, tg^{-1}\sqrt{\frac{a^2-\rho^2}{z_o^2+\rho^2}}$$

$$\rho > a \quad \phi_{sec}(\rho,0) = 0$$

and the normal derivative, which for $\rho < a$ results, as expected, in $\overline{G}(\rho)$, becomes for $\rho > a$:

$$\left(\frac{\partial\phi_{sec}}{\partial z}\right)_o = \frac{a}{(a^2+z_o^2)\sqrt{\rho^2-a^2}} - \frac{z_o}{\sqrt{z_o^2+\rho^2}^3} \, tg^{-1} \frac{a}{z_o}\sqrt{\frac{z_o^2+\rho^2}{\rho^2-a^2}}$$
$$- \frac{a\sqrt{\rho^2-a^2}}{(z_o^2+\rho^2)(z_o^2+a^2)}.$$

If one interprets this as the z component of the velocity then the first term shows the expected square root singularity at the edge, in analogy with the plane case. The fact that for $\rho > a$ $\phi(\rho,0)$ vanishes, and that for $\rho >> a$ the derivative behaves as ρ^{-3} shows again that ϕ has dipole character.

Concluding Remarks. The developments presented here can be generalized, by suitably operating upon the potentials (11) and (20), to obtain formulae for potentials depending upon an azimuth angle as well. In this way a Green's function with an arbitrary position of the source can be constructed.

It should also be mentioned that all the problems discussed here may also be written down in terms of oblate spheroidal coordinates, in terms of which Laplace's equation is separable, so that solutions in the form of infinite series proceeding in terms of products of spherical harmonics P and Q may be postulated. But it is evident that these series are very difficult and awkward to discuss if one wants to obtain an understanding of the general behavior. The solutions discussed here can ultimately all be expressed in terms of elementary functions, a fact which certainly would be difficult to see from spherical harmonics series. It is intended to apply this method to the discussion of vortex rings.

References

1) Kleinman, Arch. Ratl. Mech. 18, 205 (1966) and Kleinman and Ar, ibid. 23, 219 (1966).
2) E. Beltrami, Opere Vol. III, p. 349 (1911).
3) N. Nielsen, Cylinderfunktionen, Leipzig 1903, p. 200, eq. (9).
4) Ibidem, p. 200, eq. (9).

Received 3/1/69

Diffraction Dissociation in the Chou-Yang Model

by
N. Byers
University of California, Los Angeles
and
S. Frautschi
California Institute of Technology

Viewing the Chou-Yang model of elastic scattering as the high-energy limit of Glauber's multiple scattering theory, we show that the model predicts certain inelastic reactions. For example, $a + c \to b + c$ where b has the same intrinsic quantum numbers (isospin, strangeness, etc.) as a. The spin-parity as well as the mass of b may differ from a. As in the elastic case, as $s \to \infty$, the differential cross section $d\sigma/dt$ depends only on t. In the Chou-Yang model, $d\sigma/dt$ for $a + c \to a + c$ has the approximate form $d\sigma/dt \cong \mu_{ac}^2 \, [F_a \, F_c]^2$ where $F_a(t)$ and $F_c(t)$ are the form factors of a and c (measured by electron scattering) and μ_{ac} is a (real) constant; with the normalization $F_a(0) = F_c(0) = 1$, μ_{ac} is determined by fitting their expression to the elastic scattering data. Their assumption that F_a and F_c are the form factors measured in electron scattering is supported by the impressive experimental agreement of the t-dependence of $d\sigma/dt$ for p-p scattering with $[F_p(t)]^4$ as measured in e-p scattering. In our extension of their model, we obtain, for a certain class of states--"collective excitations"--a differential cross section for the reaction $a + c \to b + c$ which has the approximate form $d\sigma/dt \cong \mu_{ac}^2 \, |F_{a\to b} \, F_c|^2$ near $t = 0$, if μ_{ac} is not too large. Here $F_{a\to b}$ is a transition form factor, which, presumably, can be measured in electroproduction experiments. We also discuss briefly higher order corrections in μ_{ac} to this form, and possible diffraction dissociation productions of other classes of states.

In electroproduction of nuclear excited states, $e + A \rightarrow e + B$, selection rules often restrict the reaction at $\theta = 0°$ when the spin of B differs from that of A, causing the cross section to dip in the forward direction. Similar selection rules and forward dips may occur in diffraction dissociation of nuclei. We show, however, that forward dips of this type should not normally occur in hadron physics because of the large mass differences between target and excited states.

1. Introduction

Experiments[1] indicate that there exists a class of production reactions $a + c \rightarrow b + d$ which have two distinguishing features.

(1) Their differential cross sections $d\sigma/dt$ $[t = (p_a - p_b)^2]$ at high energies are energy independent.

As is common at high energies, the reactions tend to occur mainly with small t values. The other distinguishing feature which appears to characterize this class is that

(2) a and b and c and d have the same intrinsic quantum numbers (isospin, baryon number, strangeness, etc.).

Good and Walker[2] predicted the occurrence of a phenomenon they called diffraction dissociation of beam particles--a class of production reactions $a + A \rightarrow b + A$ associated with the shadow (elastic) scattering of beam particles a by a nucleus A. They suggested that the produced system b would satisfy (2) and, in addition, that the spin-parity of b would be the same as that of a. In this paper, we wish to show that models of hadron-hadron elastic scattering like the Chou-Yang model (or the quark model) yield a possible explanation of production reactions which satisfy (1) and (2). The production mechanism we propose here allows for spin-parity changes in reactions which we shall call diffraction dissociation reactions. We will, however, adopt the terminology of Good and Walker and call the class of reactions which satisfy (1) and (2) diffraction dissociation events. The mechanism which we propose to account for these reactions is, we believe, the same as was discussed by Good and Walker.

We shall adopt a Chou-Yang model of hadrons.[3] We shall show that it is very similar to Glauber's multiple scattering theory[4,5] of high-energy nucleus-nucleus scattering. Glauber's theory describes both inelastic and elastic nucleus-nucleus scatterings. Similarly, this Chou-Yang model describes inelastic as well as elastic scattering. Many of our conclusions are independent of the details of the model and are also valid for quark models of hadrons.[6] Our basic assumptions are the following:

(A) Hadrons are extended objects.

(B) Their internal spatial (or space-time) structure is described by amplitudes χ; S-matrix elements for reactions $a + c \rightarrow b + d$ are linear functionals of χ_a and χ_c (and of corresponding conjugate amplitudes $\bar{\chi}_b$ and $\bar{\chi}_d$) as, in the non-relativistic theory of atomic or nuclear reactions, they are linear functionals of the atomic or nuclear wave functions (and complex conjugate wave functions of the final states).

(c) At high energies ($s \rightarrow \infty$), elastic hadron-hadron scattering occurs owing to a local, absorptive interaction between "pieces" of hadronic matter.

More precisely, we shall assume a model for hadron-hadron elastic scattering which is formally similar to Glauber's multiple scattering theory[4,5] of high-energy nucleus-nucleus scattering.

These assumptions are motivated by: (i) the remarkable similarity of the t-dependence of proton-proton elastic scattering to the fourth power of the electric form factor of the proton, and (ii) indications of multiple scattering structure in the proton-proton data at larger t-values.[7] Glauber's multiple scattering theory of high-energy particle-nucleus scattering gives, for example, proton-nucleus scattering amplitudes in terms of measured nucleon-nucleon scattering amplitudes and nuclear wave functions. Measured and calculated cross sections agree remarkably well.[8] The theory yields a multiple scattering expansion in which the leading term is the effect of single nucleon-nucleon encounters (impulse approximation). The next term contains the effects of single nucleon scattering from two nucleons in the target; the third from three, etc. Experimental proton-nucleus scattering data clearly indicate the presence of such effects.[8] Glauber's theory can also be used with existing quark models to express hadron-hadron scattering in terms of quark-quark scattering.

Chou and Yang have shown that their model for hadron-hadron elastic scattering yields an expansion for the scattering amplitude whose terms can be put in one-to-one correspondence with Glauber's multiple scattering expansion.[9] We shall explore their model from this viewpoint. However, some of our results are more general. We shall use Glauber's theory of nucleus-nucleus scattering as a heuristic tool. In the next section we summarize briefly Glauber's multiple scattering theory for nucleus-nucleus reactions, and show that if nucleon-nucleon scattering is mainly absorptive and spin and isospin independent, inelastic nuclear reactions occur which might naturally be called diffraction dissociation reactions.

In Section 3 we show that, if certain conditions are satisfied, the Chou-Yang model may be regarded as Glauber's multiple scattering expansion for elastic hadron-hadron scattering. Here hadrons are assumed to be extended objects whose constituent matter distribution is described by amplitudes which are generalizations of nuclear wave functions. Further, it is assumed that the hadron-hadron interaction can be decomposed into constituent interactions (as in Glauber's theory one decomposes the nucleus-nucleus interaction in terms of the nucleon-nucleon interactions). In the Chou-Yang model, the constituent interactions are purely absorptive, spin and isospin independent, and highly local.

In Section 4, we argue that this model may also describe certain production reactions. If these are mainly given by the single scattering approximation, the resulting amplitude has a simple form involving transition form factors which may be compared with electroproduction measurements. Higher order corrections due to elastic rescattering are taken into account.

In Section 5 selection rules are given. In Glauber's theory with spin independent interactions, spin-parity changes may occur owing to transfer of orbital angular momentum to internal motions. This yields the selection rule that spin-parity changes are limited to those obtained by adding 0^+, 1^-, 2^+, ... to the initial state. Relativistic effects, however, may be expected to violate this selection rule for production at $\theta \neq 0°$. Additional selection rules that obtain in the limit of no mass change are also given.

In sections 6 and 7, unfamiliar effects to be expected in hadronic diffractive productions are discussed. In particular, forward dips which occur in the nuclear case, it is argued, need not occur in the hadronic case owing to the large mass changes.

2. Diffraction dissociation in Glauber's theory

In this and the following sections, we shall confine our attention to high energy, small angle scattering and use Glauber's eikonal approximations. We first summarize briefly Glauber's theory of nucleus-nucleus scattering. Consider a nucleus A incident on a nucleus C with impact parameter b (see Fig. 1). For simplicity, assume both A and C are spinless, and consider final states in which C is left in its ground state. Let ψ_C be the corresponding wave function for C at rest. The basic assumption in Glauber's theory is that the amplitude change of the wave function describing the A nucleons after the collision with C is given by

$$S^{(C)}(\vec{b}, \vec{s}_1, \vec{s}_2 \ldots, \vec{s}_A) = \int d\tau_C \, \psi_C^* \exp\,[2i \sum_{i,j} \delta_{ij}(\vec{s}_1 - \vec{b} - \vec{s}_j{}'))]\, \psi_C, \quad (1)$$

where δ_{ij} is the elastic scattering phase shift for the scattering of the i-th nucleon in A by the j-th nucleon in C; the quantity $|\vec{s}_i - \vec{b} - \vec{s}_j{}'|$ is the impact parameter of the collision (See Fig. 1)[10]; the sum is over all i in A and j in C; $\int d\tau_C$ denotes integration (and sum, if spin is taken into account) over all the internal variables which describe the nucleon configuration of C. The function

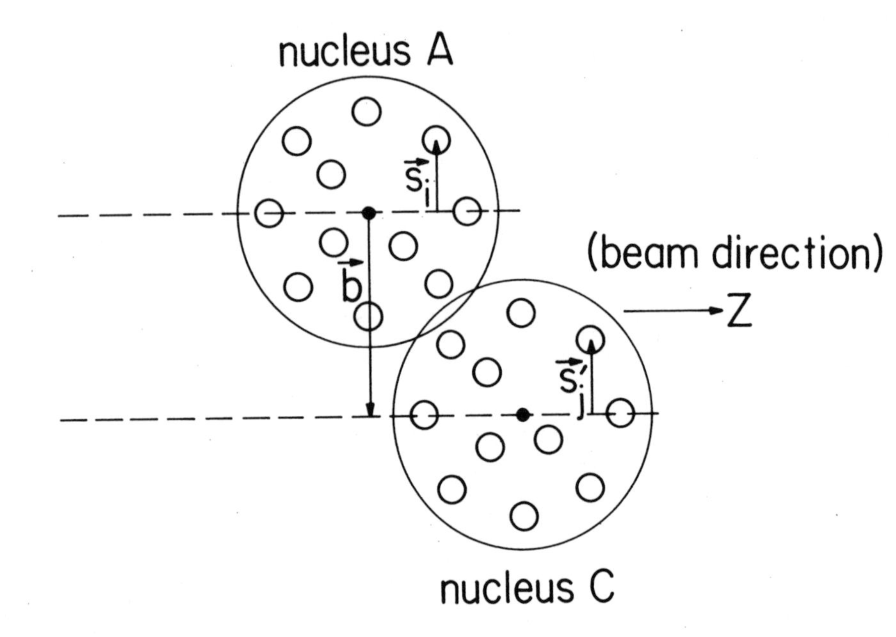

Fig. 1

Diagram which defines the variables which explicitly appear in Eq. (1); $\vec{s}_i$ is a two-dimensional vector which gives the displacement of the i-th nucleon in A from the center of mass of A in the plane perpendicular to the incident beam direction (z-axis); similarly, $\vec{s}_j'$ gives the displacement in this plane of the j-th nucleon in C from the center of mass of C; $\vec{b}$ is the distance in this plane of the center of mass of C from that of A.

$S^{(C)}$ depends upon the transverse displacements $\vec{s}_i$ of the nucleons in A and on the impact parameter vector $\vec{b}$. If the wave function describing the nucleon configuration in A is ψ_A, after the collision with C it is the product

$$S^{(C)}(\vec{b};\ \vec{s}_1,\ \ldots,\ \vec{s}_i,\ \ldots,\ \vec{s}_A)\ \psi_A(\underset{\sim}{r}_1,\ \ldots,\ \underset{\sim}{r}_i,\ \ldots,\ \underset{\sim}{r}_A). \tag{2}$$

In (2), $\underset{\sim}{r}_i$ is the coordinate vector for the i-th nucleon with components $(x_i,\ y_i,\ z_i)$ and the corresponding 2-vector $\vec{s}_i$ has components $(x_i,\ y_i)$. The coordinate vectors in (2) satisfy $\sum_{i=1}^{A} \underset{\sim}{r}_i = 0$.

If (2) is expressed as an expansion in a complete set of states ψ_B, viz.:

$$S^{(C)}\ \psi_A = \psi_A - \sum_B \alpha_{BA}^{(C)}(\vec{b})\ \psi_B\ , \tag{3}$$

the expansion coefficients $\alpha_{BA}^{(C)}$ give, in Glauber's approximation, partial wave amplitudes for the reactions A + C → B + C. The partial wave amplitudes for A + C → A + C are

$$\alpha_{AA}^{(C)}(\vec{b}) = \int d\tau_A\ \psi_A^*\ (1 - S^{(C)})\ \psi_A\ . \tag{4}$$

The elastic scattering differential cross section is

$$(d\sigma/dt)_{elastic} = \pi |A_{elastic}|^2 \tag{5}$$

where $t \cong - \vec{q}^2$ and

$$A_{elastic} \cong i \int \frac{d^2b}{2\pi}\ e^{i\vec{q}\cdot\vec{b}}\ \alpha_{AA}^{(C)}(\vec{b})\ . \tag{6}$$

In (6), the momentum transfer in the z-direction has been neglected.[4] For elastic scattering at high energies and small angles, $q_z << |\vec{q}|$.

The expansion coefficients in (3) with $B \neq A$

$$\alpha_{BA}^{(C)}(\vec{b}) = - \int d\tau\ \psi_B^*\ S^{(C)}\ \psi_A \tag{7}$$

similarly yield partial wave amplitudes for the inelastic reactions $A + C \to B + C$. Neglecting spin, the differential cross sections are given by

$$(d\sigma/dt)_{AC\to BC} = \pi |A_{AC\to BC}|^2 \tag{8}$$

with

$$A_{AC\to BC} \cong i \int \frac{d^2b}{2\pi} e^{i\vec{q}\cdot\vec{b}} \alpha_{BA}^{(C)}(\vec{b}) \ . \tag{9}$$

Here again one has assumed $|\vec{q}| >> q_z$ and neglected q_z. However, if the mass of B differs from that of A, the relation $q_z << |\vec{q}|$ may not be satisfied. In this case, q_z must be taken into account and corrections made. We shall return to this point below.

Glauber's multiple scattering expansion may be obtained for the cases above by expanding (1) in powers of the nucleon-nucleon partial wave amplitudes $\alpha_{ij} = 1 - \exp[2i\ \delta_{ij}]$. Writing (1) as

$$S^{(C)} = \int d\tau_C\ \psi_C^*\ S\ \psi_C\ , \tag{10}$$

one has

$$S = \prod_{i,j} (1 - \alpha_{ij}) = 1 - \sum_{i,j} \alpha_{ij} + \frac{1}{2!} \sum{}' \alpha_{ij}\ \alpha_{k\ell} - \ . \ . \ . \tag{11}$$

where $\sum{}'$ means sum over all indices $i = 1, 2, 3, \ldots N_A$ and $j = 1, 2, 3, \ldots N_C$ omitting terms in which the same pair occurs in more than one factor. Note that there will be precisely $N + 1$ terms in the right-hand side of (11) if $N = N_A N_C$, N_A = number of nucleons in nucleus A, and N_C = number of nucleons in nucleus C. Glauber's multiple scattering expansion yields an infinite series only in the limit $N \to \infty$. Substituting (11) into (10) and using this expression in (7) or (4), one obtains the multiple scattering expansions for inelastic and elastic reaction amplitudes. If one keeps only the leading or single scattering term in such an expansion, one obtains what is usually called the impulse approximation. This term usually dominates near $-t = 0$ if the nucleon-nucleon scattering is not too strong.

Glauber's theory also describes reactions like $A + C \to B + D$. In this case, one considers instead of (10)

$$S^{(DC)} = \int d\tau_C \, \psi_D^* \, S \, \psi_C \tag{12}$$

and obtains as above partial wave amplitudes given by

$$\alpha_{BA}^{(DC)}(\vec{b}) = \int d\tau \, \psi_B^* \, \psi_D^* \, (1 - S) \, \psi_C \, \psi_A \tag{13}$$

where, when q_z is negligible,

$$S = e^{2i \sum_{i,j} \delta_{ij}(\vec{s}_i - \vec{b} - \vec{s}_j{}')} \quad . \tag{14}$$

In this approximation, the differential cross sections are given by expressions like (8) with

$$A_{AC\to BD}(t) = i \int \frac{d^2b}{2\pi} \, e^{i\vec{q}\cdot\vec{b}} \, \alpha_{BA}^{(DC)}(\vec{b}) \quad . \tag{15}$$

As is shown in section 3, the Chou-Yang model for the high-energy limit of hadron-hadron <u>elastic</u> scattering yields amplitudes which are formally identical with those obtained from (13) and (14) with $D = C$ and $B = A$ under the following assumptions:
(1) The nucleon-nucleon partial wave scattering amplitudes $\alpha_{ij} = 1 - e^{2i\delta_{ij}}$ are spin and isospin independent; i.e., they have the form $\alpha_{ij}(\vec{b}) = f(b)$ for all i and j.
(2) The nucleon-nucleon scattering is purely absorptive (δ_{ij} pure imaginary; α_{ij} pure real).
(3) The α_{ij} are energy independent; i.e., they depend only on b.
(4) The range of impact parameters for which the α_{ij} differ appreciably from zero is small compared to the distances over which the nuclear wave functions extend so that in, e.g., (13) they may be approximated by

$$\alpha_{ij} \approx \text{constant} \times \delta^2(\vec{b}) \qquad \text{(zero range approximation).}$$

(5) Fluctuations in expectation values of products of sums $\sum \alpha_{ij}$ are small; i.e., the number of nucleons N in A and C is large and one has, for example,

$$<AC \,|\sum{}' \alpha_{ij} \alpha_{k\ell}|\, AC> \approx <AC \,|(\sum \alpha_{ij})^2|\, AC>$$

and

$$<AC \,|(\sum \alpha_{ij})^2|\, AC> \approx <AC \,|\sum \alpha_{ij}|\, AC>^2 .$$

Physically, assumption (5) may be expected to hold when the nucleon-nucleon collision time is short compared to the characteristic period of nuclear motions in the nuclei. A detailed analysis of this assumption from a viewpoint that may be more directly applied to hadrons is given in section 3.

In all that follows, we assume $N >> 1$.

Assumptions (1) - (4) may be expressed, e.g., by taking

$$\alpha_{ij}(\vec{b}) = \frac{2\pi\mu}{N} \delta^2(\vec{b}) \tag{16}$$

with μ a real number to be determined by a fit to elastic scattering data. We shall refer to (16) as a zero-range approximation which is purely absorptive and <u>locally</u> spin and isospin independent.

Using (16), $S^{(C)}$ has the form

$$S^{(C)} = 1 - (2\pi\mu/N) \sum_i D_1^{(C)}(\vec{s}_i - \vec{b}) + \frac{1}{2!} (2\pi\mu/N)^2 \sum{}' D_2^{(C)}(\vec{s}_i - \vec{b}, \vec{s}_k - \vec{b}) - \ldots \tag{17}$$

where $D_1^{(C)}(\vec{s})\, d^2s$ is the probability of finding a target nucleon in a tube of cross-sectional area d^2s whose axis is along z and displaced by $\vec{s}$ from the center of mass of C. Similarly, $D_2^{(C)}(\vec{s}, \vec{s}')\, d^2s\, d^2s'$ is the probability of finding a nucleon in two such tubes. (More precisely, for the terms $i = k$, the nucleons in the two tubes must be different nucleons; when $i \neq k$, the same nucleon might be in both tubes.) The higher order terms can similarly be interpreted. The "wave function" $S^{(C)}\ \psi_A$ obtained from (17) is then just ψ_A with the shadow of C on it; that is to say, the amplitude for finding, for example, the i-th nucleon of A displaced $\vec{s}_i$ from the center of mass of A in $S^{(C)}\ \psi_A$ is suppressed relative to that in ψ_A by the probability it has "seen" one or more nucleons in C. Thus the resulting wave function $S^{(C)}\ \psi_A$ differs from that of the original nucleus A and has components of states other than A in it. It is these states that appear in the expansion (3).

From this correspondence, it is natural to infer that if the Chou-Yang model correctly describes elastic scattering it also predicts certain production processes. These productions are, for example, reactions like $A + C \to B + C$ where the states B have non-vanishing coefficients in (3). These coefficients depend only on $\vec{b}$ and yield differential cross sections $d\sigma/dt$ which are energy independent functions of t. Thus one may obtain a description of productions which arise owing to diffraction.

3. The Chou-Yang model as a relativistic extension of Glauber's theory

The Chou-Yang model for elastic hadron-hadron scattering may be viewed as the high-energy limit of Glauber's theory as follows. Consider the expansion (11) in the limit $N = N_A N_C \to \infty$. It is then an infinite series. Assume that the eikonal approximation holds so that the elastic partial wave scattering amplitudes α_{ac} for $a + c \to a + c$ are related to the differential cross section by (6) with

$$A_{elastic} = i \int \frac{d^2b}{2\pi} e^{i\vec{q}\cdot\vec{b}} \alpha_{ac}(\vec{b}) , \tag{18}$$

and assume that the α_{ac} are given by

$$\alpha_{ac} = \langle ac \,|(1 - S)|\, ac\rangle \tag{19}$$

where S has the expansion (11). Then (19) has the form

$$\alpha_{ac} = \langle ac \,|\sum \alpha_{ij}|\, ac\rangle - \frac{1}{2!} \langle ac \,|\sum{}' \alpha_{ij}\, \alpha_{k\ell}|\, ac\rangle + \ldots \tag{20}$$

In the non-relativistic theory (Section 2 above), the contribution of the leading term in (20) to $A_{elastic}$ is easily evaluated. For simplicity, let

$$\alpha_{ij}(\vec{b}) = N^{-1}\, \tilde{f}(\vec{b}) \tag{21}$$

and assume axial symmetry; i.e., that $\tilde{f}(\vec{b}) = \tilde{f}(-\vec{b})$. In (21), N = (number of "nucleons" in a) x (number of "nucleons" in c) = $N_a \times N_c$. Then one has for

$$\alpha^{(1)}_{ac}(\vec{b}) = \langle ac \,|\sum \alpha_{ij}|\, ac\rangle \tag{22}$$

the expression

$$\alpha^{(1)}_{ac} = \int d\tau_a\, d\tau_c\, N^{-1} \sum_{i,j} \tilde{f}(\vec{s}_i - \vec{b} - s_j{}')\, |\psi_a|^2\, |\psi_c|^2, \tag{23}$$

and substituting this expression into (18), one obtains

$$A^{(1)}_{ac} = i\, f(\vec{q})\, F_a(\vec{q})\, F_c(\vec{q}) \tag{24}$$

where

$$f(\vec{q}) = \int \frac{d^2b}{2\pi}\, e^{i\vec{q}\cdot\vec{b}}\, \tilde{f}(\vec{b}) \tag{25}$$

and the form factors F_a and F_c are given by

$$F_a = N_a^{-1} \sum_{i=1}^{N_a} \int d\tau_a\, e^{i\vec{q}\cdot\vec{s}_i}\, |\psi_a|^2,$$

$$F_c = N_c^{-1} \sum_{j=1}^{N_c} \int d\tau_c\, e^{i\vec{q}\cdot\vec{s}_j{}'}\, |\psi_c|^2. \tag{26}$$

If $f(\vec{q})$ = constant [i.e., $\tilde{f}(\vec{b}) = 2\pi$ (constant) $\delta^2(\vec{b})$], (23) gives the leading term in the expansion of Chou and Yang.[9]

The remaining terms in (20) in general differ from the corresponding terms in the Chou-Yang expansion. The Chou-Yang expansion, however, may be viewed as an approximation to (20). To see what approximation this is, consider first the second term in (20). One can insert a complete set of states $|x\rangle$ and write

$$2!\ \alpha_{ac}^{(2)} = \langle ac\ |\textstyle\sum'\ \alpha_{ij}\ \alpha_{k\ell}|\ ac\rangle = \sum_{x} \langle ac\ |\textstyle\sum'\left(\alpha_{ij}|\ x\rangle\ \langle x\ |\alpha_{k\ell}\right)|ac\rangle \tag{27}$$

which also can be written as

$$2!\ \alpha_{ac}^{(2)} = \langle ac|\textstyle\sum'\left(\alpha_{ij}|ac\rangle\ \langle ac|\alpha_{k\ell}\right)|ac\rangle + \sum_{x\neq ac} \langle ac|\textstyle\sum'\left(\alpha_{ij}|x\rangle\langle x|\alpha_{k\ell}\right)|ac\rangle. \tag{28}$$

Let us assume the α_{ij} describe purely absorptive "nucleon"-"nucleon" scattering. The α_{ij} are then real and positive. It follows that their diagonal matrix elements are real and positive. In the limit $N \to \infty$, the first term in (28) contains a sum of N^2 positive terms. On the other hand, the second term in (28) for each $x \neq ac$ contains a sum of N^2 terms whose phases may be random. If indeed those phases are random, each such sum will contribute a term of order N. Let us assume this (assumption [i]). Now given that each term in the sum over $x \neq ac$ is of order 1/N compared to the diagonal term, we still face the fact that we have an infinite sum. If we assume (assumption [ii]) that this sum also vanishes in the limit $N \to \infty$ relatively to the diagonal term, we have

$$\alpha_{ac}^{(2)} \cong \frac{1}{2!} \langle ac\ |\textstyle\sum'\left(\alpha_{ij}|\ ac\rangle\ \langle ac\ |\alpha_{k\ell}|\right)ac\rangle \tag{29}$$

which, in the limit $N \to \infty$, is related to (22) by

$$\alpha_{ac}^{(2)} = \frac{1}{2!}\ (\alpha_{ac}^{(1)})^2. \tag{30}$$

The above assumptions may be similarly applied to the triple scattering term in (20) to obtain the result that it is, in the limit $N \to \infty$, given by

$$\alpha_{ac}^{(3)} = \frac{1}{3!} (\alpha_{ac}^{(1)})^3 \tag{31}$$

and, similarly, one may obtain for the n-tuple scattering term

$$\alpha_{ac}^{(n)} = \frac{1}{n!} (\alpha_{ac}^{(1)})^n. \tag{32}$$

One then obtains the result that

$$\alpha_{ac} = 1 - e^{-\langle\alpha\rangle} \tag{33}$$

where $\langle\alpha\rangle = \alpha_{ac}^{(1)}$ and has the form (23). If (zero range approximation)

$$\tilde{f}(\vec{s}_i - \vec{b} - \vec{s}_j{}') \cong 2\pi\, \mu_o\, \delta^2(\vec{s}_i - \vec{b} - \vec{s}_j{}'), \tag{34}$$

(33) yields the Chou-Yang model for elastic scattering[9]

$$A_{elastic} = i\ [\Delta - \frac{1}{2!} \Delta \otimes \Delta + \frac{1}{3!} \Delta \otimes \Delta \otimes \Delta - \ldots] \tag{35}$$

with

$$\Delta = \mu\, F_a\, F_c \tag{36}$$

and $\otimes$ denoting the convolution integral:

$$\Delta \otimes \Delta = \int \frac{d^2k}{2\pi}\, \Delta(\vec{q})\ \Delta(\vec{q}-\vec{k}).$$

We have thus shown that if one assumes that the off-diagonal matrix elements both individually and summed are relatively small, one obtains the Chou-Yang model as the relativistic generalization of Glauber's multiple scattering theory.

However, the off-diagonal matrix elements $<x\ |\sum \alpha_{k\ell}|\ ac>$ may not all be small. In particular, in the next section we discuss a class of states $|x>$ in which one has, for example, a collective excitation of a. These matrix elements may be of the same order in N as the diagonal matrix elements. If such states exist, one might expect corrections to (35).

4. Diffraction dissociation in the Chou-Yang model

If one looks at the Chou-Yang model from the above viewpoint, the considerations of Section 2 lead naturally to the expectation that diffractive dissociation production reactions should occur along with elastic scattering. The production mechanism is like that described in the last part of Section 2. In that part, we assumed the short-range approximation (16). That assumption is equivalent to taking $f(\vec{q}) = \mu$ in (24). It leads, in the present analysis, to the form (36) for the single scattering approximation to $A_{elastic}$. In this approximation, the amplitude for the diffraction dissociation production of b in

$$a + c \rightarrow b + c \tag{37}$$

would be given by

$$A_{bc;ac} \cong i\ \mu\ F_{(a \rightarrow b)}\ F_c \tag{38}$$

where $F_{(a \rightarrow b)}$ is a transition form factor. The remarkable fact that elastic proton-proton scattering data agree with the assumption that

$$A_{elastic} \cong i\ \mu\ [F_p]^2 \tag{39}$$

where F_p is the measured electric form factor of the proton[12] leads naturally to the conjecture that the form factor $F_{(a \rightarrow b)}$ in (38) might be the same as that which is measured in the electroproduction reaction

$$e^- + a \to e^- + b. \tag{40}$$

The conjecture can be checked experimentally. It gives for (37) at $t = 0$ for example, the prediction that

$$(d\sigma/dt)_{ap\to bp} \cong \mu^2\,[F_{(a\to b)}(0)]^2 \tag{41}$$

since $F_p(0) = 1$. The constant μ may be evaluated from elastic scattering data.

Some accessible reactions of type (37) and (40) are $p + p \to N^* + p$ and $e + p \to e + N^*$. Fits of the Chou-Yang model to p-p elastic scattering yield values for μ which are large.[7] In this case, multiple scattering corrections to (38) and, possibly, (41) may be appreciable. These may be approximately taken into account as follows. Consider the partial wave amplitude $\alpha_{bc;ac}$ which is related to $A_{bc;ac}$ by

$$A_{bc;ac} = i \int \frac{d^2b}{2\pi}\, e^{i\vec{q}\cdot\vec{b}}\, \alpha_{bc;ac}(\vec{b}). \tag{42}$$

It has an expansion of form (20); viz.,

$$\alpha_{bc;ac} = \langle bc\,|\textstyle\sum \alpha_{ij}|\,ac\rangle - \frac{1}{2!}\langle bc\,|\textstyle\sum' \alpha_{ij}\,\alpha_{k\ell}|\,ac\rangle + \ldots \tag{43}$$

Let the leading term in this series be denoted by $\alpha^{(1)}_{bc;ac}$ and the n-th term by $\alpha^{(n)}_{bc;ac}$ as in (22) and (27). Similarly to (27), we may write

$$2!\,\alpha^{(2)}_{bc;ac} = \sum_x \langle bc\,|\textstyle\sum'\left\{\alpha_{ij}|\,x\rangle\langle x\,|\alpha_{k\ell}\right\}|ac\rangle. \tag{44}$$

If we again make the perhaps unjustified assumption that "elastic" terms dominate, we may take

$$2!\,\alpha^{(2)}_{bc;ac} \cong \langle bc|\textstyle\sum'\left\{\alpha_{ij}|bc\rangle\langle bc|\alpha_{k\ell}\right\}|ac\rangle + \langle bc|\textstyle\sum'\left\{\alpha_{ij}|ac\rangle\langle ac|\alpha_{k\ell}\right\}|ac\rangle. \tag{45}$$

If one similarly evaluates all the higher order multiple scattering terms, using similar reasoning as given above for the derivation of (33), one may sum the series. Neglecting terms of order 1/N, one then obtains

$$\alpha_{bc;ac} \cong \alpha^{(1)}_{bc;ac}\, e^{i(\delta_{bc} + \delta_{ac})} \, [\sin(\delta_{bc} - \delta_{ac})/(\delta_{bc} - \delta_{ac})] \tag{46}$$

where δ_{bc} and δ_{ac} are the b-c and a-c elastic scattering phase shifts. If $\delta_{bc} = \delta_{ac}$, (46) has the familiar form

$$\alpha_{bc;ac} = \alpha^{(1)}_{bc;ac}\, e^{2i\delta} . \tag{47}$$

In our model [see (33)]

$$\exp\,[2i\ \delta_{ac}] = \exp\,[-\langle\alpha\rangle_{ac}], \tag{48}$$

and the result may be expressed in a form similar to (35); viz.,

$$- i\ A_{bc;ac} \cong \Delta_{bc;ac} - \Delta_{bc;ac} \otimes \Delta + \frac{1}{2!}\Delta_{bc;ac} \otimes \Delta \otimes \Delta - \frac{1}{3!}\Delta_{bc;ac} \otimes \Delta \otimes \Delta \otimes \Delta + \ldots \quad . \tag{49}$$

where Δ is given by (36) and

$$\Delta_{bc;ac} = \mu\, F_{(a \to b)}\, F_c . \tag{50}$$

We have neglected spin in the above discussion. For reactions like $p + p \to N^* + p$, effects of spin may be important. However, analysis shows (see below, in particular, Eq. (57) and Reference 25) that if the elastic scattering is spin independent and the absorption is a scalar effect the differential cross section may still have the simple form

$$d\sigma/dt = \pi\, |A_{bc;ac}|^2 \tag{51}$$

with $A_{bc;ac}$ given by (49) and (50).

It would be remarkable indeed if some diffraction dissociation differential cross sections were simply related to electroproduction form factors as indicated above. In the electroproduction case $e + p \to e + N^*$, there are three form factors.[13] If they differ significantly, we do not know which one should be used in (50), although inference from non-relativistic theory (see below) leads us to conjecture that the Coulomb form factor[13] may be the most likely candidate. However, we wish to remark here that perhaps the "scaling law" for the elastic form factors of the proton (i.e., $G_E^p \approx G_M^p$)[14] is in some way connected with the agreement of (39) with experimental data. Experimentally, separate determinations of transverse and Coulomb electroproduction form factors is under way (see Panofsky, Ref. 12).

The diffraction dissociation production mechanism we are proposing here is, in some respects, analogous to the Coulomb excitation of nuclei. Both are due to local, spin independent interactions. For heavy nuclei, first-order Coulomb excitation produces a special class of nuclear states.[15] They are collective excitations;[16] e.g., the 2^+, 4^+, etc. rotational states first discussed by Bohr and Mottelson. First-order Coulomb excitation is given by matrix elements which are similar to the single scattering approximation $\alpha^{(1)}_{bc;ac}$; i.e., neglecting the spin and charge structure of the nucleons, the nuclear matrix elements are proportional to

$$\int d\tau \, \psi_B^* \sum_{i=1}^{Z} e^{i\underset{\sim}{q}\cdot\underset{\sim}{r}_i} \psi_A \tag{52}$$

where the sum is over protons only. If Z is large, collective excited states ψ_B are dominantly produced because for these all, or nearly all, the terms in the sum contribute. Excited states B known as single particle excitations are not seen in Coulomb excitation experiments with heavy nuclei; for these only one term in the sum is appreciable.

Similarly, we would not expect that an amplitude given by (50) would describe all diffractive hadronic productions. Presumably, only a limited class of states could be produced in this "single scattering approximation". These would be "collective excitations". If some diffractive production differential cross sections are found to be simply related to electroproduction form factors as proposed here, this relation may serve as identification of such "collectively excited states".

5. Selection rules

In this section we shall use the Glauber theory of Section 2 as a heuristic device to infer properties of diffractive productions that may be expected in the Chou-Yang model. Unless otherwise stated, we confine our attention for simplicity to the case $a + c \rightarrow b + c$. We extend the model to particles with spin and isospin. We assume that the absorption is locally spin independent and locally isospin independent. In the Glauber theory, the above assumptions may be more precisely expressed by taking the "nucleon"-"nucleon" partial wave amplitudes $\alpha_{ij}(b)$ to have the form (16) or, more generally, (21). In the relativistic theory, these assumptions are taken to mean the following. One may write the matrix element for the reaction $a + c \rightarrow b + c$ as

$$\langle b \,|\, \mathcal{S}^{(c)}(0,p_c) \,|\, a \rangle \,. \tag{53}$$

where $|a\rangle$, for example, is a state of particle a with momentum p_a and helicity λ_a. If c has spin zero, the operator $\mathcal{S}^{(c)}$ is given by[17]

$$\mathcal{S}^{(c)}(0,p_c) = \int d^4y \; e^{-ip_c \cdot y} \, (\Box_y + m_c^2) \; T\left[j_c(0) \; \phi_c^{\,+}(y)\right]. \tag{54}$$

In the Glauber theory, $\mathcal{S}^{(c)}$ is related to $S^{(c)}$ of Section 2 by

$$\mathcal{S}^{(c)} = \int \frac{d^2b}{2\pi} \, e^{i\vec{q}\cdot\vec{b}} \, S^{(c)} \,. \tag{55}$$

In the relativistic theory, we assume that the high-energy limit of the piece of the operator $\mathcal{S}^{(c)}$ that connects b to a has properties similar to (55); i.e., $\mathcal{S}^{(c)}$ contains no spin or isospin dependence. Note, however, that the $\mathcal{S}^{(c)}$ in (54) depends on the momentum of c. In Glauber's theory, $\mathcal{S}^{(c)}$ depends only on the CM momentum transfer $\vec{q}$ owing to the fact that the center of mass motion may be completely factored out of nuclear wave functions. But in a relativistic theory this does not hold in general. The amplitudes describing the internal structure of composite states[18] depend on the velocity of the center of mass motion. Therefore we expect that $\mathcal{S}^{(c)}$ may depend upon the velocity of c.

Our model gives the following selection rules:

(a) *Isospin selection rule*. The isospin of b must be the same as that of a.

(b) *Spin-parity selection rules*. The produced state b need not have the same spin-parity as a. The dependence of $S^{(c)}$ on $\vec{b}$ allows orbital angular momentum transfer from the initial a + c state to internal motions. Consequently, spin-parity *changes* of 0^+, 1^-, 2^+, etc., may occur. At 0°, these would be the only spin-parity changes that could occur.[19] Even away from 0°, these are the only spin-parity changes that occur in the Glauber theory for nuclei where $S^{(C)}$ depends only on b. In the relativistic case, however, in general $\mathcal{S}^{(c)}$ depends both on $\vec{q}$ and the velocity of c. Additional spin-parity changes may occur away from the forward direction due to this velocity dependence. There is, however, one absolute selection rule at all angles, owing to the fact that one cannot construct a pseudoscalar from two vectors; no $0^- \rightarrow 0^+$ transitions can occur.

(c) *Other internal quantum numbers*. Since in our picture diffraction productions are, so to speak, purely geometrical (or spatial) effects, only spin-parity changes occur. In reactions a + c → b + d, if b is produced with small momentum transfer it should have the same strangeness, baryon number, etc., as a.

Of course, the mass of b may differ from that of a. When this mass difference is negligible (i.e., the nuclear case), the following additional selection rules hold:[20]

(d) Equal mass selection rule at 0°. If the masses of b and a are equal, non-vanishing production occurs at 0° only if their parities are equal. To see this, consider (53) in the rest frame of a. If the mass of b equals that of a, then b is produced at rest and both $|a\rangle$ and $|b\rangle$ are parity eigenstates. In our model, when $\vec{q} = 0$, $\mathcal{S}^{(c)}$ is an even operator. It depends, of course, on the incident beam direction $\underset{\sim}{p}_c$; however, it is spin independent and unchanged when $\underset{\sim}{p}_c \to -\underset{\sim}{p}_c$. It is therefore an even function of $\underset{\sim}{p}_c$. Thus, since $\mathcal{S}^{(c)}$ is diagonal in c, it can only connect states $|a\rangle$ and $|b\rangle$ of the same parity.

If the amplitudes for (37) may be obtained from an expansion like (17) and the single scattering term dominates, one has further selection rules owing to the fact that this term contains a product of form factors [as in (38)]. To illustrate this point, let us consider the equal mass case and assume that the transition form factor $F_{a\to b}$ is given by an expansion like (26) (in the rest frame of a); viz.,

$$F_{a\to b} \cong N^{-1} \sum_i \int d\tau \; e^{i\vec{q}\cdot\vec{s}_i} \; \psi_B^* \; \psi_a . \qquad (56)$$

Then as $\vec{q} \to 0$ (forward production), $F_{a\to b}$ will tend to vanish owing to the orthogonality of the states. Thus one might expect the selection rule that such productions vanish in the forward direction. Indeed, sharp dips in the forward direction are found in Coulomb excitation differential cross sections.[21]

It is of interest to explore the conjecture that at least some diffraction production amplitudes can also be written as a product of form factors. Generally this would occur over a limited range of momentum transfer owing to, e.g., multiple scattering corrections. In particular, multiple scattering presumably dominates at large t, and partially fills in any dips the single scattering may exhibit. Nevertheless, single scattering should often dominate over a range of small t, and we discuss it in the next section with special reference to the question of forward dips.

6. Hadronic diffractive productions

There is an important difference to be noted between hadronic excitations and nuclear excitations. Hadronic mass differences are of the order of magnitude of hadronic masses, whereas mass differences of nuclear excited states are small compared to nuclear masses. Owing to these large mass differences, some of the selection rules present in the nuclear case should not be expected to hold in hadronic productions. Consider, for example, a nuclear process whose matrix element is proportional to a form factor like (56); viz., (here we no longer make the approximation $q_z << |\vec{q}|$)

$$F_{A\to B}(q^2) = N^{-1} \sum_{i=1}^{N} \int d\tau \; e^{i q\cdot r_i} \; \psi_B^* \; \psi_A . \tag{56'}$$

Let us consider (56') in the rest frame of A. At $\theta = 0°$, $\vec{q} = 0$ but energy-momentum conservation requires that in the rest frame of A, q_z = momentum of B $\sim (M_B - M_A)$. If $M_B = M_A$, $F_{A\to B}$ vanishes at $\theta = 0°$ owing to the orthogonality of ψ_B and ψ_A. If $M_B - M_A \neq 0$, the magnitude of $F_{A\to B}$ is controlled by the dimensionless parameter $q_z R$ where R is the radius of A or B (whichever is smaller). If $q_z R << 1$, the differential cross section may be expected to show a dip in the forward direction. In the hadronic case, $q_z R$ is large. Indeed, it is so large that one cannot expect the simple expression (56') to apply. If one considers $a + c \to b + c$, in the limit $s \to \infty$, one finds that at $\theta = 0°$, in the rest frame of a, $q_z = (m_b^2 - m_a^2)/2m_a$. In most cases (see Table I), b is produced with relativistic velocities. The relativistic analogue of (56') may still contain a factor like ψ_a which describes the internal structure of a; however, the factor describing the internal structure of b will now be a function which depends on the center of mass motion of b.[18] Thus an expression like (56') evaluated in the rest frame of a will have an unsymmetrical dependence on the internal structure of a and b; e.g., a will be spherically symmetric but b will be Lorentz contracted and appear like a disc. To restore the symmetry, consider the expression in the Breit frame where $p_a + p_b = 0$. In this frame, as $s \to \infty$,

$$q_z^{Breit} = (m_b^2 - m_a^2)\ [2(m_b^2 + m_a^2)]^{-1/2}. \tag{57}$$

For small mass differences, one might expect that $F_{a \to b}$ might be controlled for forward productions by the dimensionless parameter $q_z^{Breit} R$ where R is a measure of the size of b (or a, whichever is smaller); i.e., when $q_z R \ll 1$. Since all hadrons appear to have approximately equal sizes, one might take $R \approx 4.5\ (\text{GeV}/c)^{-1}$ as obtained from elastic scattering diffraction peaks.[22] In Table I, the q_z^{Breit} is tabulated for various possible diffractive production reactions and one sees that $q_z R > 1$.

We therefore see no reason to expect that corresponding transition form factors for forward production are small.[23] This is consistent with present experimental information.[1] The information is incomplete, but where production has been followed into very small t -- notably in production of $N^*(1688)$[24]--there is no sign of a forward dip in the cross section.

More generally, for the relativistic case let us consider $\langle b\ |\$^{(c)}|\ a\rangle$ as in (53) and (54) above. If we neglect p_c dependence, $\$^{(c)}$ is a scalar operator that connects the one particle states a and b. By known techniques,[25] one may show that matrix elements $\langle b\ |\$^{(c)}|\ a\rangle$ with $|a\rangle$ and $|b\rangle$ states of definite momenta and helicities p_a, λ, and p_b, λ', respectively, have the form

$$\langle p_b\ \lambda'\ |\$^{(c)}|\ p_a\ \lambda\rangle = \sum_{\ell=|s'-s|}^{s'+s} f_{\lambda'\lambda}^{s',s;\ell}\ \ (p'')^{\ell}\ g_{\ell}(t) \tag{58}$$

where s and s' are the spins of a and b, respectively, the $f_{\lambda'\lambda}^{s',s;}$ are non-singular functions of p_a and p_b;

$$p'' = \sqrt{\frac{[(m_a + m_b)^2 - t]\ [m_a - m_b)^2 - t]}{2(m_a^2 + m_b^2)}}\ ; \tag{59}$$

and the $g_\ell(t)$ are invariant form factors. If (58) is calculated from a simple model, in lowest order perturbation theory the $g_\ell(t)$ are coupling constants. Note that

$$p'' = \sqrt{(q_z^{\text{Breit}})^2 - t + t^2/2(m_a^2 + m_b^2)} \ . \tag{59'}$$

At $t = 0$, the factor $(p'')^\ell$ in (58) is simply the familiar centrifugal barrier penetration factor. Parity conservation requires that when $\mathcal{S}^{(c)}$ is a scalar operator, $g_\ell \neq 0$ only when

$$(-)^\ell = \eta_a \eta_b \tag{60}$$

where η_a and η_b are the intrinsic parities.[26]

If the lower of the two spins s and s' is 0 or 1/2, the sum over ℓ in (58) reduces to a single term. If the spin and parity of b and a are the same, $\ell = 0$ and, when b and a are complex structures, g(t) may vanish as $p'' \to 0$ $[t \to (m_b - m_a)^2]$ owing to orthogonality.

The effect of these mass differences can be seen directly from (59). When $m_a \neq m_b$, the physical region in t for the reaction $a + b \to b + c$, as $s \to \infty$, begins at a distance $(m_a - m_b)^2$ from the threshold for $a \to b$. If p'' were small, one would have a suppression controlled by some dimensionless parameter $p''R$. It is clear from Table I that even when R is as small as 1 $(\text{GeV}/c)^{-1}$, this parameter is not small compared to one and one need not expect suppression of forward production.

In the above, we have neglected any $\underset{\sim}{p}_c$ dependence in $\mathcal{S}^{(c)}$ and treated it as a scalar operator. This is the case in Glauber's theory which was derived using non-relativistic wave mechanics. However, we expect this is not the case relativistically. The $\mathcal{S}^{(c)}$ in a fully relativistic theory will generally depend upon $\underset{\sim}{p}_c$. If this dependence is not negligible, the $\mathcal{S}^{(c)}$ in $\langle b|\mathcal{S}^{(c)}|a\rangle$ may not behave like a scalar. Then (58) no longer applies and the threshold behavior of $\langle b|\mathcal{S}^{(c)}|a\rangle$ may be quite different. However, the important point we wish to emphasize here is that in hadronic productions the threshold p'' = 0 is, in most cases, far from the physical threshold $\theta = 0°$.

This conclusion can be restated in the language of "kinematic constraints". The point $t = (m_b - m_a)^2$ is a "pseudothreshold" where kinematic constraints apply, which can be interpreted physically in terms of centrifugal barrier effects in the t-channel. In particular cases (most familiarly for one photon exchange--i.e., J = 1 in the t-channel), the constraints produce zeros at the pseudothreshold. When $(m_b - m_a)^2$ is large, however, these zeros have relatively little effect on the physical region at $t \leq 0$.

7. A Speculation

A speculation one might make from the above considerations is that production of "collective states" in the single scattering approximation may become small for high mass states. This would be the case if, near $q^2 = 0$, form factors are rapidly decreasing functions of q^{T^2} where $q^{T^2} = -(q^{Breit})^2$ (q^T_μ = four-momentum transfer transverse to the average momentum $p_\mu = p^a_\mu + p^b_\mu$). The speculation is motivated by the facts that we found it useful to think of form factors as functions of $\tilde{q}^{Breit}$, and that (apart from centrifugal barrier effects) hadron form factors appear to fall rapidly as their arguments take on large values. Note that

$$q^{T^2} = q^2 - \frac{(m_b^2 - m_a^2)^2}{2(m_b^2 + m_a^2) - q^2} \tag{61}$$

so that even at $q^2 = 0$, q^{T^2} becomes large as $|m_b^2 - m_a^2|$ grows. In this case, the contribution (49) to $A_{bc;ac}$ is small and, presumably, the dominant term corresponds to multiple scattering which takes a to b through a succession of lower mass intermediate states. Estimations of multiple scattering effects are difficult because they are very model dependent. Estimation of terms involving a succession of inelastic excitations yield results which are sensitive to the internal structure of these states. However, a qualitative remark might be made. The slope of the contribution to $d\sigma/dt$ vs. t from a given order of multiple scattering is generally smaller the larger the number of multiple scatterings. This might account for the general tendency of the slope to decrease as m_b increases, which is noted clearly in continuum production $\pi + N \rightarrow (MM) + N$,[27] and also occurs in production of specific isobars[24,28] with the exception of $N^*(1400)$.

Acknowledgments

One of us (N.B.) wishes to acknowledge with pleasure stimulating discussions with Professors W. Slater and W. N. Cottingham. We would both like to thank Professors J. Prentki, L. Van Hove, and R. E. Peierls for the hospitality and facilities that were afforded us during our work at CERN and the Department of Theoretical Physics in Oxford.

TABLE I

The values of q_z in the Breit frame for reactions at $\theta = 0°$ in the high energy limit $s \rightarrow \infty$; $q_z = [-qT^2]^{1/2}$.

$a + c \rightarrow b + c$	$q^{Breit} = (m_b^2 - m_a^2)/[2(m_b^2 + m_a^2)]^{1/2}$ in GeV/c
$\pi\ p \rightarrow A_1\ (1058)p$	0.73
$K\ p \rightarrow K^*\ (1320)p$	0.75
$K\ p \rightarrow K^*\ (1790)p$	0.79
$p\pi \rightarrow N^*(1470)\pi$	0.51
$pp \rightarrow N^*(1690)p$	0.71
$pp \rightarrow N^*(2190)p$	1.2

References

1. See, e.g., D. R. O. Morrison, Phys. Rev. 165, 1699 (1968).
2. M. L. Good and W. D. Walker, Phys. Rev. 120, 1857 (1960). A more recent work, with many references, is M. Ross and L. Stodolsky, Phys. Rev. 149, 1172 (1966).
3. T. T. Chou and C. N. Yang, Proceedings of Conference on High Energy Physics and Nuclear Structure, ed. G. Alexander, North-Holland Publishing Co., Amsterdam (1967), pp. 348-360.
4. R. J. Glauber, Proceedings of Conference on High Energy Physics and Nuclear Structure, ed. G. Alexander, North-Holland Publishing Co., Amsterdam (1967), pp. 311-338; also Lectures in Theoretical Physics, Vol. I, ed. W. E. Brittin and L. G. Dunham, Interscience Publishers, Inc., New York (1959), pp. 315-414.
5. See, e.g., V. Franco (to be published) for a detailed analysis of deuteron-nucleus scattering in the Glauber approximation.
6. D. R. Harrington and A. Pagnamenta, Phys. Rev. Letters 18, 1147 (1967
7. T. T. Chou and C. N. Yang, Phys. Rev. Letters 20, 1213 (1968); L. Durand and R. Lipes, Phys. Rev. Letters 20, 637 (1968).
8. H. Palevsky et al., Phys. Rev. Letters 18, 1200 (1967); G. Igo et al., Nucl. Phys. B3, 181 (1967); and R. H. Bassel and C. Wilkin, Phys. Rev. 174, 1179 (1968).
9. T. T. Chou and C. N. Yang, Phys. Rev. 170, 1591 (1968) and 175, 1832 (1968).
10. Throughout this paper we use the notation $\vec{s}$ for two-dimensional vectors and the usual notation $\underset{\sim}{q}$ and q_μ for three and four-dimensional vectors.
11. Such a mechanism for deuteron disintegration and the disintegration of light nuclei has been extensively discussed. It was first proposed by R. J. Glauber, Phys. Rev. 99, 1515 (1955) and E. L. Feinberg, J. Exptl. Theoret. Phys. (USSR) 29, 115 (1955) (translation: Soviet Phys. JETP 2, 58 (1956)). Other references are given in Ref. 2 above.

12. For $s \to \infty$, this was conjectured by T. T. Wu and C. N. Yang, Phys. Rev. 137, B708 (1965). R. P. Feynman remarked that experimental data support this relation for $-t \lesssim 1$ $(\mathrm{GeV/c})^2$ (XIIIth International Conference on High Energy Physics, Berkeley, 1966). Elastic p-p scattering at high energies now support this conjecture over the range $0 \leq -t \leq 10$ $(\mathrm{GeV/c})^2$; see Rapporteur's Report of W.K.H. Panofsky, XIVth International Conference on High Energy Physics, Vienna (1968).
13. J. Bjorken and J. D. Walecka, Ann. Phys. (New York) 38, 35 (1966).
14. See, e.g., H. H. Joos, report to Heidelberg Conference (1967).
15. See, e.g., P. H. Stelson and F. K. McGowan, Ann. Rev. Nucl. Sci. 13, 163 (1963).
16. For extensive discussion of collective nuclear states, see D. J. Rowe in Fundamentals of Nuclear Theory, International Course on Nuclear Physics, Trieste (1966) (published by IAEA, Vienna); pp. 93-160 and 531-621.
17. J. Bjorken and S. Drell, Relativistic Quantum Fields, McGraw-Hill, Inc., New York (1965).
18. See, e.g., the Bethe-Salpeter amplitude given in G. C. Wick, Phys. Rev. 96, 1131 (1954).
19. In more complicated models where, for example, space parts of currents play a role, one may have additional spin-parity changes. For the purposes of our present discussion, however, we take the simplest point of view and assume that effects other than the purely geometrical are negligible.
20. Rules applicable to the equal mass case have already been discussed in T. T. Chou and C. N. Yang, Phys. Rev. 175, 1832 (1968).
21. See, e.g., T. A. Griffy and L. C. Biedenharn, Nucl Phys. 36, 452 (1962).
22. N. Byers and C. N. Yang, Phys. Rev. 142, 976 (1966).
23. J. Bjorken and J. D. Walecka (Ref. 13) have explicitly shown this q_z^{Breit} dependence for electroproduction form factors and have similarly remarked that, owing to the large mass differences, these need not suppress forward productions.

24. K. J. Foley et al., Phys. Rev. Letters 19, 397 (1967).
25. H. Joos, Fortsch. Physik 10, 65 (1962).
26. M. Jacob and G. C. Wick, Ann. Phys. (New York) 17, 404 (1959).
27. Aachen-Berlin-CERN Collaboration and Aachen-Berlin-CERN-London (I.C.)-Vienna Collaboration, Physics Letters 27B, 336 (1968).
28. E. W. Anderson et al., Phys. Rev. Letters 16, 855 (1966).

Received 5/6/69

On the Nature of Diffraction Scattering

by

Peter G. O. Freund
The University of Chicago

1. Introduction

At high energies the elastic scattering of hadrons is primarily diffractive. A full relativistic quantum theory of diffraction phenomena is still missing.[1] We want to point out below some features of diffraction scattering that can be derived from rather general considerations. This should be useful in guiding work towards a consistent dynamical model of diffraction. Some of these general features have been noted before though maybe in a less systematic way. Others, such as the relation between the energy dependences of the width of the diffraction peak and of a certain combination of total cross-sections (see section 5), are new.

I presented the contents of this paper in the "Wentzel Seminar" at the Enrico Fermi Institute in November 1968. The comments received from the participants in this seminar have been very useful for the clarification of my ideas. It is with great pleasure that I dedicate this "Quantum" to Gregor Wentzel.

2. The Pomeranchuk-singularity and its known physical attributes

As we are interested in genuine diffraction processes we shall confine our attention to elastic scattering and so-called diffraction dissociation processes. The common feature of these is that they exhibit cross-sections that are very nearly constant

$$\sigma(E) \gtrsim \text{const } E^{-0.15} + O(E^{-0.5}). \quad (1)$$

The smaller terms are due to lower lying Regge-poles. We shall discuss here the Pomeranchuk amplitude responsible for the leading (diffractive) term in (1).

Let us parametize the Pomeranchuk amplitude $A_P(s,t)$ of a process in the form

$$A_P(s,t) \approx \beta(t)\, s^{\alpha_P(t)}\, (\ln s)^{\lambda_1(t)} (\ln \ln s)^{\lambda_2(t)} \ \ldots \qquad (2)$$

A full theory of diffraction should yield unique forms for the functions β, α_P, λ_1, λ_2, ... For the time being all that is known is that

$$\alpha_P(0) \approx 1 \qquad (3)$$

and very likely[1]

$$\alpha'_P(0) < \alpha'_{Regge} \qquad (4)$$

where $\alpha'_{Regge} \approx 1 GeV^{-2}$ is the value characteristic of the lower Regge trajectories. In terms of the language of complex angular momenta, the set of functions $\beta(t)$, $\alpha(t)$, $\lambda_1(t)$, $\lambda_2(t)$, ... uniquely identifies the nature of a singularity P in the complex (t-channel) angular momentum plane. We shall call this the Pomeranchuk singularity.[2] Its signature is of course positive.

From the point of view of symmetries P has the following quantum numbers: isospin I = 0, hypercharge Y = 0, charge conjugation C = +1. If SU(3) were an exact symmetry then P ought to be a unitary singlet. This can be most easily seen by observing that if it were not the case, then processes other than of elastic scattering or diffraction dissociation would receive a contribution from P. In reality P may also exhibit a small SU(3) octet (and possibly even smaller 27-plet, etc. ...) component. If it were a pure SU(3) singlet then one should find[3]

$$\sigma_{\pi^- p}(\infty) - \sigma_{K^- p}(\infty) = 0 \,. \qquad (5)$$

Present fits to total cross-section data[4] indicate that the left hand side of eq. (5) is of the order of 3 mb. This can be viewed as evidence for an octet part in P.

3. Factorization of the Pomeranchuk singularity

Regge poles are well known to have the property that their residua factorize. As we pointed out above, the nature of the Pomeranchuk singularity is not known at present. As such, it is not obvious whether P factorizes. If P would factorize and lower Regge contribution to elastic and diffraction dissociation processes would be negligible, then one would find that[5]

$$\sigma_{\pi\pi} = \sigma^2_{\pi N}/\sigma_{NN} \quad (6$$

and[6]

$$\frac{d\sigma}{dt}_{pp\to pN^*} \bigg/ \frac{d\sigma}{dt}_{\pi p\to\pi N^*} = \frac{d\sigma}{dt}_{pp\to pp} \bigg/ \frac{d\sigma}{dt}_{\pi p\to\pi p} \quad (7$$

for all N^* such that the processes $pp\to pN^*$ and $\pi p\to\pi N^*$ can proceed through P-exchange. While the left hand side of eq. (6) is not easily measurable, both sides of eq. (7) have been measured[7] for $N^* = N^*_{1/2}+(140$ $N^*_{5/2}+(1688)$. The equations (7) have been found[6] to hold within 20%, a degree of accuracy characteristic of the approximations used in obtaining them. It therefore appears that to a good approximation P is a factorizable singularity.

4. The s-channel build-up of the Pomeranchuk singularity

It has become clear over the past year that Regge-pole exchange is an average representation of the contributions of many direct channel resonances.[8] The question then is what s-channel intermediate states is the Pomeranchuk contribution built of. It is clear that the P-term cannot be built of s-channel resonances alone. Indeed, there is a P-term in $\pi^+\pi^+$-scattering and there are no known $\pi^+\pi^+$ resonances. In $\pi^+\pi^+$ scattering therefore, the P-term must be built completely out of nonresonant s-channel "background." The question now is which part of this nonresonant background builds up the P. An attractive possibility is that the whole nonresonant background is "used up" in building the P. This offers an interesting picture in which for any process[9]

Sum of t-channel Regge exchanges $\sim$ average of s-channel resonances (8)

and

t-channel P-exchange $\sim$ s-channel nonresonant background (9)

This picture can be easily shown to have many implications on the coupling of Regge-poles. In particular it leads to an extended form of exchange degeneracy.[9,10] It also might require the existence of certain "exotic" mesons with baryon-antibaryon as their only two-body decay mode.[11] The duality relation (8) is embodied in the Veneziano[12]-model. The relation (9) has to be embodied in any serious model of P. It is maybe the strongest piece of dynamical information about P that we dispose of at present.

5. Pomeranchuk singularity and unitarity

It is well known that the successive exchange of two or more Regge-poles gives rise to cuts in the angular momentum plane. These may be viewed as unitarity corrections to the Regge pole amplitude. In the same spirit, one may consider the possibility of unitarizing an approximate "bare" P-term. There are various methods for doing this: the rescattering model, the absorption model and the eikonal model. To first order, for an elastic process, they all give the same result[13]

$$\sigma^P_{tot}(s) = \frac{1}{s}\,\mathrm{Im}\,A_{P\ corrected}(s,o) = \frac{1}{s}\,\mathrm{Im}\,A_{P\ bare}(s,o)\left(1 - \frac{\sigma_{el}(s)}{\sigma^P_{tot}(s)}\right) \tag{10}$$

where $\sigma^P_{tot}(s)$ is the part of the total cross-section contributed by the P-term through the optical theorem. Eq. (10) may be regarded as essentially model independent.

For the purpose of the following discussion let us replace eq. (3) by the stronger requirements

$$\begin{aligned} &\alpha_P(0) = 1 \\ &\lambda_1(0) = \lambda_2(0) = \ldots = 0\,. \end{aligned} \tag{11}$$

Then $\sigma^{P}_{tot}(s)$ approaches a constant for $s \to \infty$. As is done in all the models mentioned above[13,14] let us assume

$$\frac{1}{s} \operatorname{Im} A_{P\ bare}(s,o) \approx \sigma^{P}_{tot\ bare}(s) = \sigma^{P}_{bare}(\infty) = \text{const.} \tag{12}$$

for large s. Now, there is an essential qualtitative difference between the cases $\alpha'_P(0) = 0$ and $\alpha'_P(0) \neq 0$. If $\alpha'_P(0) = 0$ then σ_{el}/σ_{tot} = const so that

$$\sigma^{P}_{tot}(s) = \text{const.} \tag{13}$$

If $\alpha'_P(0) \neq 0$ then

$$1 - \frac{\sigma_{el}(s)}{\sigma_{tot(s)}} \approx 1 - \frac{a}{\ln s} \tag{14}$$

with

$$a \approx \frac{\sigma_{tot}(\infty)}{32\pi\alpha'_P(0)} \tag{15}$$

(s in units of GeV^2) so that

$$\sigma^{P}_{tot}(s) = \text{const}\ (1 - a/\ln s) \tag{16}$$

which because of $a > 0$ (see eq. (15)) <u>increases</u> towards its asymptotic value.[14]

Using either universal Regge couplings or the quark model it is easy to show[15] that in the combination of total cross-sections

$$\bar{\sigma}(s) = \frac{1}{2}\left(\sigma_{K^-p} + \sigma_{K^+p} + \sigma_{K^-n} + \sigma_{K^+n} - \sigma_{\pi^-p} - \sigma_{\pi^+p}\right) \tag{17}$$

all Regge pole contributions cancel out and only the Pomeranchuk contribution is left over. $\bar{\sigma}(s)$ is thus precisely $\sigma^{P}_{tot}(s)$ discussed above. If $\sigma_{\pi N}(\infty)$ were equal to $\sigma_{KN}(\infty)$, then we could apply eqs. (15) and (16) directly to $\bar{\sigma}(s)$. In reality $\sigma_{\pi N}(\infty) > \sigma_{KN}(\infty)$. Therefore we have to make a slight correction to eq. (16). Moreover eq. (16) is only valid for $\ln s \gg 1$. For moderate values of $\ln s$, one has to replace $\alpha_p'(0)\ln s$ by $L(s) = b + \alpha_p'(0)\ln s$ in Eq. (16). Here $b \approx (4\ \text{GeV}^{-2} - 3\alpha_p'(0))$ is chosen to give the correct width of the diffraction peak. Finally then

$$\sigma^p_{tot}(s) = \bar{\sigma}(s) = \bar{\sigma}(\infty)\ [1 - \frac{2(\sigma_{KN}(\infty))^2 - (\sigma_{\pi N}(\infty))^2}{32\pi\alpha_p'(0)\ \bar{\sigma}(\infty)\ L(s)}] \tag{18}$$

Experimentally[15] at $P_{lab} = 6, 8, 10, 12, 14, 16$ GeV/c $\bar{\sigma}(s) = 12.9 \pm 0.4;\ 12.8 \pm 0.4;\ 13.3 \pm 0.4;\ 13.3 \pm 0.4;\ 13.5 \pm 0.4;$ 13.8 ± 0.5 mb respectively. These values are consistent within one standard deviation with $\bar{\sigma}(s)$ = const and therefore $\alpha_p'(0) = 0$. With present experimental accuracies, Eq. (18) sets a rather weak limit on $\alpha_p'(0)$. The combination $\bar{\sigma}(s)$ of the total cross-sections contains precious information about the slope of the Pomeranchuk trajectory. We believe that it would be worthwhile to remeasure $\bar{\sigma}(s)$ with higher accuracy. At the same time it is important to note that the amplitude for ϕ-photoproduction has been shown to be proportional to $A_p(s,t)$.[16] Thus $(d\sigma/dt)_{\gamma p\to\phi p}$ contains a great deal of information about $\alpha_p(t)$. Determining experimentally the shape of the angular distribution in $\gamma p\to\phi p$ is the optimum method for unraveling the nature of the Pomeranchuk singularity.

6. Conclusions

We have tried in this paper to put together a list of properties of the Pomeranchuk singularity which in our opinion rest on a solid foundation. In particular in section 5 some new ways of exploring the nature of this singularity have been pointed out. The ultimate purpose of our model independent discussion is to stimulate model building for the Pomeranchuk singularity.[17] Any acceptable model of this singularity must be able to accommodate the results presented above.

References

1) H. M. Chan, 14th International Conference on High Energy Physics, Proceedings edited by J. Prentki and J. Steinberger, CERN, Geneva 1968, p. 391 ff.

2) This singularity is often called the Pomeranchukon or Pomeron. We find both these terms misleading because their ending suggests a particle spectrum connected with the singularity. There is at present no evidence whatsoever in favor of such a suggestion.

3) P. G. O. Freund, H. Ruegg, D. Speiser and A. Morales, Nuovo Cimento. 25, 307 (1962).

4) See e.g. M. Davier, Phys. Rev. Letters 20, 952 (1968).

5) M. Gell-Mann, Phys. Rev. Letters 8, 263 (1962); V. N. Gribov and I. Ya. Pomeranchuk, Phys. Rev. Letters 8, 343 (1962).

6) Peter G. O. Freund, Phys. Rev. Letters 21, 1375 (1968).

7) K. Foley, S. Lindenbaum et. al. Phys. Rev. Letters 11, 425 (1963); ibid. 19, 397 (1967).

8) R. Dolen, D. Horn and C. Schmid, Phys. Rev. 166, 1768 (1968).

9) P. G. O. Freund, Phys. Rev. Letters 20, 235 (1968). A restatement of this result is contained in H. Harari, Phys. Rev. Letters 20, 1385 (1968).

10) C. Schmid, CERN preprint TH 960 (1968).

11) J. Rosner, Phys. Rev. Letters 21, 950 (1968).

12) G. Veneziano, Nuovo Cim. 57A, 190 (1968).

13) P. G. O. Freund and P. J. O'Donovan, Phys. Rev. Letters 20, 1329 (1968); J. Finkelstein and M. Jacob, Nuovo Cim. 56A, 681 (1968); P. J. O'Donovan, to be published.

14) This is a slightly generalized version of an observation first made by S. Frautschi and B. Margolis, Nuovo Cim. 56A, 1155 (1968) and by V. N. Gribov and A. A. Migdal, to be published.

15) P. G. O. Freund, Nuovo Cim. 46A, 563 (1966).

16) P. G. O. Freund, Nuovo Cim. 48A, 541 (1967); see also Ref. 3.

17) For work in this direction see G. F. Chew and A. Pignotti, Phys. Rev. 176, 2112 (1968) and P. G. O. Freund, Phys. Rev. Letters 22, 565 (1969).

Received 3/15/69

Asymptotic Behavior of Partial Widths in the Veneziano Model of Scattering Amplitudes

by

Yoichiro Nambu and Paul Frampton
The Enrico Fermi Institute and the Department of Physics
The University of Chicago

A general procedure is given for determining the partial wave residues of the Veneziano amplitudes at sufficiently high resonance energies. This is explicitly worked out for π-π scattering, and it is found that the residues are all asymptotically positive if the Regge intercept $\alpha(0) > 1/3$.

1. Introduction

Recently Veneziano[1] has constructed a simple and elegant mathematical model for scattering amplitudes which manifestly exhibits the duality[2], or equivalence, of direct channel resonance amplitudes and the usual Regge behavior coming from crossed channel resonances, a concept which has proven successful in the analysis of hadron scattering. It has also been pointed out[3] that this model is consistent with the requirements of chiral dynamics. An important prediction of the model is that there exist an infinite number of parallel and linearly rising Regge trajectories. They represent a sequence of regularly spaced resonances (in the mass squared variable), each level comprising degenerate states of different angular momenta in a way reminiscent of the hydrogenic spectrum.

The present article is devoted to the study of an elementary problem: to determine the residues C_n^ℓ when a Veneziano amplitude is decomposed into a sum of Breit-Wigner terms with various resonance energies (m_n) and spins (ℓ). This is interesting for two reasons. Firstly, for an elastic scattering (such as $\pi^+ + \pi^- \to \pi^+ + \pi^-$), C_n^ℓ must be ≥ 0 as it is equal to the sum of the partial widths (apart from kinematical factors), or the squares of the coupling parameters, of the states (m_n, ℓ) with respect to the pion channel. (These states themselves may have internal degeneracy.) This would be necessary even though the Veneziano model completely ignores the width in the resonance denominator, and thus the unitarity is not incorporated. If the positivity condition were not met in general, the model could not be said to be physically satisfactory. Such a suspicion is not entirely without a basis in view of the previous difficulties[4] encountered in the saturation of superconvergence relations (which the Veneziano model nicely satisfies).

The second motivation for the present work is to see if it is possible to understand the basic features of the model in terms of the quark model and the like, in conjunction with some dynamical symmetry. The linear mass spectrum and degeneracy structure of the resonances were forced by the duality principle, but they bear at least a formal resemblance to what one expects from a simple dynamical model of bound states based on SU(3) (harmonic oscillator) or O(4) (hydrogen atom) symmetry. The partial width, which is determined by the overlap of a resonant state and the incident channel, may give us a clue to the possible connection between these two seemingly unrelated pictures of the resonances.

The partial wave analysis of a Veneziano amplitude is actually not as trivial a problem as it may appear. Although it requires only elementary algebra, no simple and practical general formula seems to exist. Some numerical calculations have been done for π-π scattering[5,6]. One can also derive certain necessary conditions for the positivity of the residues[5,7]. We will give here a brief account of the basics of the problem.

Taking π^+-π^- elastic scattering for definiteness, a Veneziano amplitude is represented by the function

$$A(s,t) = \frac{\Gamma(1-\alpha(s))\ \Gamma(1-\alpha(t))}{\Gamma(1-\alpha(s) - \alpha(t))} \qquad (1)$$

Here s and t are the usual channel energy variables, $\alpha(s)$ the degenerate ρ-meson and f-meson trajectory function $\alpha(s) = s + a$, $a = \alpha(0) \approx 1/2$, measuring the masses in units (≈ 1 Gev) so as to make the slope equal to unity. The residue of $A(s,t)$ at the nth level $\alpha(s) = n$ is given by

$$\begin{aligned} F_n(t) &= \alpha(t)\ [\alpha(t) + 1]\ \ldots\ [\alpha(t) + n-1]/(n-1)! \\ &= (t + a)\ (t + a + 1)\ \ldots\ (t + a + n-1)/(n-1)! \end{aligned} \tag{2}$$

Since t is related to the center of mass angle θ by

$$\cos\theta = Z = 1 + 2t/(s-4\mu^2),\ s = n-a,\ \mu^2 \approx 0.02\ \text{(pion mass)}, \tag{3}$$

$F_n(t)$ is an nth order polynomial with equally spaced zeros in the variable t as well as Z, distributed over their physical range. The ℓ-wave projection of $F_n(t)$ is then

$$C_n^\ell = (\ell + \tfrac{1}{2}) \int_{-1}^{1} F_n(t(Z))\ P_\ell(Z)dZ, \tag{4}$$

which obviously vanishes for $\ell > n$.

The positivity condition for C_n^ℓ can be easily worked out in some simple cases[7]. For example, $C_n^n \geq 0$ for all n if $C_1^1 \geq 0$, which means

$$a \leq 1 - 4\mu^2 \approx 1 \tag{5}$$

Another obvious criterion is that $F_n(Z=1) \geq 0$ and $F_n(Z=1) \geq F_n(Z=-1)$, i.e. the forward amplitude is larger than the backward. In view of the equal spacing of zeros, this will be the case if the largest zero Z_1 and the smallest zero Z_n are so situated that $1 - Z_1 \geq Z_n - (-1)$, which leads to

$$a \geq (1 - 4\mu^2)/3 \approx 1/3. \tag{6}$$

We see that C_n^ℓ is a rather sensitive function of the Regge intercept a. From other low lying levels we obtain further conditions.

In fact $C_2^0 \geq 0$ gives

$$a \geq 1/2, \tag{7}$$

and so far as is known, this seems to make all coefficients positive, at least for low lying levels ($n \lesssim 50$).[6]

In the following, we are going to determine C_n^ℓ for large $n >> 1$, and arbitrary $n \geq \ell \geq 1$ in the sense of an asymptotic formula. We will find that no new conditions on a are required; indeed Eq. (6) is already sufficient to make all C_n^ℓ asymptotically positive.

2. Asymptotic formulas for C_n^ℓ

Let us introduce a generating function for $F_n(t)$:

$$\begin{aligned} f(x,t) &= 1 + \sum_{n=1}^{\infty} F_n(t)\ x^n/n \\ &= (1 - x)^{-\alpha(t)} \\ &= \exp\ [\alpha(t)\zeta] \end{aligned} \tag{8}$$

where

$$\zeta \equiv -\ln\ (1-x),$$

$$\begin{aligned} \alpha(t) &= (Z-1)\ (n-a-4\mu^2)/2 + a \\ & (Z-1)\ (n-a)/2 + a. \end{aligned}$$

We have ignored μ^2 for the sake of simplicity, but this will cause no loss of generality. The function (8) can be decomposed into Legendre polynomials with the aid of the formula

$$\begin{aligned} &(\ell + \tfrac{1}{2}) \int_{-1}^{1} e^{\lambda Z}\ P_\ell(Z)\ dZ \\ &= [(\ell + \tfrac{1}{2})/2^\ell \ell!] \int_{-1}^{1} e^{\lambda Z} (d/dZ)^\ell\ (Z^2-1)^\ell\ dZ \\ &= [(\ell + \tfrac{1}{2})\ \lambda^\ell/2^\ell \ell!] \int_{-1}^{1} e^{\lambda Z}\ (1-Z^2)^\ell\ dZ \\ &= [(\ell + \tfrac{1}{2})\ (2\pi)^{1/2}/\lambda^{\ell+1/2}]\ I_{\ell+1/2}(\lambda), \end{aligned} \tag{9}$$

where $I_{\ell+1/2}(\lambda)$ is a modified Bessel function. To get C_n^ℓ, we have only to extract the nth power of x by means of a Cauchy integral. Using ζ as the integration variable, we obtain the basic formula

$$C_n^\ell = n(\ell + \tfrac{1}{2})\ (2\pi)^{1/2}\ [(n-a)/2]^{-1/2}$$
$$\times \frac{1}{2\pi i} \oint \frac{d\zeta}{\zeta^{1/2}} \exp\ [(3a-1)\zeta/2]\ I_{\ell+1/2}\big((n-a)\zeta/2\big)/[2\ \sinh(\zeta/2)]^{n+1} \tag{10}$$

where the integration is about the origin. Eq. (10) is general and exact, but not easy to evaluate. For large values of n, however, we can make use of the asymptotic expansion of Bessel functions, and then further evaluate the integral by the method of saddle points.

We observe first that the integrand of Eq. (10) has a string of poles, beside the origin, on the imaginary axis at $\zeta = \pm 2n\pi i$, which we must avoid. As $\mathrm{Re}\zeta \to \pm\infty$, on the other hand, it behaves like

$$\begin{aligned} &\sim \exp\ [(a-1)\zeta]/\zeta, && \zeta \to +\infty \\ &\sim \exp\ [2a\zeta]/\zeta, && \zeta \to -\infty. \end{aligned} \tag{11}$$

For finite $|\zeta|$, the asymptotic formula (for large ν)

$$\begin{aligned} &I_\nu(\nu Z) \sim \exp\ [\nu f(Z)]/(2\pi\nu)^{1/2}(1+Z^2)^{1/4}, \\ &f(Z) = (1+Z^2)^{1/2} - \sinh^{-1}(1/Z), \quad \mathrm{Re}Z \neq 0, \end{aligned} \tag{12}$$

enables one to split the integrand into a slowly varying factor ϕ and a fast varying one χ in the limit of large n and ℓ with a fixed ratio

$$0 \leq k \equiv (\ell+1/2)/(n-a) \leq 1. \tag{13}$$

Here

$$\phi \equiv \exp[(3a-1)\zeta/2] \; / \; \zeta^{1/2}[2\pi(\ell+1/2)]^{1/2}(1+Z^2)^{1/4}(2 \sinh \zeta/2)^{1+a},$$

$$\chi \equiv \left(\exp[kf(Z)] \; / \; 2 \sinh(\zeta/2)\right)^{n-a} \equiv \exp[(n-a)g(\zeta)], \tag{14}$$

$$Z \equiv \zeta/2k, \quad g(\zeta) = kf(Z) - \ln[2 \sinh(\zeta/2)].$$

The saddle points will be mainly controlled by χ. Taking the logarithmic derivatives of χ, then, we find

$$\begin{aligned} d \ln \chi/d\zeta &= (n-a) \; dg/d\zeta \\ &= \frac{1}{2} (n-a)[(1+Z^2)^{1/2}/Z - \coth(\zeta/2)], \\ d^2 \ln \chi/d\zeta^2 &= (n-a) \; d^2g/d\zeta^2 \\ &= \frac{1}{4} (n-a)[1/\sinh^2(\zeta/2) - 1/kZ^2(1+Z^2)^{1/2}]. \end{aligned} \tag{15}$$

The condition $dg/d\zeta = 0$ yields two real solutions $\pm\zeta_o$ satisfying

$$\sinh(\zeta_o/2) = \zeta_o/2k \tag{16}$$

so that

$$\begin{aligned} d^2g/d\zeta_o^2 &= [1/4 \sinh^2(\zeta_o/2)][1 - (2/\zeta_o) \tanh(\zeta_o/2)] \geq 0, \\ f(\zeta_o) &= \cosh(\zeta_o/2) + \ln \tanh(\zeta_o/4), \\ g(\zeta_o) &= kf(\zeta_o) - \ln\left(2 \sinh(\zeta_o/2)\right). \end{aligned} \tag{17}$$

As k varies from 0 to 1, ζ_o varies from ∞ to 0, $g(\zeta_o)$ varies from 0 to $\ln(e/4)$, and the second derivative from 0 to $(n-a)/24$. This justifies a posteriori the approximations we have made, except near $k = 0$ and $k = 1$. We will come back to these exceptional cases later.

The contour integral in Eq. (10) is now reduced to a sum of two Gaussian integrals running through the points $\pm\zeta_o$ in the vertical direction. The two contributions differ by a factor $(-1)^{n-\ell}\exp[i(3a-1)\zeta_o]$ as is clear from Eq. (10). The final result is

$$
\begin{aligned}
C_n^\ell &\sim n(\ell + 1/2)\ [2/g''(\zeta_o)(n-a)]^{1/2} \\
&\quad \times \sum_{\pm} \phi(\pm\zeta_o)\ \chi(\pm\zeta_o) \\
&\sim \left(\frac{\ell + 1/2}{2}\right)^{1/2} / [(\zeta_o/2)\cosh(\zeta_o/2) - \sinh(\zeta_o/2)]^{1/2} \\
&\quad \times [\exp\left(\cosh(\zeta_o/2)\right)\ \tanh(\zeta_o/4)]^{\ell+1/2} / [2\sinh(\zeta_o/2)]^n \\
&\quad \times \begin{cases} 2\cosh[(3a-1)\zeta_o/2] & n-\ell \text{ even} \\ 2\sinh[(3a-1)\zeta_o/2] & n-\ell \text{ odd} \end{cases}
\end{aligned}
\tag{18}
$$

This shows immediately that the positivity of C_n^ℓ for both even and odd $n-\ell$ requires $a = \alpha(0) > 1/3$, the same condition as Eq. (6). (In fact it is exactly the same even when the external mass is not ignored: In the last factor of Eq. (18), a will be replaced by $a + 4\mu^2/3$.) The dependence of C_n^ℓ on n and ℓ (or k), on the other hand, is not so transparent since ζ_o is an implicit function of k. But the main controlling factor is

$$
\begin{aligned}
&[\exp\left(\cosh(\zeta_o/2)\right)\ \tanh(\zeta_o/4)]^{\ell+1/2} / [2\sinh(\zeta_o/2)]^n \\
&= \exp[(n-a)g(\zeta_o)]\ /\ [2\sinh(\zeta_o/2)]^a .
\end{aligned}
\tag{19}
$$

Since $\exp[g(\zeta_o)]$ decreases from 1 to $e/4$ with increasing k, C_n^ℓ will decrease very rapidly for large n, as ℓ increases toward n. The remaining slowly varying factor tends to counteract this. The function $g(\zeta_o)$ is plotted in Fig. 1.

Let us next turn to the extreme cases $k \approx 1$ and $k \approx 0$, which need special care.

a) $k \simeq 1$, or $\ell \simeq n$. In this limit, the saddle points $\pm\zeta_o$ approach zero like $1/n^{1/2}$, and we must take into account both χ and ϕ from the beginning. However, it turns out to be sufficient to make an expansion

$$\begin{aligned} \ln \phi\, \chi &\simeq (\ell + 1/2)\, f(\zeta/2k) - (n+1) \ln \sinh(\zeta/2) - 1/2 \ln\zeta \\ &\quad + (3a-1)\zeta/2 \\ &\simeq (\ell + 1/2)\, [(1 - \ln 2) + \ln(\zeta/2k) + \tfrac{1}{4} (\zeta/2k)^2] \\ &\quad - (n+1)\, [\ln\zeta + \zeta^2/24] - 1/2 \ln\zeta \\ &\quad + (3a-1)\zeta/2. \\ &\simeq -(m+1) \ln\zeta + (3a-1)\zeta/2 + \zeta^2/48 + \\ &\quad + (n-a) - (n + 1/2 - m) \ln 4, \quad m = \ell - n = 0, 1, 2, \ldots \end{aligned} \tag{20}$$

which is correct to order $1/n^{1/2}$ for large n, $\zeta^2 \sim 1/n$, and fixed m. In fact from Eq. (20) we obtain

$$\zeta_o^2 = (m+1)/24n + O(n^{-3/2}). \tag{21}$$

Instead of using the saddle point method, however, we can evaluate the integral rigorously with the aid of the formula

$$\begin{aligned} &\frac{1}{2\pi i} \oint d\zeta\, \zeta^{-(1+m)} \exp\, [\alpha\zeta - \beta^2\zeta^2/2] \\ &\quad = H_m(\alpha/\beta)\, \beta^m/m! \end{aligned} \tag{22}$$

where $H_m(x)$ is the Hermite polynomial of order m, and $\alpha = (3a-1)/2$, $\beta^2 = -n/24$ in the present case. Since $\alpha/\beta = O(n^{-1/2})$, we may actually keep only the leading term in $H_m(x)$:

$$H_m(x) = \begin{cases} (-1)^{m/2}\, (m-1)!! & m \text{ even} \\ (-1)^{(m-1)/2}\, m!!x & m \text{ odd} \end{cases} \tag{23}$$

In this way we obtain

$$C_n^{n-m} \sim e^{n-a}/4^{n-m+1/4}$$

$$\times \begin{cases} (n/24)^{m/2}/m!!, & m \text{ even} \\ \frac{(3a-1)}{2}(n/24)^{(m-1)/2}/(m-1)!!, & m \text{ odd}, \end{cases} \tag{24}$$

which is > 0 if 3a-1 > 0. We also observe the pairwise degeneracy

$$C_n^{n-2m}/C_n^{n-2m-1} \sim 1/2(3a-1) = 1 \text{ for } a = 1/2. \tag{25}$$

b) $k \approx 0$, $\ell << n$. As $k \to 0$, ζ_o and the width of the Gaussian peak around ζ_o grow according to Eqs. (15) and (16). Since the original contour had to avoid the zeros at $\zeta = 2n\pi i$, the saddle point approximation will begin to fail when the width becomes comparable to 2π. This means

$$n/8 \sinh(\zeta_o/2) \approx \pi,$$

$$\text{or} \quad k_1 = (\zeta_o/2)/\sinh(\zeta_o/2) \tag{26}$$

$$\approx \sinh^{-1}(n/8\pi)/(n/8\pi)$$

$$\approx (8\pi/n)^{1/2} \ln 2(n/8\pi)^{1/2}.$$

Around this value k_1, C_n^ℓ should reach a maximum and then begin to decrease. We are now concerned with this region $0 \leq k \leq k_1$.

To handle the situation, we change the original contour to two straight lines running parallel with the real axis through $\pm\pi i$, taking account of Eq. (11) and the asymptotic form of $I_{\ell+1/2}((n-a)/2)$ for large arguments and fixed ℓ. We then find that the integral in Eq. (9) reduces to

$$\int_0^\infty \frac{\exp[(a-1)\zeta]}{[1+\exp(-\zeta)]^{n+1}} \frac{2}{\zeta^2+\pi^2} [\zeta\sin(a\pi) - \pi\cos(a\pi)]\, d\zeta$$
$$+ (-1)^{n-\ell} \int_0^\infty \frac{\exp[-2a\zeta]}{[1+\exp(-\zeta)]^{n+1}} \frac{2}{\zeta^2+\pi^2} [\zeta\sin(2a\pi) + \pi\cos(2a\pi)]\, d\zeta. \tag{27}$$

The first denominator causes a strong suppression of the integrand for $\zeta \lesssim \ln n$, so we extend the integration from $-\infty$ to $+\infty$, observing that

$$\int_{-\infty}^{\infty} \exp\,[(a-1)\zeta]/[1+\exp(-\zeta)]^{n+1} d\zeta$$
$$= \int_{0}^{\infty} t^{a}/(1+t)^{n+1}\, dt = \Gamma(1-a)\Gamma(n+a)/\Gamma(n+1) \sim \Gamma(1-a)n^{a-1}. \tag{28}$$

We can therefore write

$$\int_{0}^{\infty} \frac{\exp[(a-1)\zeta]}{[1+\exp(-\zeta)]^{n+1}} \frac{\zeta}{\zeta^2+\pi^2} d\zeta \sim \Gamma(1-a)n^{a-1} < \frac{\zeta}{\zeta^2+\pi^2} >, \text{ etc.} \tag{29}$$

where the average may be estimated by taking $\zeta \sim \ln n$. Collecting other factors, we obtain

$$c_n^{\ell} \sim \frac{(2\ell+1)}{\pi} [\Gamma(1-a)n^{a-1}\left(\frac{\sin a\pi}{\ln n} - \frac{\pi\cos a\pi}{(\ln n)^2}\right)$$
$$+ (-1)^{n-\ell}\Gamma(2a)n^{-2a}\left(\frac{\sin 2a\pi}{\ln n} + \frac{\pi\cos 2a\pi}{(\ln n)^2}\right)], \tag{30}$$

$$\ln n >> 1,$$

which is again > 0 if $a > 1/3$. For $a = 1/2$, we have

$$c_n^{\ell} \sim (2\ell+1)/\pi^{1/2}n^{1/2}\ln n. \tag{31}$$

3. Remarks

It is obvious that our procedure is applicable to any function of the type

$$\Gamma\big(n-\alpha(s)\big)\ \Gamma\big(m-\alpha(t)\big)\ /\ \Gamma\big(\ell-\alpha(s)-\alpha(t)\big),$$

$$\max\ (m,n) \leq \ell \leq m + n;\ m,\ n \geq 0.$$

The general amplitude may be a superposition of such terms with different n, m, and ℓ. However, each term individually is known to have an infinite number of negative residues except when $n = m = \ell = 1$.[7] In this last case, our result shows that the residues are ≥ 0 for sufficiently high levels provided that $a \gtrsim 1/3$. On the other hand, the low lying levels demand that $1 \gtrsim a \gtrsim 1/2$, which seems then sufficient to guarantee the positivity except at most a finite number of states. Quite probably there are no exceptions.

The behavior of C_n^ℓ as a function of ℓ and n is rather complicated, but its gross feature is what one expects from the Regge behavior. For low values of ℓ, it grows like $(2\ell+1)/n^{1/2}\ln n$ up to a maximum $\ell_1 \sim n^{1/2}\ln n$, and then decreases more or less exponentially, as is characteristic of an expanding interaction radius $R \sim \ell_1/\sqrt{n} \sim \ln n$. The strong suppression of the residues for $n \gtrsim \ell \gtrsim n^{1/2}\ln n$ is an inevitable consequence of the linearly rising trajectory. The decay of these levels to the incident channel is highly forbidden by the centrifugal barrier. It may be, however, that the coupling between nearby levels is strong. We also note the fine structure in C_n^ℓ between even and odd trajectories, which is not obvious from the beginning. These points should be kept in mind when we try to construct a dynamical theory of hadrons, whether in terms of bootstrapping or in terms of fundamental constituents.

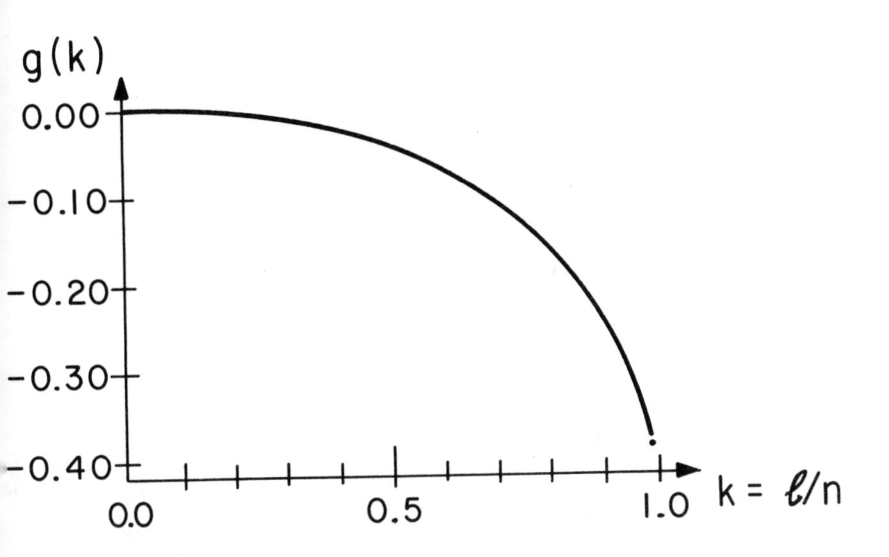

References

1) G. Veneziano, Nuovo Cimento 57A, 190 (1968).

2) R. Dolen, D. Horn and C. Schmid, Phys. Rev. 166, 1768 (1968); C. Schmid, Phys. Rev. Letters 20, 689 (1968).

3) C. Lovelace, Physics Letters 28B, 265 (1968); M. Ademollo, G. Veneziano, and S. Weinberg, Phys. Rev. Letters 22, 83 (1969).

4) E.g., R. Oehme, Ann. Phys. (New York) 46, 347 (1968).

5) J. Shapiro and J. Yellin, Preprint UCRL 18500 (1968), and further preprints by J. Yellin.

6) F. Wagner, Preprint CERN TH 978 (Dec. 1968).

7) R. Oehme, Preprint EFI 69-12 (Jan. 1969).

Received 5/1/69